NEW TRENDS IN THE PHOTOCHEMISTRY OF POLYMERS

The proceedings of an international symposium held in honour of Professor Bengt Rånby at The Royal Institute of Technology, Stockholm, Sweden, 26–29 August 1985.

PROFESSOR BENGT RÅNBY

NEW TRENDS IN THE PHOTOCHEMISTRY OF POLYMERS

Edited by

NORMAN S. ALLEN

Department of Chemistry, John Dalton Faculty of Technology,
Manchester Polytechnic, UK

and

JAN F. RABEK

Department of Polymer Technology, The Royal Institute of Technology,
Stockholm, Sweden

ELSEVIER APPLIED SCIENCE PUBLISHERS
LONDON and NEW YORK

ELSEVIER APPLIED SCIENCE PUBLISHERS LTD
Crown House, Linton Road, Barking, Essex IG11 8JU, England

Sole Distributor in the USA and Canada
ELSEVIER SCIENCE PUBLISHING CO., INC.
52 Vanderbilt Avenue, New York, NY 10017, USA

British Library Cataloguing in Publication Data

New trends in the photochemistry of polymers
1. Polymers and polymerization 2. Photochemistry
I. Rånby, Bengt II. Royal Institute of Technology, (*Stockholm*) III. Allen, Norman S.
IV. Rabek, J. F.
547.7'0455 QD381

ISBN 0-85334-365-9

WITH 36 TABLES AND 75 ILLUSTRATIONS

Printed in Great Britain by Galliard (Printers) Ltd, Great Yarmouth

Preface

During the past decade, the *photochemistry of polymers* has become a field of central importance in polymer science and technology. Such applications as photopolymerisation, photocrosslinking, photostabilisation and solar energy devices have evolved from esoteric basic research specialities into industrial products. The use of photocurable coatings and printing inks is steadily gaining ground in many industries. Research on photoinitiated polymerisation has given rise to entirely new applications in microelectronics, e.g. in resists, barrier coatings, encapsulants, and printing wiring board technologies. Without the use of photoresists it would not be possible to develop modern electronic and computer industries.

In many of the above-mentioned subjects research has been carried out in the Department of Polymer Technology, The Royal Institute of Technology, Stockholm, Sweden, directed by Professor Bengt Rånby. On the occasion of his retirement in 1985 the International Symposium on 'New Trends in the Photochemistry of Polymers' was organised and held in Stockholm, 26–29 August 1985. We have the honour of editing the proceedings of this auspicious occasion which deals essentially with new developments in the field of polymer photochemistry. It was impossible to invite all the prominent specialists working in the photochemistry and photophysics of polymers and in the practical applications of these problems in industry. On this occasion friends and colleagues of Professor Rånby directly involved in cooperation with his department were invited to present their contribution. It is gratifying that, considering the short time schedule, most of those invited to contribute have completed their papers.

The papers presented in this book cover a very wide range of topics important in the photochemistry and photophysics of polymers. Some lecturers review specified areas in the field while others concentrate more on the specified interest of the speakers.

Professor Webber (USA) reviews intramolecular energy transfer processes in polymers with particular emphasis on excimer kinetics, while Professor Guillet (Canada) presents a critical authoritative overview of the antenna effect in macromolecules which provides an important approach for developing polymers to harvest solar energy. Some polymers are shown here to have an energy conversion efficiency as good as chloroplasts. Professor Morawetz (USA) discusses molecular dynamics as studied by fluorescence spectroscopy for both tagged and doped polymers as well as polymer blends.

Professor Schnabel (Federal Republic of Germany) presents experimental results on the initiation of free radical polymerisation by acylphosphine oxides and derivatives, whereas Professor Wu (China) presents results on the photochemical behaviour of some monomers which contain electron donor and electron acceptor pairs.

Professor Pappas (USA) reviews the importance of radical versus cationic photopolymerisation, whereas Professor Smets (Belgium) discusses the photochemical synthesis of block copolymers with particular emphasis on efficiencies, mechanism and advantages. Professor Bamford (UK) discusses the importance of transition metal derivatives in the modification of polymers by photochemical methods and their importance in biomedical applications.

Dr Wiles (Canada) provides an overview of polyolefin photo-oxidation processes and places particular importance on the key role of hydroperoxides in photo-initiation and their removal/destruction during photostabilisation with emphasis on hindered amine light stabilisers.

Dr Winslow (USA) and Professor Weir (Canada) present results on the photochemistry of poly(ethylene-co-carbon monoxide) and polystyrene, respectively. Dr Hansen (Denmark) describes in depth unwanted photochemical reactions during the preparation and use of polymeric coatings such as delamination, chalking, erosion and colour changes.

One of the editors, Dr Allen (UK), reviews our current understanding of the mode of action of orthohydroxyaromatic and hindered piperidine light stabilisers with highlights of some of his recent work, while Professor Scott (UK) reviews the importance of antioxidant processes in polymer photostabilisation with particular emphasis on the role of the redox chain breaking donor/acceptor mechanism for most classes of light stabilisers. Professor Vogl (USA) describes his latest work on the synthesis of novel polymerisable dibenzotriazole absorbers with particular emphasis on mechanism, efficiencies, properties and applications.

Professor Rabek (Sweden) presents a review on the applications of polymer and model systems as a means of harvesting solar energy. Professor Kagiya (Japan) presents results on the TiO_2 catalysed photolysis of water in the presence of water-soluble polymers and discusses ways of improving the efficiency of the process.

Lastly, Professor Miyama (Japan) reviews the use of photochemistry in preparing biomedical polymers with particular emphasis on photografting.

We hope that the contents of the lectures will be of value to all who attend the meeting as well as polymer scientists and technologists, chemists and biologists worldwide with an interest in this important developing field.

It is with great pleasure that we, the editors and publishers (Elsevier Applied Science Publishers Ltd), present this book on behalf of the organisers at the Royal Institute of Technology (Stockholm) to Professor Bengt Rånby in honour on his retirement and in recognition of his work. We wish him well in future years with health and happiness and hope that he will treasure this book with fond memories of us all.

NORMAN S. ALLEN
JAN F. RABEK

Contents*

* The names of invited speakers are printed in capitals. Only these speakers are included in the List of Contributors on pp. xiii–xv.

List of Contributors

N. S. Allen

Department of Chemistry, John Dalton Faculty of Technology, Manchester Polytechnic, Chester Street, Manchester M1 5GD, UK

C. H. Bamford

Bioengineering and Medical Physics Unit, University of Liverpool, PO Box 147, Liverpool L69 3BX, UK

J. E. Guillet

Department of Chemistry, University of Toronto, Toronto, Canada M5S 1A1

C. H. Hansen

Scandinavian Paint and Printing Ink Research Institute, Agern Allé 3, DK-2970 Hørsholm, Denmark

V. T. Kagiya

Department of Hydrocarbon Chemistry, Faculty of Engineering, Kyoto University, Sakyo-ku, Kyoto 606, Japan

H. Mark

Polytechnic Institute of New York, 333 Jay Street, Brooklyn, New York 11201, USA

H. Miyama

Department of Materials Science and Technology, The Technological University of Nagaoka, Nagaoka, Niigata, 949-54, Japan

H. MORAWETZ

Department of Chemistry, Polytechnic Institute of New York, 333 Jay Street, Brooklyn, New York 11201, USA

S. P. PAPPAS

Polymers and Coatings Department, North Dakota State University, Fargo, North Dakota 58105, USA

J. F. RABEK

Department of Polymer Technology, The Royal Institute of Technology, Teknikringen 50, S-100 44, Stockholm, Sweden

W. SCHNABEL

Hahn-Meitner-Institut für Kernforschung Berlin, Bereich Strahlenchemie, Postfach 39 01 28, D-1000 Berlin 39, Federal Republic of Germany

G. SCOTT

Department of Chemistry, University of Aston in Birmingham, Gosta Green, Birmingham B4 7ET, UK

G. J. SMETS

Laboratory of Macromolecular and Organic Chemistry, Katholieke Universiteit Leuven, Celestijnenlaan 200F, B-3030 Leuven (Heverlee), Belgium

O. VOGL

Polytechnic Institute of New York, 333 Jay Street, Brooklyn, New York 11201, USA

S. E. WEBBER

Department of Chemistry and Center for Polymer Research, University of Texas at Austin, Austin, Texas 78712, USA

N. A. WEIR

Chemistry Department, Lakehead University, Thunder Bay, Ontario, Canada PB7 5E1

D. M. WILES

Chemistry Division, National Research Council of Canada, Ottawa, Canada K1A 0R9

F. H. WINSLOW
AT&T Bell Laboratories, 600 Mountain Avenue, Murray Hill, New Jersey 07974, USA

S. K. WU
Institute of Photographic Chemistry, Academia Sinica, Beijing, China

SYMPOSIUM CHAIRMAN

Professor J. F. Rabek	*The Royal Institute of Technology, Stockholm, Sweden*

ORGANISING COMMITTEE CHAIRMAN

Professor J. F. Jansson	*The Royal Institute of Technology, Stockholm, Sweden*

ORGANISING COMMITTEE

Professor K. Abbås	*Bofors Plast AB, Tidaholm, Sweden*
Professor B. Åkermark	*The Royal Institute of Technology, Stockholm, Sweden*
Professor G. Ahlgren	*The Royal Institute of Technology, Stockholm, Sweden*
Professor A. C. Albertsson	*The Royal Institute of Technology, Stockholm, Sweden*
Professor P. Carstensen	*ASEA Kabel AB, Stockholm, Sweden*
Professor P. Flodin	*Chalmers University of Technology, Sweden*
Dr P. Forsgren	*National Swedish Board for Technical Development, Stockholm, Sweden*
Dr S. Göthe	*AB Wilh. Becker, Stockholm, Sweden*
Dr A. Hult	*The Royal Institute of Technology, Stockholm, Sweden*
Professor P. O. Kinell	*Umeå University, Umeå, Sweden*
Dr J. Lucki	*The Royal Institute of Technology, Stockholm, Sweden*
Professor B. Stenberg	*The Royal Institute of Technology, Stockholm, Sweden*
Professor B. Törnell	*Kemicentrum, Lund University, Lund, Sweden*

Professor Bengt Rånby: Contributions to Science and Technology of Polymers

HERMAN MARK
Polytechnic Institute of New York, New York, USA

In early Spring 1920—far north in Sweden near the Arctic Circle—a baby boy Bengt was born to the family on the Västaback farm; two years later the father took over the management of a cooperative dairy farm near by. The mother was a schoolteacher in the village.

1922—Sweden, the land of forests and lakes, of great natural resources of scientists and pioneers; 1922—Meeting of the German Scientists and Physicians in Leipzig; General Chairman Svante Arrhenius, major topics of discussion: Organic Chemistry, Willstaetter and von Euler; Colloid Chemistry, Svedberg; X-ray Physics, Manne Siegbahn.

Wood—the wealth of the land—was the prime target for intense activities, particularly how to isolate and purify its value-dominating component: *cellulose.* Klason, Hägglund, Erdtman and Adler and later Stockman and Gierer provided important fundamentals; tentative but very attractive results of M. Polanyi in Berlin and K. Freudenberg in Heidelberg permitted the suggestion of a long chain structure of cellulose. H. Staudinger, coming from rubber and polystyrene—not from cellulose—dispelled, not too easily, all doubts with the concept of *macromolecules.*

When this all started, Bengt was only two years old but the waves of this *revolutionary* aspect and its enormous *practical* consequences are even now producing innovations far beyond their original limits—and Bengt is still happily and successfully floating on them.

After a remarkably accelerated public and intermediate education Bengt entered Uppsala University—Sweden's scientific Mecca—and began, surprisingly but very interestingly, with the most basic disciplines: Astronomy; Mathematics and Theoretical Physics. B.Sc. in 1940 at the age 20. Then came Military Service and—in 1943—the first 'job', Assistant in Svedberg's Institute of the Uppsala University—hired by Sven Brohult,

Svedberg's right hand at the time. Nils Gralén, a rising star of these days and later for many years the leader of Sweden's modern Textile Research Institute in Göteborg, was in charge as Bengt's first supervisor. And the main topic was *cellulose*. All aspects on a broad front: on molecular and supermolecular level; configuration, conformation, hydrogen bonding, steric hindrance, thermodynamics and others. In 1952, PhD and successor to Gralén.

It was at that time that the Polytechnic Institute of Brooklyn had received from American Companies a number of fellowships to study new aspects of Polymer Science on a broad basis. Bengt Rånby got one of them—from Du Pont and in 1952 came to Brooklyn. Later, he and also others very kindly evaluated their stay at Poly as 'formative and of immense value'. Let me try today—with the hindsight of more than 30 years—to explain the impact of our 'way of work' on so different visitors from abroad: it was an unusual and delicate blend of discipline and freedom: hard and good work had to be done at any time. There were specialists available to insure that: Charlie Overberger, Bruno Zimm, Paul Doty, Turner Alfrey, Isidor Fankuchen and others, but beyond that: the sky was the limit. And what a sky! Debye and Flory at Cornell, Onsager and Fuoss at Yale and Tobolsky at Princeton. Short junkets in crowded cars to Wilmington, Bell Telephone, Allied Chemical, Cyanamide and even to Dow in Michigan. And on top of it, all summer long, the fabulous Gordon Research Conferences. There: exchange of experience, approaches and results in polymer research went on freely for a whole week in the invigorating air of New Hampshire: confidence was established and friendships were formed.

Rånby returned to Sweden in 1953 as a 'Docent of Physical Chemistry' at the University of Uppsala. But not for too long—somewhere on the lawns of Colby College in New Hampshire a bait had been prepared and in 1956 he was back in the USA. This time it was *cellulose* again. R and D Project Manager at the American Viscose Corporation in Markus Hook, PA, invited by Dr H. Cudd, Vice President, who created this new R and D Division. Rayon was in danger—nylon, the acrylics and polyesters were building up. A new basic understanding of the structure and texture of cellulosics was needed and its conversion into useful properties: action of modifiers, rayon tyre cord with improved 'fatigue', high wet modulus and 'polynosic' types. For Rånby these were just the two sides of one coin: fundamental knowledge made it *exist*—practical results let it *persist*.

But to make sure, during his years at American Viscose, Bengt studied also the new competing materials: *polyolefins* for fiber and film.

Within two years his personality, his working style and efficiency became known. There is always a need for good people in a large country like the USA, and in 1959 a prestigious dual position was offered to him by the farsighted Professor Edwin Jahn, Dean of Science: Research Professor of Pulp and Paper Technology at the State University of New York and Director of the Empire State Paper Research Institute at the College of Forestry in Syracuse, N.Y.

Apparently the background was still *wood* and *cellulose*; why not? After all he was a Swede and a leader in the field. But Rånby had been looking around carefully during these years: a wave of new *concepts*—chain folding—new *tools*—Ziegler–Natta catalysts—and new *procedures*—Michael Szwarc's 'living polymers' (right on his doorstep) had introduced new and fascinating approaches: it was not only cellulose any more, it was polymers; *all* polymers. At the same time Bengt went through the traditional transition of all of us: we develop from the *things* to the *people*; manipulate first with chemicals, cooperate then with chemists and, finally, have to deal with *executives*. He met no difficulties on any level.

One of his earlier supervisors, a keen observer of the American scene with inherent linguistic sense told me one day in 1960: Bengt is 'smart'.

Interesting and stimulating publications, well organized and conducted symposia, remarkable successes in the organization of teaching and research did not remain unnoticed in Sweden. The pressing need for a modern center to cultivate this new discipline condensed in a call of significance: to organize and to run the first and only *Department of Polymer Technology* in Stockholm and, in fact, in Scandinavia.

1961—home is the sailor, home from the sea and his ship is full of valuable cargo: patience, experience and optimism. The new responsibility is heavy but the ingredients for success are there: leadership, people, space and money.

Of prime importance was the decision which main path the new Department should take? Evidently there existed several options: concentrate on certain materials, or on certain methods of preparation, characterization and processing. After careful deliberation Rånby decided to study, on a broad front, the interaction of light with polymers. There was a truly wide range: new polymers can be prepared by photosynthesis and photopolymerization; existing systems may be modified by controlled degradation or crosslinking, they may be stabilized by reactive additives, by the removal of sensitive groups, ring closure or other similar reactions. Quantitative studies on free radical formation with UV and ESR spectroscopy had started already in 1963 with several associates such as Hiroshi Yoshida, Göran Mählhammar and Peter Carstensen and led to a series of publications; they were followed (since 1970) by the investigation of free radical polymerization with Koichi Takakura and Zenzi Izumi; of photosensitized reactions of various dienes, polystyrene, PVC and polyester together with Jan Rabek, Zenon Joffe, Julia Lucki and others. The amount of new information grew and its incorporation into the existing art had to be done on a larger scale. That means nothing else but writing or—at least editing—monographs and comprehensive volumes. Hence:

1972: Kinell, Rånby and Reio; *ESR Applications to Polymer Research.*
1975: Rånby and Rabek; *Photodegradation, Photooxidation and Photostabilization of Polymers.*
1976: Same authors; *Singlet Oxygen Reactions with Organic Compounds and Polymers* and
1977: Same authors; *ESR Spectroscopy in Polymer Research.*
1979: Same authors; *Long Term Properties of Polymers and Polymeric Materials*

The work in the laboratory continued along different lines—A-C Albertsson, G. Canbäck, S. Paul, T. Skrowronski, G. Arct, S. Göthe, A. Hult and others; the number of publications swelled to more than 300.

Still Rånby did not forget his first love, cellulose, and developed with R. Mehrotra, L. Gädda, C. Rodehed, J. Persson, D. Zuchowska and others new methods to modify starch and cellulose by graft copolymerization.

The time had come to delegate responsibility, to form working groups and teams in order to gain time for the only activity, which Rånby alone could perform: high level international representation in Western Europe, Japan, The Americas, China and all over the world. We all know and dislike the hectic days of such visits but for the Department of Polymer Technology at the Royal Institute in Stockholm they became important and, very fortunately, Bengt is an agile and indefatigable traveller: airport, plane, taxi, lecture hall, taxi, hotel, airport and so on. Within a few years the Department was a shining star in the Galaxy of Polymer Science Centers.

Very difficult to achieve and equally difficult to maintain!

Here we are: I told you what he did and how it was done. But, now that others will soon—January 1986—take over responsibility and representation; what is Rånby going to do?

In Brooklyn we have a saying: the first 30 years of your life you are doing what you are told to do, the next 30 years you are doing what you are paid for and—if there is anything left after that—then you can do what you really want. Bengt, the young man, has many such years left. What will he do? Keep himself busy—to be sure! Maintain contact with all friends all over the world, think about new polymers and new applications, new ideas and new dreams. And then the sky only is the limit beyond that. And—Bengt—what a sky!

Twenty Years of Polymer Photochemistry in the Department of Polymer Technology, The Royal Institute of Technology, Stockholm, Sweden

JAN F. RABEK

Research on polymer photochemistry in the Department of Polymer Technology, The Royal Institute of Technology, Stockholm, Sweden, was started by B. Rånby at the beginning of 1963. The first publication by H. Yoshida and B. Rånby appeared on the study of ESR spectra of free radicals formed and trapped in polypropylene[1,2] and polyethylene[2] after ultra-violet radiation. Further work with P. Carstensen was extended to the ESR study of free radicals formed during the photodegradation of polyisobutylene[3,4] and other elastomers (cis-1,4-polybutadiene, cis-1,4-polyisoprene and cis-1,4-polypiperylene).[5-10] The photodegradation of polyethylene was then studied by G. Mälhammar in a dissertation for Tekn.lic. degree. To make polymeric materials more acceptable environmentally, paraffins and carbonyl compounds were added to polyolefins and PVC, which made them more susceptible to photo-oxidation.

The photochemical programme was then extended to sensitised polymer reactions in cooperation with J. F. Rabek who joined the department in 1971. The mechanisms of polymer photodegradation, photo-oxidation and photostabilisation were studied in a programme, directed by B. Rånby and J. F. Rabek, and carried out by J. Lucki, Z. Joffe, G. Canbäck, D. Lala and visiting scientists from Poland (J. Arct, T. A. Skowronski) and from China (S. K. Wu, Y. Y. Yang, S. Z. Jian, C. S. Dai, R. Liu and others).

Studies on the degradation of commercially available polymers have concentrated on polyolefins,[11,12] polystyrene,[16,17,31,38,41-43,49,53] poly(vinyl chloride),[18,25,26,30,36,48,50,52,54,59,74,80,81,99,100] polybutadiene,[8,10,27,39,40,44-47,55,56,58-61,76] polynorbornene,[65,66] unsaturated polyesters,[62,68,77,89] copolymers of butadiene/acrylonitrile[73] and ethylene/vinyl acetate[75] and polymer blends.[59,75]

Photosensitised degradation and/or crosslinking studies on polymers have also formed part of these researches.[14,15,18,24,28,37,51,72] The reactions of several sensitisers such as benzophenone,[13] 4-chlorobenzophenone,[81] hexachlorobenzene,[81] quinones,[17,20,81] tetrahydrofuran,[25,54] anthracene,[31,58] N-methyl-2-benzoyl-beta-naphthiazoline,[35,40] 1,3-diphenylisobenzofuran,[56] various dyes[27,39,45] and metal salts[11,50,52] have also been examined and their modes of action established.

Special attention has been given to the effect of commercial additives such as thermostabilisers,[30,74] pigments[80] and lubricants[99] on the photo-oxidative degradation of poly(vinyl chloride).

A great part of the research programme has been devoted to singlet oxygen oxidation of unsaturated polymers such as polybutadiene,[19,21–23,27,29,32–34,45–47,58] polynorbornene[66] and polyene structures in partially dehydrochlorinated poly(vinyl chloride).[48]

Stemming from this research, studies of the stabilisation against photochemical[60,63,64,69,71,76–78,82,83,89–91,98] and singlet oxygen[44,45,60,70,71,76,84] oxidative degradation have been further developed. Some work focused on the antagonistic/synergistic effects between photostabilisers, antioxidants and other additives.[82–84,91,98]

In addition to the references quoted, parts of the results from these researches were presented at three international meetings organised by the department:

1. 22nd Nobel Symposium on *ESR Applications to Polymer Research*, Stockholm, 1972.[10–12]
2. IUPAC Symposium on *Long Term Properties of Polymers and Polymeric Materials*, Stockholm, 1976.[49–52]
3. EUCHEM Conference on *Singlet Oxygen Reactions with Organic Compounds and Polymers*, Stockholm, 1976.[32,33]

Starting in 1977, B. Rånby has initiated and developed a new research programme on the photopolymerisation, photocuring and photocrosslinking of polymers[85–88,92–97] in cooperation with S. Göthe, A. Hult, G. Broodh, J. Hilborn, K. Allmér and visiting scientists from China, Y. L. Chen, P. Y. Zhang, Z. M. Gao and B. J. Qu.

In order to present the most recent achievements in coatings technology an *International EUCHEM Conference on Photopolymerization and Photocrosslinking of Organic Coatings* was organised in Stockholm in 1980.

Studies of the photoinitiated crosslinking and surface modifications of polyesters, polyolefins and rubber polymers, are currently in progress.

From the beginning of this programme the department has had broad international cooperation with the USA in the field of polymer photochemistry:

1. US–Swedish Workshop on *Photodegradation and Photostabilization of Polymers* was held in Stockholm in 1981[67] and
2. US–Swedish Workshop on *Photochemistry of Polymers* was held in Pasadena, California in 1984 (both workshops were sponsored by National Science Foundation (NSF), USA, and National Swedish Board for Technical Development (STU), Sweden),

with Japan (Professor V. T. Kagiya, Department of Hydrocarbon Chemistry, University of Kyoto),[35,40] with China (Professor S. K. Wu, Institute of Photographic Chemistry, Academia Sinica, Beijing),[57,65,66,69,70,77,83,84] with Poland (Institute of Organic and Polymer Technology, Technical University of Wroclaw)[54,63,64,73–75,79–81,99,100] and lately with Japan, Czechoslovakia and Hungary. More than 20 scientists from these countries have worked from a few months up to two years in our department. Particularly important is the continuous and very successful cooperation with Professor S. K. Wu at the Institute of Photographic Chemistry, Academia Sinica, Beijing, China, in the field of polymer degradation,[65,66] photostabilisation,[69,70,83,84] study of excimers[57] and lately of photocuring.

B. Rånby and J. F. Rabek have on numerous occasions been visiting professors in the USA, Japan, China and many European countries where they have done research and presented lectures and courses on the photodegradation, photostabilisation and photocrosslinking of polymers and related problems.

The major objective of the department is the education of professional chemists and chemical engineers to prepare them for industry and the academic world. Research-oriented programmes of study leading to the degrees of Master and 'Licentiat' of Technology and Doctor of Technology are offered. Several doctors' degrees are conferred every year.

Several contributions in books[101,107] and as monographs[104–106] in the field of photochemistry of polymers,[101,102,105] ESR spectroscopy,[103] singlet oxygen[104,107] and experimental methods[106] have been published by B. Rånby and/or J. F. Rabek.

Modern research in photochemistry requires major instrumentation. The department is very well equipped with many different light sources, weatherometers, photoreactors, special apparatuses for the measuring of oxygen uptake, microwave generators for singlet oxygen, high pressure optical cell (up to 10 000 bars), IR and UV–VIS and emission spectroscopic instruments (with computer station on line), ESR and ESCA computerised spectrometers, electron microscopes and a complete set of equipment for measuring mechanical and other physical properties of polymers. An environmental photoreactor, probably the biggest in Europe, to study the photodegradation of polymers in an atmosphere containing ozone and other environmental pollutants (e.g. automobile exhaust) is currently under construction.

Research on the following photochemical projects is currently in progress:

photocuring of paints and lacquers (B. Rånby),
photocrosslinking of unsaturated polyesters without styrene (B. Rånby),
photocrosslinking of polyolefins (B. Rånby),
photovulcanisation of rubber (B. Rånby),
photomodification of polymer surfaces (B. Rånby),
photochemistry as a basis for photoresist technology (B. Rånby),
reversible photo-initiated networks (A. Hult),
photostabilisation of polymers and polymer blends (J. F. Rabek),
environmental accelerated degradation of polymers (J. F. Rabek),
singlet oxygen oxidation of polymers (J. F. Rabek),
high-pressure spectroscopy for the study of oxygen–polymer complexes (B. Rånby and J. F. Rabek),
excimer photophysics (J. F. Rabek).

In addition to these photochemical researches, A. C. Albertsson from our department in cooperation with Professor O. Vogl from the Polytechnic Institute of New York, USA, is involved in the synthesis of polymerisable and polymer bound photostabilisers.

A new building for the department will be completed in 1985 which will provide improved facilities for experimental research and education programmes, e.g. special laboratories for organic coatings, rubber technology, polymer photochemistry and photophysics and advanced spectroscopy for energy transfer and nanosecond events.

Most of the research projects in our department in the field of photochemistry of polymers have been financially supported by the National Swedish Board for Technical Development (STU) and lately by the Swedish Natural Science Research Council (NFR) and several public and private foundations, which is gratefully acknowledged.

In honour of Professor B. Rånby and in recognition of his achievements in the field of polymer photochemistry and on the occasion of his retirement in 1985 an *International Symposium on New Trends in the*

Photochemistry of Polymers was organised on August 26–29, 1985, in Stockholm, Sweden. The lectures given are published in this Symposium Volume.

List of publications from the photochemistry field published in the Department of Polymer Technology, The Royal Institute of Technology, Stockholm, Sweden

1. Yoshida, H. and Rånby, B., ESR spectra of methyl radicals trapped in polypropylene after ultraviolet irradiation. *J. Polym. Sci.*, **B2**, 1155 (1964).
2. Rånby, B. and Yoshida, H., ESR studies of polyethylene and polypropylene irradiated by UV light. *J. Polym. Sci.*, **C12**, 263 (1966).
3. Carstensen, P. and Rånby, B., Free radicals in polyisobutylene induced by ultraviolet radiation. In: *Radiation Research*, North-Holland, Amsterdam, 1967, p. 297.
4. Rånby, B. and Carstensen, P., Free radicals in polyolefins initiated with ultraviolet and ionizing radiation. *Advances in Chemistry Series*, no. 66, 256 (1967).
5. Carstensen, P., Et ESR-Studium af Frie Radikaler i UV bestrålet Poly(isobutylene) (in Danish). *Dansk Kemi*, **49**, 97 (1968).
6. Carstensen, P., Electron spin resonance studies of free radicals formed in elastomers by ultraviolet radiation. *Acta Polytech. Scand.*, No. 97,3 (1970).
7. Carstensen, P., Free radicals in diene polymers induced by ultraviolet radiation. I. An ESR study of cis-1,4-poly(isoprene). *Makromol. Chem.*, **135**, 219 (1970).
8. Carstensen, P., Free radicals in diene polymers induced by ultraviolet irradiation. II. An ESR study of cis-1,4-poly(butadiene). *Makromol. Chem.*, **142**, 131 (1971).
9. Carstensen, P., Free radicals in diene polymers induced by ultraviolet irradiation. III. An ESR study of cis-1,4-poly(piperylene). *Makromol. Chem.*, **141**, 145 (1971).
10. Carstensen, P., Degradation of 1,4-polydienes induced by ultraviolet irradiation. In: *ESR Applications to Polymer Research* (Ed. P. O. Kinell, B. Rånby and V. Runnström-Reio), Almqvist-Wiksell, Stockholm, 1972, p. 159.
11. Joffe, Z. and Rånby, B., ESR studies of UV-induced degradation of polyolefins. In: *ESR Applications to Polymer Research* (Ed. P. O. Kinell, B. Rånby and V. Runnström-Reio), Almqvist-Wiksell, Stockholm, 1972, p. 171.
12. Rabek, J. F. and Rånby, B., Photochemical oxidation reactions of synthetic polymers. In: *ESR Applications to Polymer Research* (Ed. P. O. Kinell, B. Rånby and V. Runnström-Reio), Almqvist-Wiksell, Stockholm, 1972, p. 201.
13. Rabek, J. F., The sensitization effect of benzophenone and Michler's ketone in photochemistry of polymers. *Materials of 3rd Techn. Conference on Photopolymers*, Soc. Plast. Eng. Mid. Hudson Sect., Ellenville, 1973, p. 27.
14. Rånby, B., Rabek, J. F. and Joffe, Z., Photochemically induced reactions. *Materials of Conference on the Degradability of Polymers and Plastics*, London, 1973, p. 3.1.

15. Rånby, B., Fotokemisk Nedbrytning och Oxidation av Alifatiska och Aromatiska Polymerer (in Swedish). *Kemia-kemi*, **1–8**, 477 (1974).
16. Rabek, J. F. and Rånby, B., Studies of the photooxidation mechanism of polymers. I. Singlet oxygen mechanism for the photooxidation of polystyrene. *J. Polym. Sci.*, A1, **12**, 273 (1974).
17. Rabek, J. F. and Rånby, B., Studies of the photooxidation mechanism of polymers. II. The role of quinones as sensitizers in the photooxidation degradation of polystyrene. *J. Polym. Sci.*, A1, **12**, 295 (1974).
18. Rånby, B., Rabek, J. F., Shur, Y. J., Joffe, Z. and Wikström, K., Photo-oxidative degradation of poly(vinyl chloride) in the presence of various kinds of additives. *Proc. Intern. Symp. Degradation and Stabilization of Polymers*, Brussels, Sept. 11–13, 1974, p. 35.
19. Rabek, J. F. and Rånby, B., Role of singlet oxygen in photo-oxidative degradation and photostabilization of polymers. *Proc. Intern. Symp. Degradation and Stabilization of Polymers*, Brussels, Sept. 11–13, 1974, p. 257.
20. Yoshida, H., Kambara, Y. and Rånby, B., ESR spectra of radicals during photolysis of chloranil in solution. *Bull. Chem. Soc. Japan*, **47**, 2599 (1974).
21. Rabek, J. F. and Rånby, B., Oxidation of polymers by singlet oxygen. *Proc. XXIVth Intern. Symp., Macromolecules*, IUPAC, Jerusalem, July 13–18, 1975, p. 255.
22. Rabek, J. F. and Rånby, B., Role of singlet oxygen in photo-oxidative degradation and photostabilization of polymers. *Polym. Engng Sci.*, **15**, 40 (1975).
23. Rånby, B. and Rabek, J. F., Some problems of the photo-oxidative degradation of polymers by singlet oxygen. *Proc. Intern. Symp. Ultraviolet Light Induced Reactions in Polymers*, ACS, Philadelphia, Penn., April 1975, p. 148.
24. Rabek, J. F., Primary reactions in the sensitized photodegradation of polymers. *Proc. Intern. Symp. Ultraviolet Light Induced Reactions in Polymers*, ACS, Philadelphia, Penn., April 6–11, 1975, p. 190.
25. Rabek, J. F., Shur, Y. J. and Rånby, B., Studies of the photooxidative mechanism of polymers. III. Role of tetrahydrofuran in the photooxidative degradation of poly(vinyl chloride). *J. Polym. Sci.*, A1, **13**, 1285 (1975).
26. Rabek, J. F., Canbäck, G., Lucki, J. and Rånby, B., Studies on the photooxidation mechanism of polymers. IV. Effect of ultraviolet light (2537A) on solid PVC particles in different liquids. *J. Polym. Sci.*, A1, **14**, 1447 (1976).
27. Rabek, J. F. and Rånby, B., Studies on the photooxidation mechanism of polymers. V. Oxidation of polybutadiene by singlet oxygen from microwave discharge and dye-photosensitized reactions. *J. Polym. Sci.*, A1, **14**, 1463 (1976).
28. Rabek, J. F., Photosensitized degradation of polymers. In: *Ultraviolet Light Induced Reactions in Polymers* (Ed. S. S. Labana), ACS Symp. Series, No. 25, Am. Chem. Soc., Washington, D.C., 1976, p. 255.
29. Rånby, B. and Rabek, J. F., Photooxidative degradation of polymers by singlet oxygen. In: *Ultraviolet Light Induced Reactions in Polymers* (Ed. S. S. Labana), ACS Symp. Series, No. 25, Am. Chem. Soc., Washington, D.C., 1976, p. 391.

30. Rabek, J. F., Canbäck, G. and Rånby, B., Studies of the photooxidation mechanism of polymers. IV. The role of commercial thermostabilizers in the photostability of poly(vinyl chloride). *J. appl. Polym. Sci.*, **21**, 2211 (1977).
31. Rämme, G. and Rabek, J. F., A flash photolysis study of the triplet state quenching of anthracene by oxygen in the presence of polystyrene. *Europ. Polym. J.*, **13**, 855 (1977).
32. Rånby, B. and Rabek, J. F., Singlet oxygen reactions with synthetic polymers. In: *Singlet Oxygen* (Eds B. Rånby and J. F. Rabek), Wiley, Chichester, 1978, p. 211.
33. Rabek, J. F., Shur, Y. J. and Rånby, B., Photosensitized singlet oxygen oxidation of polydienes. In: *Singlet Oxygen* (Eds B. Rånby and J. F. Rabek), Wiley, Chichester, 1978, p. 264.
34. Rabek, J. F. and Rånby, B., The role of singlet oxygen in the photooxidation of polymers. *Photochem. Photobiol.*, **28**, 557 (1978).
35. Kagiya, V. T., Takemoto, K., Shi, H. and Rabek, J. F., Sensitized photo-oxidation of polyethylene film by addition of N-methyl-2-benzoyl-beta-naphthiazoline. *J. Polym. Sci.*, **B16**, 619 (1978).
36. Rånby, B., Rabek, J. F. and Canbäck, G., Investigation of PVC degradation mechanism by ESR spectroscopy. *J. Macromol. Sci. Chem.*, **A12**, 587 (1978).
37. Rånby, B., Photodegradation and photo-oxidation of polymer surfaces. In: *Polymer Surfaces* (Eds E. T. Clark and W. J. West), Wiley, Chichester, 1978, p. 381.
38. Rånby, B. and Lucki, J., Photo-oxidation of polystyrene studied by model compounds. Radiation research. *Proc. 6th Intern. Congress Radiation Research*, Japan Assoc. Radiat. Res. Tokyo, 1979, p. 374.
39. Rabek, J. F. and Rånby, B., Studies on the photo-oxidative mechanism of polymers. VIII. The role of singlet oxygen in the dye-photosensitized oxidation of cis-1,4- and 1,2-polybutadiene and butadiene–styrene copolymers. *J. appl. Polym. Sci.*, **23**, 2481 (1979).
40. Rabek, J. F., Rämme, G., Canbäck, G., Rånby, B. and Kagiya, V. T., Studies on the photo-oxidation mechanism of polymers. VIII. The photo-sensitized oxidation mechanism of cis-1,4-polybutadiene by N-methyl-2-benzoyl-beta-naphthiazoline. *Europ. Polym. J.*, **15**, 339 (1979).
41. Lucki, J. and Rånby, B., Photo-oxidation of polystyrene. I. Formation of hydroperoxide groups in photo-oxidised polystyrene and 2-phenyl butane. *Polym. Degrad. Stab.*, **1**, 1 (1979).
42. Lucki, J. and Rånby, B., Photo-oxidation of polystyrene. II. Formation of carbonyl groups in photo-oxidized polystyrene. *Polym. Degrad. Stab.*, **1**, 159 (1979).
43. Lucki, J. and Rånby, B., Photo-oxidation of polystyrene. III. Photo-oxidation of 2-phenyl butane as model compound for polystyrene. *Polym. Degrad. Stab.*, **1**, 251 (1979).
44. Lucki, J., Rabek, J. F. and Rånby, B., Stabilizing effect of 4-(1-imidazolyl)-phenol against the oxidation of polybutadiene by molecular and singlet oxygen. *Polym. Bull.*, **1**, 563 (1979).
45. Rabek, J. F. and Rånby, B., Dye-sensitized photooxidation of 1,4-polydienes. *Photochem. Photobiol.*, **30**, 133 (1979).

46. Rabek, J. F., Lucki, J. and Rånby, B., Comparative studies of reactions of commercial polymers with molecular oxygen, singlet oxygen and ozone. I. Reactions with cis-1,4-polybutadiene. *Europ. Polym. J.*, **15**, 1089 (1979).
47. Lucki, J., Rånby, B. and Rabek, J. F., Comparative studies of reactions of commercial polymers with molecular oxygen, singlet oxygen and ozone. II. Reactions with 1,2-polybutadiene. *Europ. Polym. J.*, **15**, 1101 (1979).
48. Rabek, J. F., Rånby, B., Östensson, B. and Flodin, P., Oxidation of polyene structures in poly(vinyl chloride) by molecular oxygen and singlet oxygen. *J. appl. Polym. Sci.*, **24**, 2407 (1979).
49. Lucki, J., Rabek, J. F. and Rånby, B., Study of photo-oxidation of polystyrene using model compounds. *J. appl. Polym. Sci., Polym. Symp.*, No. 35, 275 (1979).
50. Rabek, J. F., Canbäck, G. and Rånby, B., Some problems in the mechanism of photosensitized degradation of poly(vinyl chloride). *J. appl. Polym. Sci., appl. Polym. Symp.*, No. 35,299 (1979).
51. Rånby, B. and Rabek, J. F., Accelerated reactions in photodegradation of polymers. *J. appl. Polym. Sci., appl. Polym. Symp.*, No. 35,243 (1979).
52. Joffe, Z. and Rånby, B., Some weathering and UV ageing problems of poly(vinyl chloride) contaminated with Fe-complexes and organic additives. *J. appl. Polym. Sci., appl. Polym. Symp.*, No. 35,307 (1979).
53. Rånby, B. and Lucki, J., New aspects of photodegradation and photo-oxidation of polystyrene. *Pure appl. Chem.*, **52**, 295 (1980).
54. Rabek, J. F., Skowronski, T. A. and Rånby, B., Studies on photodegradation of poly(vinyl chloride) accelerated by alpha-hydroperoxy-tetrahydrofuran. *Polymer*, **21**, 226 (1980).
55. Rabek, J. F. and Lala, D., Carotenoids as effective stabilizers against photooxidation of cis-1,4-polybutadiene. *J. Polym. Sci.*, **B18**, 427 (1980).
56. Lala, D., Rabek, J. F. and Rånby, B., The effect of 1,3-diphenylisobenzofuran on the photo-oxidative degradation of cis-1,4-polybutadiene. *Europ. Polym. J.*, **16**, 735 (1980).
57. Wu, S. K., Jiang, Y. C. and Rabek, J. F., The temperature effect on the formation of excimer in siloxanes and polystyrene solutions. *Polym. Bull.*, **3**, 319 (1980).
58. Rabek, J. F. and Rånby, B., Polynuclear aromatic hydrocarbons as sensitizers of photooxidative degradation of cis-1,4-polybutadiene. *Rev. Roumanie Chim.*, **25**, 1045 (1980).
59. Lala, D., Rabek, J. F. and Rånby, B., Photodegradation and photo-stabilization of the two-phase system poly(vinyl chloride)–polybutadiene. *Polym. Degrad. Stab.*, **3**, 307 (1980–81).
60. Lala, D. and Rabek, J. F., Photostabilizing effect of Ni compounds in photo-oxidative degradation of cis-1,4-polybutadiene. *Polym. Degrad. Stab.*, **3**, 383 (1980–81).
61. Lala, D. and Rabek, J. F., The role of hydroperoxides in photo-oxidative degradation of cis-1,4-polybutadiene. *Europ. Polym. J.*, **17**, 7 (1981).
62. Lucki, J., Rabek, J. F., Rånby, B. and Ekström, C., Photolysis of polyesters, poly(propylene-1,2-maleate) and poly(propylene-1,2-ortho-phthalate). *Europ. Polym. J.*, **17**, 933 (1981).
63. Arct, J., Dul, M., Rabek, J. F. and Rånby, B., Studies on modified

benzotriazoles as photostabilizers for poly(vinyl chloride). *Europ. Polym. J.*, **17**, 1041 (1981).

64. Rabek, J. F., Rånby, B., Arct, J. and Golubski, Z., Metal salts of 2-((1-hydroxy-2-naphthalenyl)carbonyl)-benzoic acid as a new class of photostabilizers. *Europ. Polym. J.*, **18**, 87 (1982).
65. Wu, S. K., Lucki, J., Rabek, J. F. and Rånby, B., Photo-oxidative degradation of polynorbornene. *Polym. Photochem.*, **2**, 73 (1982).
66. Wu, S. K., Lucki, J., Rabek, J. F. and Rånby, B., Singlet oxygen oxidation of polynorbornene. *Polym. Photochem.*, **2**, 125 (1982).
67. Rånby, B. and Vogl, O., U.S.–Swedish workshop on photodegradation and photostabilization of polymers. *Polym. News*, **8**, 188 (1982).
68. Rånby, B., Rabek, J. F. and Lucki, J., Surface oxidation reactions of unsaturated polymers. In: *Physicochemical Aspects of Polymer Surfaces* (Ed. K. L. Mittal), Plenum Press, New York, 1983, Vol. 1, p. 283.
69. Yang, Y. Y., Lucki, J., Rabek, J. F. and Rånby, B., Photostabilizing effect of bis(hindered piperidine) compounds: I. In photo-oxidative degradation of cis-1,4-polybutadiene. *Polym. Photochem.*, **3**, 47 (1983).
70. Yang, Y. Y., Lucki, J., Rabek, J. F. and Rånby, B., Photostabilizing effect of bis(hindered piperidine) compounds: II. In singlet oxygen degradation of cis-1,4-polybutadiene. *Polym. Photochem.*, **3**, 97 (1983).
71. Rabek, J. F., Rånby, B. and Arct, J., Stabilization of cis-1,4-polybutadiene against photo-oxidation and singlet oxygen oxidation by hindered phenols. *Polym. Degrad. Stab.*, **5**, 65 (1983).
72. Rånby, B. and Rabek, J. F., Environmental corrosion of polymers. In: *The Effects of Hostile Environment on Coatings and Plastics* (Ed. D. P. Garner and G. A. Stahl), ACS Symp. Ser. No. 299, Am. Chem. Soc., Washington, D.C., 1983, p. 291.
73. Skowronski, T. A., Rabek, J. F. and Rånby, B., Photo-oxidation of copolymers of butadiene and acrylonitrile. *Polym. Degrad. Stab.*, **5**, 173 (1983).
74. Skowronski, T. A., Rabek, J. F. and Rånby, B., Effect of thermal stabilizers on the photo-oxidative degradation of solid poly(vinyl chloride). *Polymer*, **24**, 1189 (1983).
75. Skowronski, T. A., Rabek, J. F. and Rånby, B., Photo-oxidation of ethylene–vinyl acetate copolymers. *Polym. Photochem.*, **3**, 341 (1983).
76. Rabek, J. F., Recent development in photodegradation and photostabilization of polydienes. In: *Polymer Additives* (Ed. J. E. Kresta), Plenum Press, New York, 1984, p. 1.
77. Jian, S. Z., Lucki, J., Rabek, J. F. and Rånby, B., Photooxidation and photostabilization of unsaturated crosslinked polymers. *Polymer Preprints*, **25**, 56 (1984).
78. Lucki, J., Rabek, J. F. and Rånby, B., ESR study of interaction between hindered piperidine stabilizers and antioxidants or photostabilizers. *Polymer Preprints*, **25**, 38 (1984).
79. Skowronski, T. A., Rabek, J. F. and Rånby, B., Photodegradation of some poly(vinyl chloride) (PVC) blends: PVC/ethylene–vinyl acetate (EVA) copolymers and PVC/butadiene–acrylonitrile (NBR) copolymers. *Polym. Engng Sci.*, **24**, 278 (1984).

80. Skowronski, T. A., Rabek, J. F. and Rånby, B., The role of commercial pigments on the photodegradation of poly(vinyl chloride). *Polym. Degrad. Stab.*, **8**, 37 (1984).
81. Skowronski, T. A., Rabek, J. F. and Rånby, B., Effects of additives on photodegradation of polymers: photosensitized crosslinking of poly(vinyl chloride). *Polym. Photochem.*, **5**, 77 (1984).
82. Lucki, J., Rabek, J. F. and Rånby, B., Photostabilizing effect of hindered piperidine compounds: interaction between hindered phenols and hindered piperidines. *Polym. Photochem.*, **5**, 351 (1984).
83. Lucki, J., Rabek, J. F., Rånby, B. and Dai, G. S., ESR study of interaction between hindered piperidine stabilizers and low molecular weight ketones, hydroperoxides and peroxides. *Polym. Photochem.*, **5**, 385 (1984).
84. Rabek, J. F., Rånby, B., Arct, J. and Liu, R., Singlet oxygen and free radical oxidation of polydienes and related problems with stabilization: synergistic and antagonistic effects. *J. Photochem.*, **25**, 519 (1984).
85. Hult, A. and Rånby, B., Photostability of photocured organic coatings. I. Secondary reactions in thioxanthane/amine photocured organic coatings. *Polym. Degrad. Stab.*, **8**, 75 (1984).
86. Hult, A. and Rånby, B., Photostability of photocured organic coatings. II. Yellowing and photooxidation of thioxanthone/amine photocured organic coatings. *Polym. Degrad. Stab.*, **8**, 89 (1984).
87. Hult, A. and Rånby, B., Photostability of photocured organic coatings. III. Photodegradation of organic coatings, photocured with alkoxy acetophenone and hydroxy alkylacetophenone photoinitiators. *Polym. Degrad. Stab.*, **8**, 241 (1984).
88. Hult, A. and Rånby, B., Photostability of photocured organic coatings. IV. Interaction between photoinitiators and photostabilizers in photocuring reactions. *Polym. Degrad. Stab.*, **9**, 1 (1984).
89. Song, Z., Rånby, B., Gupta, A., Borsig, E. and Vogl, O., ESCA spectroscopy of polyesters stabilized with polymer bound ultraviolet stabilizers. *Polym. Bull.*, **12**, 245 (1984).
90. Lucki, J., Photostabilization of cis-1,4-polybutadiene with phenylene sulphide oligomers. *Polym. Degrad. Stab.*, **11**, 75 (1985).
91. Lucki, J., The interaction of UV-absorbers with hindered piperidines in photostabilization of cis-1,4-polybutadiene. *Polym. Photochem.* (in press).
92. Göthe, S. and Rånby, B., Photocuring of organic coatings. I. Absorption spectra and photo-reduction of thioxanthone/amine systems. *J. appl. Polym. Sci.* (in press).
93. Göthe, S. and Rånby, B., Photocuring of organic coatings. II. Primary radical formation in thioxanthone/amine systems. *J. appl. Polym. Sci.* (in press).
94. Göthe, S. and Rånby, B., Photocuring of organic coatings. III. Polymerization of methyl methacrylate photoinitiated by thioxanthone/amine systems. *J. appl. Polym. Sci.* (in press).
95. Göthe, S. and Rånby, B., Photocuring of organic coatings. IV. Application using thioxanthone/amine systems as initiator. *J. appl. Polym. Sci.* (in press).
96. Rånby, B. and Göthe, S., Free radical initiation of photopolymerization and photocrosslinking. IUPAC Symposium, Bucharest, Sept. 1983.

97. Rånby, B. and Hult, A., Photodegradation and photostabilization of organic surface coatings. *Intern. Conf. Organic Coatings, Sci. Technol.*, Athens, July, 1983 (in press).
98. Lucki, J., Jian, S. Z., Rabek, J. F. and Rånby, B., Antagonistic effects of hindered piperidine and organic sulphides in photostabilization of cis-1,4-polybutadiene. *Polym. Photochem.* (in press).
99. Skowronski, T. A., Rabek, J. F. and Rånby, B., Effect of lubricants on the photooxidative degradation of solid poly(vinyl chloride). *Polym. Degrad. Stab.* (in press).
100. Rabek, J. F., Rånby, B. and Skowronski, T. A., Photo-thermal dehydrochlorination of poly(vinyl chloride). *Macromolecules* (in press).

BOOKS

101. Rabek, J. F., Oxidative degradation of polymers, In: *Degradation of Polymers, Comprehensive Chemical Kinetics*, Chapter 4 (Ed. C. H. Bamford and C. F. Tipper), Vol. 14, Elsevier, Amsterdam, 1975, pp. 425–538.
102. Rånby, B. and Rabek, J. F., *Photodegradation, Photooxidation and Photostabilization of Polymers*, Wiley, London, 1975, pp. 1–573. Translated to Russian by MIR, Moskva, 1978.
103. Rånby, B. and Rabek, J. F., *ESR Spectroscopy in Polymer Research*, Springer Verlag, Berlin, 1977, pp. 1–410.
104. Rånby, B. and Rabek, J. F. (Eds) *Singlet Oxygen*, Wiley, Chichester, 1978, pp. 1–331.
105. Rånby, B. and Rabek, J. F. (Eds) *Long-term Properties of Polymers and Polymeric Materials*, Applied Polymer Symposium No. 35, Wiley, New York, 1979.
106. Rabek, J. F., *Experimental Methods in Photochemistry and Photophysics*, Wiley, Chichester, 1982, pp. 1–1098. Translated into Russian by MIR, Moskva, 1985.
107. Rabek, J. F., Singlet oxygen oxidation of polymers and their stabilization. In *Singlet Oxygen* (Ed. A. Frimer), Chapter 1, Vol. 4, CRC Press, USA, 1985, in press.

1

Intramolecular Energy Transfer in Polymers

S. E. WEBBER

Department of Chemistry and Center for Polymer Research, University of Texas at Austin, USA

1. INTRODUCTION

Much of the early interest in the photochemistry of polymers was directed toward the prevention of photodegradation. Photodegradation of polymers clearly remains of paramount importance as polymeric material is used more and more frequently in severe weathering environments (e.g. solar collectors). In more recent years polymer photochemistry has found applications in the electronics industry via photoresists, in which case a photoreactive polymer is desirable.[1]

In many polymers (e.g. polystyrene) the absorbing chromophore is present at each repeating unit, or in certain copolymers there exists a significant mole fraction of the absorbing unit on the chain. If certain criteria are met (which will be discussed in the next section) the energy deposited in the initially excited chromophore may be transferred to a neighboring or nearby chromophore such that the excitation is delocalized over all or some fraction of the polymer chain. Because there is no driving force for the energy to migrate in any particular direction it is generally best to think of this process as a 'random walk' of the excitation on a disordered, nearly one-dimensional lattice. There may exist polymer-bound chromophores that act as energy traps, or the majority chromophores may self-trap. Thus the fraction of the polymer coil over which the excitation migrates depends on: (1) the average 'hopping' time (τ_h) between chromophores relative to the lifetime of the excited state (τ_E) since the approximate number of transfer 'steps' is τ_E/τ_h; (2) the density of trapping groups and/or the rate of self-trapping. In general the exact relationship between these characteristics and the average energy migration length can be expected to be quite complex and is the subject of active theoretical investigation.

In this chapter we are going to consider energy transfer between polymer-bound chromophores for synthetic polymers only. We are not going to discuss energy transfer between chromophores on biopolymers,[2] nor the use of polymer films or solids to disperse small molecule chromophores that are involved in energy transfer. Thus the processes we are going to review are 'intrinsic' to synthetic polymers and have relevance to photodegradation and/or photo-active polymers and 'photon harvesting' by polymers with ultimate deposition of the energy into an energy trap which can be a reactive site.

2. GENERAL PRINCIPLES

2.1. Mechanisms of Energy Transfer

The process of energy transfer can be represented by the 'chemical reaction'

$$A^* + B \rightarrow A + B^* \tag{1}$$

In eqn (1) species A and B may be chemically and energetically equivalent but correspond to different locations in a crystal lattice, or as is germane to this review, sites on a polymer chain. The process represented in eqn (1) is equivalent to a simultaneous downward transition from the $A^* \rightarrow A$ state and an upward transition $B \rightarrow B^*$. Thus it is not surprising that the overall probability of this energy transfer step is proportional to the *spectral overlap integral*

$$S_n = \int \varepsilon_B(\bar{\nu}) I_{A^*}(\bar{\nu})(1/\bar{\nu})^n \, d\bar{\nu} \tag{2}$$

where $\varepsilon_B(\bar{\nu})$ is the molar absorption coefficient (in $cm^{-1}\,M^{-1}$) of B and $I_{A^*}(\bar{\nu})$ is the emission spectrum of A* normalized to unit area. The factor of $(1/\bar{\nu})^n$ takes care of dielectric effects within the medium and depends on the exact coupling mechanism. The S_n integral is extremely important in the molecular design of polymers for energy transfer because the choice of chromophores can be based on the magnitude of this quantity. In the case of energy transfer between equivalent molecules the S_n integral represents the extent of overlap of the emission and absorption of molecule A. In general this overlap is most favorable for chromophores that undergo minor geometric change in the excited state.

First order time-dependent perturbation theory can be used to treat process (1). The full discussion is beyond the scope of this review and appears in many texts.[3] There are two limiting cases that are normally treated.

2.1.1. *Förster Dipole–Dipole Coupling*

If the excited state in (1) is a singlet state the $^1A^* \rightarrow A$ transition is spin allowed. If it is assumed that the transition dipole moment for this transition is large (i.e. the transition is symmetry-allowed as well) and it is furthermore assumed that the A and B moieties are not in close contact (i.e. separated by at least a few collision diameters) then Förster derived the following[3,4]

$$k_{ET} = k_{^1A^*}(R_0/R)^6 |\boldsymbol{\mu}_A \cdot \boldsymbol{\mu}_B - 3(\boldsymbol{\mu}_A \cdot \hat{\mathbf{R}})(\boldsymbol{\mu}_B \cdot \hat{\mathbf{R}})|^2 \tag{3}$$

where

k_{ET} = the rate of energy transfer from $^1A^*$ to B;
R = the separation of the two transition dipoles on A and B;
$\boldsymbol{\mu}_x \cdot \hat{\mathbf{R}}$ = the scalar product between the unit vector of the transition moment of molecule X (i.e. A or B) (μ_x) and the unit vector between the two transition dipoles ($\mathbf{R}$);
$\boldsymbol{\mu}_A \cdot \boldsymbol{\mu}_B$ = the scalar product between the unit vectors of the two transition dipoles;
$k_{^1A^*}$ = the unimolecular decay rate of $^1A^*$ in the absence of energy transfer;
R_0 = critical transfer distance, given by

$$R_0^6 = [(9000 \ln 10)\phi_{fl}^A / 128\pi^5 n^4 N] \,.\, S_4 \tag{4}$$

where

ϕ_{fl}^A = quantum yield of fluorescence from $^1A^*$ (the energy donor);
S_4 = spectral overlap integral with $n = 4$ (see eqn (2));
n = refractive index of medium (assumed constant over the integral);
N = Avogadro's number.

The R_0 parameter is characteristic for a given donor–acceptor pair and has been measured for a large number of organic chromophores. The angularly dependent term in (3) (in brackets) can vary from 0 to 4. It is typically assumed that this term may be averaged over all possible orientations of the transition dipole moment and their orientation with respect to R. In this case

$$\langle |\boldsymbol{\mu}_A \cdot \boldsymbol{\mu}_B - 3(\boldsymbol{\mu}_A \cdot \hat{\mathbf{R}})(\boldsymbol{\mu}_B \cdot \hat{\mathbf{R}})|^2 \rangle_{random} = 2/3 \tag{5}$$

Unless the system has a known ordering of the molecular species it is not easy to avoid making the assumption embodied in eqn (5). In tabulations[5] this random averaged angular term is sometimes included in the R_0 value. In such a case one writes

$$\langle k_{ET} \rangle_{random} = k_{1_{A^*}} (\langle R_0 \rangle / R)^6 \tag{6}$$

($\langle R_0 \rangle$ notes the use of eqn (5) in eqn (3)).

There are two important features of this mechanism:

(1) The R_0 parameter can be large (> 5 nm in favorable cases) such that energy transfer is feasible between non-adjacent chromophores on the polymer chain.
(2) R_0 can be estimated accurately (except for the troublesome angular term in eqn (3), for which the assumption in eqn (5) is frequently satisfactory). Thus a polymer system can be 'designed' to enhance singlet energy transfer to a first approximation (but see sub-section 2.2).

We note for comparison with the next section that the Förster mechanism also applies to the following case

$$^3A^* + B \rightarrow A + {}^1B^* \tag{7}$$

In this case the transition $A \rightarrow {}^3A^*$ is spin-forbidden and therefore very weak, but *all* processes involving $^3A^*$ relaxation are spin-forbidden such that (7) can be a significant pathway for $^3A^*$ energy transfer.[6]

2.1.2. Dexter Exchange Coupling[3,7]

This mechanism is operative when the excited states involved in (1) both have triplet multiplicity and thus are spin-forbidden transitions, i.e.

$$^3A^* + B \rightarrow A + {}^3B^* \tag{8}$$

In this case

$$k_{ET} = (2\pi/\hbar) K^2 S_0 \tag{9}$$

where K is an exchange integral

$$K = \langle \phi_{A^*}(1) \phi_B(1) | e^2/r_{12} | \phi_{B^*}(2) \phi_A(2) \rangle \tag{10}$$

where ϕ represents a molecular orbital on either species A or B. $\phi_{A^*}(i)$ and $\phi_A(i)$ are an excited state and ground state MO for the ith electron respectively with an analogous notation for the MO's of species B. In general K is difficult to evaluate because one would have to know the exact geometric relationship between the two molecular species. However, since the K integral involves products of MO's such as $\phi_{A^*}(1)\phi_B(1)$ one can be sure that K decreases rapidly with distance. Dexter writes

$$K^2 = Qe^{-2\alpha R} \quad (11)$$

where α is an effective nuclear charge for the atoms involved, R is the approximate separation of species A and B (essentially equivalent to R in eqn (3)), and Q is a constant that is generally unknown and which would be very difficult to calculate. The important feature of this mechanism is that k_{ET} decreases very rapidly with distance such that chromophores A and B must be separated by approximately a collision distance. Thus in polymers triplet energy transfer along the polymer chain is expected to be important only if there is a high mole fraction of the chromophore present.

As was pointed out in the preceding sub-section, the Förster mechanism applies to $^3A^*$ *if* the energy acceptor B can undergo a spin-allowed transition (see eqn (6)). Thus only triplet-to-triplet energy transfer is of the short-range Dexter type.

2.2. Energy Trapping

Energy trapping refers to the localization of the excitation at some point in space, i.e. somewhere along the polymer chain. The energy trap may consist of a chemically distinct species, such that energy transfer to it is an exothermic process:

$$A^* + B \Leftrightarrow A + B^* + \text{heat} \quad (12)$$

In eqn (12) the endothermic back-reaction has been written explicitly; energy trapping is not necessarily irreversible.

In the case that the excited state of the trap can be observed (e.g. emission spectroscopy, transient absorption, etc.) the quantum efficiency of energy transfer (χ) can be determined. This quantity can be defined as

$$\chi = (\#\ \text{excited states of trap})/(\#\ \text{photons absorbed by donor}) \quad (13)$$

In a number of cases this quantity approaches unity.

Self-quenching is also possible (this is sometimes referred to as 'concentration quenching' in the photophysical literature). There are two major mechanisms:

$$A^* + A \rightarrow (A)_2^*,\ \text{excimer formation} \quad (14)$$

$$A^* + A \rightarrow A^{+\cdot} + A^{-\cdot},\ \text{disproportionation} \quad (15)$$

Either of these two mechanisms can be reversible, although (15) usually yields the lowest excited state regardless of multiplicity (usually this state is a triplet).

Excimer formation has received much attention in the polymer photophysical literature because excimer fluorescence is one of the most striking features of polymer fluorescence spectra.[8] There are two types of excimer-forming mechanisms that have been considered in detail.

2.2.1. Energy Transfer to Excimer-forming Sites (efs)

$$A^* + (A)_2 \Leftrightarrow A + (A)_2^* \tag{16}$$

In this case $(A)_2$ represents a pair of 'crowded' A chromophores in a configuration that is favorable for excimer formation.

2.2.2. Dynamic Formation of an Excimer

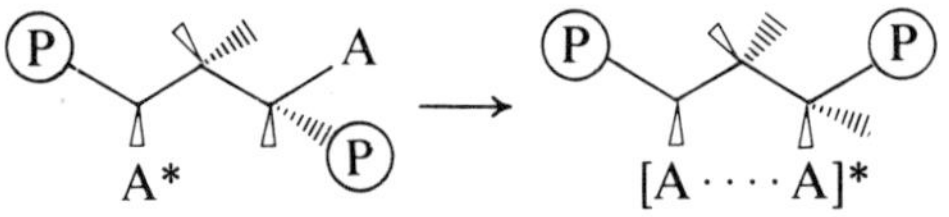

$$\text{Ⓟ} \cdots A^* \cdots A \cdots \text{Ⓟ} \longrightarrow \text{Ⓟ} \cdots [A \cdots A]^* \cdots \text{Ⓟ} \tag{17}$$

In this case excimer formation requires some intrachain rotation (in (17) Ⓟ represents the rest of the polymer chain). The process as represented in (17) could not be literally correct since the whole polymer chain would have to rotate. However a concerted rotational motion is possible that results in the juxtaposition of the two A species as represented in (17).

Processes (16) and (17) cannot be uniquely distinguished in general. It is certainly true that excimer fluorescence is observed in films at low temperatures, implying the existence of mechanism (16). Many polymers in the solution phase do not retain excimer fluorescence at lower temperatures, especially in the case that the solution is frozen.

There has been considerable effort in the past few years by Frank and coworkers to put excimer dynamics on firm theoretical ground. Historically excimer dynamics have been treated by Birk's scheme, originally derived for solutions of small molecules:

$$A^* + A \Leftrightarrow (A)_{2*} \tag{18}$$

In this scheme the monomer and excimer populations (and hence their respective fluorescence) obey the following simple kinetics

$$[A^*] = a_1 \exp(-\lambda_1 t) + a_2 \exp(-\lambda_2 t) \tag{19a}$$

$$[(A)_{2*}] = a_3[\exp(-\lambda_1 t) - \exp(-\lambda_2 t)] \tag{19b}$$

For polymeric systems this simple decay law is essentially never observed, and a number of authors have discussed at length the need to use a three-exponential fit to experimental fluorescence data.[9] Obviously a three-exponential fitting function suggests a more complex group of species than embodied in eqn (18), and consequently a variety of configurational states have been postulated. However Frank and coworkers,[8,10,11] and also Itagaki, Horie and Mita,[12] have emphasized that an excimer formation scheme like that discussed in eqns (16) and (17) will inevitably lead to a more complex decay than given in eqn (19) without the need to introduce new configurational states. The treatment given by Frank *et al.*[11b,c] is generally applicable to any energy trapping phenomenon like eqn (12) in which energy migration plays a role. In this treatment one differentiates between the energy donor (D) and the trap (T). It is assumed that only D is directly excited. We denote the probability that state D is occupied at time t by $H_D(t)$, which is postulated to obey the equation

$$(\mathrm{d}/\mathrm{d}t)H_D(t) = -(k_D + k_{DT}(t))H_D(t) \tag{20}$$

where k_D is the unimolecular decay rate of D, which includes 'static' energy transfer to the trap (i.e. not requiring any stepwise energy migration), and $k_{DT}(t)$ is a time-dependent term that describes the trapping rate 'constant' from energy migration between equivalent D species only. It is expected that $H_D(t)$ will be of the form

$$H_D(t) = \mathrm{e}^{-k_D t}G_D(t) \tag{21}$$

where $G_D(t)$ describes the loss of probability of the D state via energy migration to the trap. One recognizes that eqn (21) for $H_D(t)$ is important because $G_D(t)$ can be calculated by a variety of theoretical approaches or for specific models for the donor-trap system. Thus one might hope to include effects such as the chromophore density, which in the context of polymers would reflect the polymer coil density, etc. Note that eqns (20) and (21) together require

$$k_{DT}(t) = -(\mathrm{d}/\mathrm{d}t)(\ln G_D(t)) \tag{22}$$

As noted above the rate constant k_D includes a contribution from time-independent trapping. In Frank's treatment for excimer formation this arises from the rotational formation of the excimer state (eqn (17)). In the case of a chemically distinct trap this contribution might reflect the conformationally averaged rate of Förster single-step energy transfer.

The rate equation for $H_T(t)$ is written:

$$\mathrm{d}H_T(t)/\mathrm{d}t = -k_T H_T(t) + k_{DT}(t)H_D(t) \tag{23}$$

from which it follows:

$$H_T(t) = e^{k_T t} \int_0^t e^{-k_D \tau}(-dG_D(t)/dt)\,d\tau \tag{24}$$

The equation for $H_T(t)$ implies a build-up and decay for the trap population which also depends on the properties of $G_D(t)$. Additionally the steady-state fluorescence of donor or trap states can be calculated from

$$I_x = k_x q_{Fx} \int_0^\infty H_x(t)\,dt' \tag{25}$$

where x = D or T and q_{Fx} is the appropriate quantum yield of fluorescence. The ratio of trap and donor fluorescence is given by

$$I_T/I_D = (q_{FT}/q_{FD})[(1-M)/M] \tag{26}$$

where M is the following Laplace transform of $G_D(t)$:

$$M = k_D \int_0^\infty e^{-k_D t} G_D(t)\,dt \tag{27}$$

Thus from eqns (25) and (26) models for $G_D(t)$ can be tested against experimental results. Frank and coworkers have explored the form of $G_D(t)$ for several different models. These include one-dimensional nearest neighbor transfer only and Förster transfer under conditions of restricted dimensionality. Unfortunately calculations of $G_D(t)$ based on some measured polymer configuration and/or energy transfer mechanism is quite difficult. The most detailed application of this theoretical approach has been to excimer formation. It remains to be seen if this theory can be applied successfully to a wider range of intrapolymer energy transfer phenomena. It should also be noted that there has been considerable theoretical work on energy migration and trapping on 'fractals'[13] (lattices of fractional dimensionality) that almost certainly will be applicable to polymer photophysics.

2.3. Excited State Annihilation Processes

Excited state annihilation processes like (28), (29) are well-known:

heterogeneous annihilation

$$A^* + B^* \begin{cases} \rightarrow A^{**} + B \\ \rightarrow A + B^{**} \end{cases} \tag{28}$$

homogeneous annihilation

$$A^* + A^* \longrightarrow A^{**} + A \tag{29}$$

In these processes the highly excited state (A** or B**) may rapidly degrade to the lowest excited state of that multiplicity, or may undergo a chemical reaction such as ionization or dissociation. Note that in (28) energy transfer to the lowest excited state of the A,B pair does not necessarily occur with unit quantum yield. This is somewhat surprising because the mechanism of annihilation requires an overlap of molecular orbitals and hence a close approach by the annihilating pair of excited state molecules.

Excited state annihilation can occur if the excited state molecules diffuse together in solution, or if the excitations can themselves migrate within a crystal lattice or along a polymer chain. (In fact one of the strongest evidences for energy migration is the observation of excited state annihilation.) Thus polymer systems provide a possibility for two-photon photochemistry that would be completely absent in the small molecule model compound except at very high concentrations in solution.

In general triplet–triplet (T–T) and singlet–singlet (S–S) annihilation must be considered, shown for homogeneous annihilation in the following:[14]

T–T annihilation

$$^3A^* + {}^3A^* \longrightarrow {}^1A^* + A \longrightarrow {}^1A^* \text{ fluorescence} \qquad (30)$$
$$\searrow {}^3A^{**} + A$$

S–S annihilation

$$^1A^* + {}^1A^* \longrightarrow {}^1A^{**} + A \qquad (31)$$

In (30) production of both $^1A^*$ and $^3A^{**}$ is spin-allowed. The former can fluoresce and this emission is referred to as 'delayed fluorescence'; while it is spectrally equivalent to normal fluorescence from $^1A^*$, its decay time is typical of the metastable triplet precursor (with lifetimes typically in the 10 μs to 50 ms range for well-purified, deaerated solutions). The $^3A^{**}$ species may either radiationlessly degrade back to $^3A^*$ (in which case any photophysical property that requires $^3A^*$ will be non-linear in excitation intensity) or react. If S–S annihilation (31) is significant one may observe either a strongly non-linear excitation intensity dependence of $^1A^*$ fluorescence, transient absorption or photochemical product.

Energy transfer can be treated as a diffusive process with a 'diffusion constant' Λ_E (units: m^2/s). From 1 – D random walk theory Λ_E is given by $a^2/2\tau_h$ where a is the average chromophore separation and τ_h is the average excitation 'hopping time' for nearest neighbor transfer. The mean free distance of energy migration (L_E) can be estimated by

$$L_E = (2\Lambda_E\tau_E)^{1/2} = a(\tau_E/\tau_h)^{1/2} \qquad (32)$$

where τ_E is the excited state lifetime and the average number of energy transfer steps (S_E) is given by

$$S_E = (\tau_E/\tau_h)^{1/2} \quad (33)$$

Thus on the average excited states must be created on the order of every S_E chromophore sites for annihilation to be important, or the excited states must represent a mole fraction of S_E^{-1}. Using pulsed lasers (or even powerful steady state sources) it is not difficult to excite 1–10% of all chromophores, so S_E values of 10–100 can easily lead to facile two-photon chemistry.

The theory of intracoil excited state annihilation has not been developed to the same extent as intracoil energy transfer to traps. Building on ideas developed for excitonic processes in chloroplasts and mixed molecular crystals Webber and Swenberg[15] presented a master equation based theory in 1980. In this approach the following equation was written for $p_n(t)$, the probability of observing a coil with n excitations at time t:

$$\begin{aligned} \mathrm{d}p_n(t)/\mathrm{d}t = {} & I_{ex}p_{n-1}(t) - I_{ex}p_n(t) + k(n+1)p_{n+1}(t) \\ & - knp_n(t) - \gamma_n p_n(t) + \gamma_{n+1}p_{n+1}(t) \end{aligned} \quad (34)$$

where I_{ex} is the excitation rate in s^{-1} and k is the unimolecular decay rate for the excited state (note that the factor of n is required for $p_n(t)$ because of the simultaneous decay of n excited states). The annihilation rate constant γ_n depends on the number of excited state pairs on the polymer coil. The following proportionality was assumed, based on computer simulations:

$$\gamma_n \propto [n(n-1)]^{\alpha} \quad (35)$$

where α depends on the dimensionality of the lattice (i.e. 1-, 2- or 3-dimensional). Not written explicitly in eqn (34) is the dependence on the size of the lattice, which corresponds to the degree of polymerization for polymers.

If exciton annihilation can be neglected then the steady state solution of eqn (34) yields the Poisson distribution:

$$\bar{p}_n = (\bar{n}^n/n!)\,e^{-\bar{n}} = (f_E\bar{P})^n e^{-f_E\bar{P}}/n! \quad (36)$$

where $\bar{p}_n$ denotes the steady state probability and $\bar{n}$ is the average number of excitations per coil, given by I_{ex}/k. Note that since the rate of excitation per coil depends on the number of chromophores per coil, $\bar{n}$ is expected to be proportional to the degree of polymerization $\bar{P}$, i.e. $\bar{n} = f_E\bar{P}$ where f_E is the mole fraction of excited chromophores. If excited state annihilation cannot be neglected a more complicated expression is obtained

$$\bar{p}_n = (f_E \bar{P})^n \bar{p}_0 \prod_{i=1}^{n} (1/K_i) \tag{37}$$

(unnormalized)

where

$$K_i = n + \gamma_n/k \tag{38}$$

The value of $\bar{p}_0$ is chosen such that $\bar{p}_n$ is normalized, i.e.

$$\sum_{n=0} \bar{p}_n = 1 \tag{39}$$

The main effect of the annihilation process is to decrease the average number of excitations per coil. This effect is illustrated in Fig. 1.

The decay of an ensemble of excited state following a δ-function excitation would be expected to be a sum of exponentials from eqn (34)

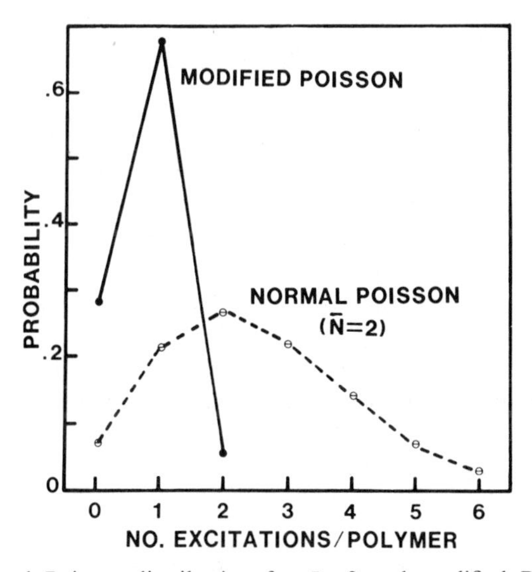

FIG. 1. Standard Poisson distribution for $\bar{n} = 2$ and modified Poisson distribution that includes the effect of exciton annihilation. A shift of $(p_n)_{max}$ to lower n values is typical.

except in the case of no annihilation. In this latter case it can be easily shown that

$$\langle n(t)\rangle = \sum_{n=0} np_n(t) = \langle n(0)\rangle \, e^{-kt} \tag{40}$$

As has been discussed in the previous section simple exponential decay is often not observed in polymers because of energy trapping at excimer forming sites. Such trapping is not considered in eqn (34).

Equation (34) also predicts that if one could initially create a state with $p_{n'}(0) = 1$ for some n' state (i.e. $p_n(0) = 0$ for $n \neq n'$) then the decay of $p_n(t)$ would be given by a single exponential. This is clearly not the case because there are many possible separations of excited states at $t = 0$, and each initial state will lead to a different average time to undergo a transition from the n' state to the $n' - 1$ state. This problem was studied by computer simulations, with a typical result presented in Fig. 2 (calculated for $k = 0$). This decay is highly non-exponential, and as expected, depends strongly on the number of excitations per coil. Note also that this decay is a function of the excitation hopping time, τ_h. An estimate of an 'effective annihilation rate constant' for state $p_n(t)$ can be made as follows:

$$k_n^x = (-\ln x)/t_x^n \tag{41}$$

where at time t_x^n

$$p_n(t_x^n) = x \tag{42}$$

i.e. $k_n^{0\cdot 5}$ corresponds to the inverse of the half-life of $p_n(t)$. Obviously the value of k_n^x will depend on the value of x as well as n. However for any choice of x the dependence of k_n^x on the dimensionality of the lattice is very strong. The results for $k_n^{0\cdot 5}$ imply that α in eqn (35) is ca. 1·8 for a 1 – D lattice and 0·7 for a 2 – D lattice. Thus in the context of polymer photophysics the important result is that exciton annihilation will become more important at high excitation rates than will be the case for corresponding two-dimensional (e.g. chloroplasts) or three-dimensional (e.g. molecular crystals) lattices. Offsetting this 'polymer effect' are two polymer properties that would tend to increase the excitation hopping time (τ_h): (1) a tendency for excitation to be trapped (perhaps irreversibly) at excimer forming sites); (2) larger average separation of chromophores than is the case for molecular crystals.

2.4. Special Features of Energy Transfer in Polymers

Every phenomenon discussed up to this point has a counterpart in standard molecular photophysics either in the solution or solid phase. There are a

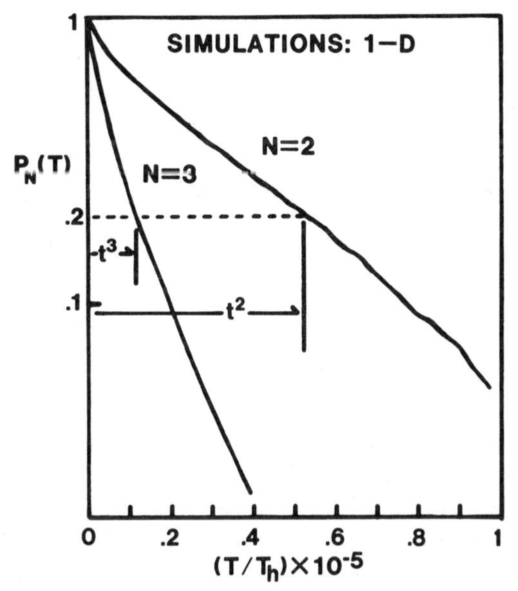

FIG. 2. Decay of probability of state $p_n(t)$ for $n=2$ and 3 from exciton annihilation only. Also illustrated is $t^n_{0\cdot 2}$ for each decay curve.

number of features in polymer photophysics that are essentially unique to this field.

2.4.1. Tacticity and/or Chain Characteristics

The rate of energy transfer between polymer-bound chromophores is obviously governed by their average separation and average mutual orientation. The distribution of chromophore separations is also important since occasional 'breaks' in a sequence of chromophores can impose a *de facto* barrier to energy transfer such that the excitation is confined to a relatively small segment of the polymer chain. (In such a case an expression like (33) correctly calculates the number of 'steps' taken but (32) does not since it is implicit in this expression that the whole chain is accessible to the excitation. Furthermore, if S_E is much greater than the degree of polymerization then once again (32) cannot be correct.[16]) The ability to form excimers is also a function of the spatial relationships that exist between adjacent (and sometimes non-neighboring) chromophores. Consequently it is not surprising that energy transfer (and all other polymer photophysical processes) are related to the type of polymer backbone, the type of bonding sequence by which the chromophore is attached to the backbone, and while not frequently tested, the tacticity of the polymer. In a sense this is one of the strengths of polymer photophysics. Since many structural modifications are possible one may hope to tailor a given polymer system to the desired use. On the other hand ready generalizations are not yet possible. Cases have been found in which inhibited or facile excimer formation both led to efficient singlet energy transfer while for the intermediate case excimer formation reduces energy transfer. One does expect in general that a more 'ordered' polymer will enhance energy transfer.

2.4.2. *Copolymers*

One can readily produce copolymers in which one chromophore acts as an energy trap or one of the components does not have an absorbing chromophore in the spectral region of interest and hence acts as an inert 'spacer' group. Obviously the degree of randomness in the copolymer will be critical to the energy transfer properties. Random copolymers have been successfully analyzed in terms of the mean sequence length of the energy transfer chromophore.[17,18] In some early work it was shown that an alternating copolymer of this type had considerably different properties than the normal homopolymer or random copolymer.[19,20]

Block copolymers are another interesting possibility. In this case the juxtaposition of the energy transfer chromophores is assured but the sequence length of these chromophores can be varied. Many effects of sequence length in such copolymers would be expected to be analogous to molecular weights, which will be discussed in the next sub-section.

2.4.3. *Effect of Molecular Weight*

The molecular weight of a polymer affects both the coil density (for a given solvent) and the number of chromophores per coil. It has been shown in a few cases that the relative intensity of excimer fluorescence is dependent on the molecular weight,[8] which obviously can change the energy transfer rate. Also the coil density would be expected to affect the extent of 'cross-chain' energy transfer between non-adjacent chromophores that are on different segments of the polymer chain. However it has not been possible to clearly distinguish cross-chain energy transfer from 'down-chain' energy transfer (i.e. between adjacent chromophores).

Obviously an increased molecular weight means that the probability of multiple photon absorption on the same coil is increased. Since intracoil excited state annihilation requires at least two excitations per coil, then one expects, and it has been found, that increasing the molecular weight tends to enhance annihilation processes.[21-23] This factor is very important in potential practical applications, since a high molecular weight polymer would be more susceptible to multiple photon chemistry.

While a full theory of these molecular weight effects is not available, it has been suggested that the probability of n excitations existing simultaneously on a polymer of degree of polymerization $\bar{P}$ should be given by a Poisson distribution (eqn (36)) (but see eqn (37) and Fig. 1).

2.4.4. Solvent Effects

The solvent used for a polymer can affect its photophysics in two ways: (1) excited state quenching, (2) coil density.

Excited state quenching is usually obvious if the fluorescence intensity for a given polymer is compared for several solvents. There are various mechanisms for excited state quenching: (1) external heavy atom effect, (2) charge-transfer or exciplex formation, and (3) energy transfer. The external heavy atom effect is important primarily for iodo- or bromo-solvents although CCl_4 may also quench in this way (note however that CH_2Cl_2 does not quench excited states). Solvents that contain cyano, amino or even aromatic ester groups may form exciplexes with excited aromatic chromophores, which may give rise to a new emission feature (exciplex fluorescence) or simply quench. Solvent quenching by energy transfer would not be encountered typically, because that would imply that the excited states of the solvent lie below those of the polymer chromophore. In such a case it would be very difficult to excite the polymer-bound chromophore. Nevertheless this possibility should be kept in mind in experiment or application design. It is most likely to be a problem for chromophores with high energy excited states such as polystyrene or polymethylacrylate.

The thermodynamic quality of the solvent affects the coil density which in turn affects the average chromophore separation. It has been demonstrated for a number of polymer systems that a poorer solvent enhances the relative intensity of excimer fluorescence. Consequently, if the excimer is an energy trap then one expects energy transfer to become less efficient. Contrarily if cross-chain energy transfer is an important mechanism then the efficiency will increase in a poor solvent. Examples of this case are found when energy transfer occurs between a small number of chromophores on the polymer chain via the Förster mechanism.[24,25]

2.4.5. Films

The majority of photophysical studies of polymers have been carried out in the solution phase on low temperature glasses. However most polymer applications involve the solid phase, including thin films (usually cast from solution). All energy transfer phenomena discussed in the previous sections do occur in the polymer films, and some of the special polymer features discussed in this section are also observed in films. The fluorescence of polymer films is typically red-shifted and structureless relative to the solution spectra, implying extensive excimer formation (primarily by the energy transfer mechanism, see preceding section on energy trapping). Quite often the film spectrum is red-shifted relative to the normal solution excimer spectrum and it is likely that in films the excimer spectrum is composed of multiple excimer types that correspond to various orientations of the overlapping chromophores.

Polymer films can either be neat, composed of only a single polymer type, or they can be mixed, in which case one must be concerned with the compatibility of the two (or more) polymer types. If there exists relatively poor compatibility then the situation is analogous to a poor polymer solvent as discussed in the preceding sub-section. In fact the appearance of mixed polymer fluorescence has been used as a probe of polymer compatibility.[8]

Most energy transfer work in films has been of two types: (1) energy transfer from the main chromophore component to a low energy trap, which may be either a small molecule species dissolved in the film or covalently bonded to a polymer, or (2) T–T annihilation. For case (1) both the triplet and singlet states have been characterized.

Energy transfer has been carefully characterized for a relatively small number of polymer films. A notable exception is poly(N-vinylcarbazole) (and derivatives) which has been heavily studied, primarily because this material is photoconducting.

3. EXAMPLES OF POLYMERIC ENERGY TRANSFER

3.1. Singlet States: Energy Transfer to Traps or Excimer Forming Sites

There are two types of situations that arise:

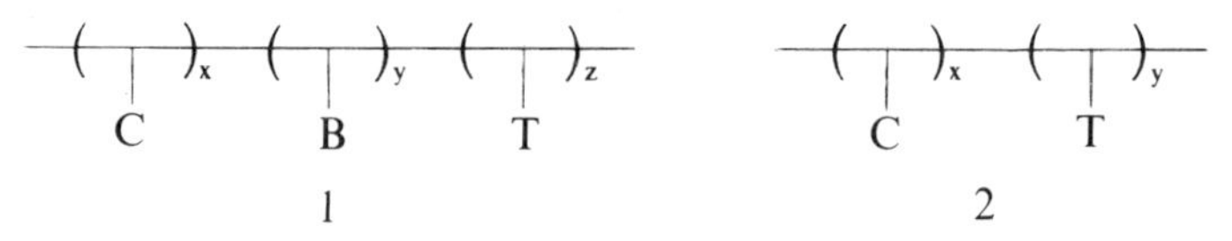

In polymer 1 there are chromophores (C), traps (T) and inert 'spacers' (B). In most cases these polymers have been random copolymers. In polymer 2

TABLE 1
Singlet Energy Transfer to Traps

Chromophore (C)	*Trap (T)*	*Spacer (B)*[a]	*Ref.*
1-naphthylmethacrylate	anthracene		30
2-vinylnaphthalene	pyrene	—	31
2-vinylnaphthalene	efs[b]	—	32
9,10-diphenylanthracene	fluorescein	styrene	25
9-10-diphenylanthracene	fluorescein	pyrollidinone	24
(9-phenanthryl)methylmethacrylate	anthracene	—	33
2-(1-naphthyl)ethylmethacrylate	efs[b]	—	34
Styrene	efs[b]	—	10

[a] If no spacer species is indicated then the polymer is like 2.
[b] efs = excimer forming site.

the inert spacer has been eliminated. In both cases the trap may be incorporated within the main chain or at chain ends, if a chemical route to accomplish this is available. In the case of excimer forming sites the T species is intrinsic to the polymer and the mole fraction of these sites is expected to be a function of chromophore structure, polymer backbone, temperature and solvent. An interesting application of these phenomena is to characterize polymer compatibility in mixed films. For incompatible combinations energy transfer to traps tends to be enhanced because of decreased coil size.

The basic observation in systems of this type is that when chromophore C is excited, energy is transferred to species T, which may emit or undergo some other photophysical or photochemical process. Examples of this kind of process are gathered together in Table 1.

3.2. Singlet States: S–S Annihilation

As was discussed in the preceding section on excited state annihilation, excited states on the same polymer coil can annihilate to produce photoproducts and/or to provide an additional radiationless deactivation pathway. These studies are less common than those described in the previous sub-section since relatively high-powered laser excitation is usually required.

Some examples are collected in Table 2. These are all for homopolymers. Heterogeneous intracoil S–S annihilation has not been reported to the author's knowledge. Also this phenomenon has not been studied for copolymers with inert spacer groups (analogous to polymer 1 in the previous sub-section).

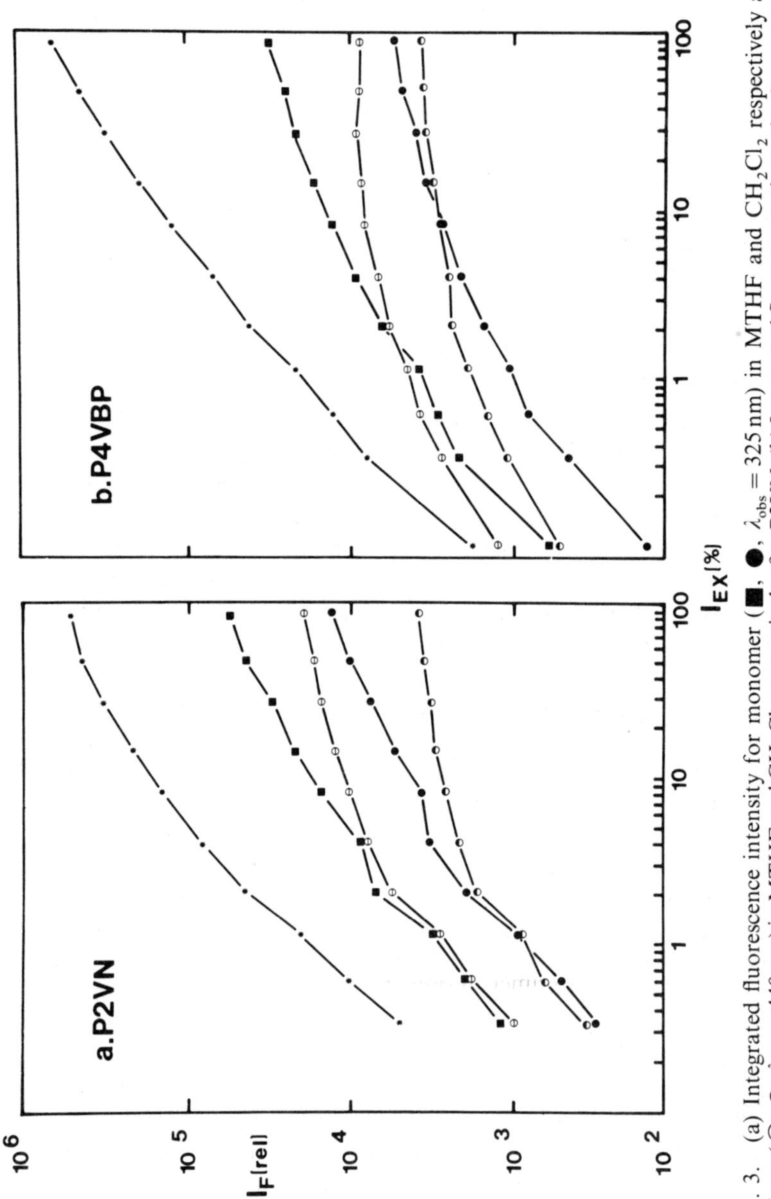

FIG. 3. (a) Integrated fluorescence intensity for monomer (■, ●, $\lambda_{obs} = 325$ nm) in MTHF and CH_2Cl_2 respectively and excimer (⊖, ◓, $\lambda_{obs} = 440$ nm) in MTHF and CH_2Cl_2 respectively; for P2VN. (b) Integrated fluorescence intensity for monomer (■, ●, $\lambda_{obs} = 315$ nm) in MTHF and CH_2Cl_2 respectively and excimer (⊖, ◓, $\lambda_{obs} = 400$ nm) in MTHF and CH_2Cl_2 respectively; for P4VBP. For both plots * is the integrated fluorescence intensity of naphthalene in cyclohexane. There was no significant difference between different P2VN or P4VBP samples.

TABLE 2
S–S Annihilation

Chromophore	*Polymer*	*Ref.*
Biphenyl	poly(4-vinylbiphenyl)	23
Naphthalene	poly(2-vinylnaphthalene)	23
Carbazole	poly(N-vinylcarbazole)	35

In the cases of S–S annihilation reported to date the primary observation has been that the fluorescence intensity or triplet yield is sublinear in the excitation flux (the polymer is usually compared to a monomeric model compound, which tends to remove artifacts such as ground state bleaching). A recent example of the former is presented in Fig. 3.[23] However what is not generally determined is if this strong non-linearity results fron new excited state absorptions in the polymer (e.g. excimer) that compete with the ground state for photons. Thus in all studies of excited state annihilation one should determine the strength of $S_1 \rightarrow S_n$ transitions and the number of photons actually adsorbed by the polymer.

It should be emphasized that excited state annihilation can lead to new photoproducts. We have recently presented an example of this for poly(2-vinylnaphthalene) in CH_2Cl_2,[23] in which a naphthalene cation radical is produced, presumably following electron transfer to the solvent, i.e.

$$^1A^* + (^1A^* \text{ or } ^1A_{2^*}) \longrightarrow {}^1A^{**} \xrightarrow[\text{RCl}]{} {}^2A^{+\cdot} + R^{\cdot} + Cl^- \qquad (43)$$

3.3. Triplet States: T–T Annihilation

This process has been studied primarily in low temperature glasses or films. Quite often this process leads to easily observed delayed fluorescence (eqn (30)). Some typical examples are compiled in Table 3. Note that these are all homopolymers. Only one case of heterogeneous polymeric T–T annihilation is known at present[26a] although Klöpffer has found T–T annihilation to enhance excimer fluorescence which was interpreted via a heterogeneous annihilation mechanism.[26b] These processes in low temperature solids and films have recently been reviewed by the present author.[27]

A more recent observation involves T–T annihilation in poly(2-vinylnaphthalene) (P2VN)[28] and poly(4-vinylbiphenyl) (P4VBP)[29] in fluid solution, leading to delayed fluorescence. In these cases the intersystem crossing yield from the monomer and excimer singlet state is not very high

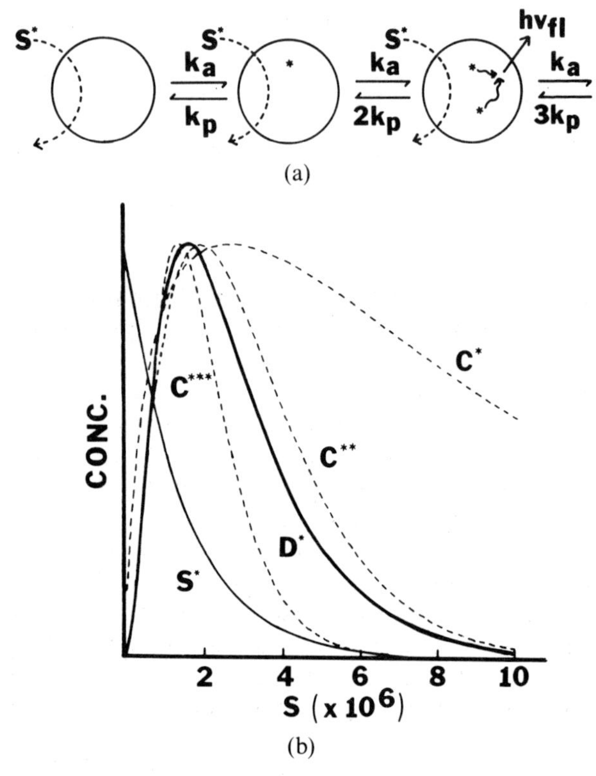

FIG. 4. (a) Representation of kinetic model in eqn (44) of text in which the excited states on a coil are represented by *. (b) Result of a numerical solution of the system of equations (44). Note that the rate of decay of an excited coil depends strongly on the number of excitations per coil because of exciton annihilation (all concentrations scaled to same maximum for ease of comparison).

so a photosensitizer was used (benzophenone, excited at 337 nm by a N_2 laser). The following kinetic equations were used:

$$\begin{aligned} C + {}^3\text{Sens}^* &\longrightarrow {}^3C^* + \text{Sens} \\ {}^3C^* + {}^3\text{Sens}^* &\longrightarrow {}^3C^{**} + \text{Sens} \\ {}^3C^{**} + {}^3\text{Sens}^* &\longrightarrow {}^3C^{***} + \text{Sens} \end{aligned} \tag{44}$$

in which ${}^3\text{Sens}^*$ is the excited triplet state of the sensitizer and ${}^3C^*$, ${}^3C^{**}$, etc., represent polymer coils with one, two, etc., excitations. The equations used to analyze these systems are similar to eqn (34), in which the delayed excimer fluorescence is assumed to arise from intracoil T–T annihilation.

TABLE 3
T–T Annihilation

Chromophore	*Polymer*	*Phase*	*Ref.*
Biphenyl	poly(4-vinylbiphenyl)	benzene, 25 °C	29
Naphthalene	poly(2-vinylnaphthalene)	benzene, 25 °C	28
Naphthalene	poly(1-vinylnaphthalene)	films	36
Naphthalene	poly(1-vinylnaphthalene)	THF, 77 K	37
Naphthalene	poly(2-vinylnaphthalene)	MTHF, 77 K	21
Carbazole	poly(N-vinylcarbazole)	MTHF, 77 K	22
Carbazole	poly(N-vinylcarbazole)	films	38, 39
Naphthalene	poly(2-naphthylmethacrylate)	MTHF, 77 K	40
Benzophenone	poly(vinylbenzophenone)	—	41

The model and the results of the solution of the kinetic model are presented in Fig. 4. The analysis is complicated but the fundamental concept is that only those coils with two or more triplet excitations can contribute to formation of $^1D^*$, the singlet excimer state (for experimental reasons the monomer excited singlet could not be observed). The kinetic model embodied in eqn (44) is only a rough approximation, but served to rationalize the main experimental observations.

ACKNOWLEDGEMENT

We would like to acknowledge the efforts of Ms Rayna Kolb in preparing this article.

REFERENCES

1. See the recent monograph, Thompson, L. F., Willson, C. G. and Bowden, M. J. (Eds), *Introduction to Microlithography: Theory, Materials, Testing*, A.C.S. Symposium Series 219, 1983.
2. For a recent book that includes a discussion of these phenomena see Steiner, R. F. (Ed.), *Excited States of Biopolymers*, Plenum Press, New York, 1983.
3. Birks, J. B., *Photophysics of Aromatic Molecules*, Wiley-Interscience, London, 1970.
4. Förster, Th., *Ann. Phys.*, **2**, 55 (1948).
5. See Berlman, I. B., *Energy Transfer Parameters of Aromatic Molecules*, Academic Press, New York, 1973.
6. Kellogg, R. E., *J. chem. Phys.*, **47**, 3403 (1967).

7. Dexter, D. L., *J. chem. Phys.*, **21**, 836 (1953).
8. Semerak, S. N. and Frank, C. W., Photophysics of excimer formation in aryl vinyl polymers, *Adv. Polym. Sci.*, **54**, 33 (1983).
9. Ghiggino, K. P., Roberts, A. J. and Phillips, D., *Adv. Polym. Sci.*, **40** (1981).
10. (a) Gelles, R. and Frank, C. W., *Macromolecules*, **15**, 741 (1982); (b) *ibid.*, 747.
11. (a) Fitzgibbon, P. D. and Frank, C. W., *Macromolecules*, **15**, 733 (1982); (b) Fredrickson, G. H. and Frank, C. W., *Macromolecules*, **16**, 572 (1983); (c) Fredrickson, G. H. and Frank, C. W., *Macromolecules*, **16**, 1198 (1983); (d) Fredrickson, G. H., Andersen, H. C. and Frank, C. W., *Macromolecules*, **16**, 1456 (1983); (e) Fredrickson, G. H., Andersen, H. C. and Frank, C. W., *J. phys. Chem.*, **79**, 3572 (1983).
12. Itagaki, H., Horie, K. and Mita, I., *Macromolecules*, **16**, 1395 (1983).
13. (a) Klafter, J. and Blumen, A., *J. chem. Phys.*, **80**, 875 (1984); (b) Klafter, J. and Zumofen, G., *Phys. Rev.*, *B*, **28**, 6112 (1983).
14. Excited state annihilation has been considered for other possibilities (e.g. singlet–triplet, singlet–doublet, etc.). For a review of these processes in molecular crystals see Swenberg, C. E. and Geacintov, N. E. In: *Organic Molecular Photophysics* (Ed. J. B. Birks), John Wiley, London, Vol. 1, 1973, p. 489; *ibid.*, Vol. 2, 1975, p. 409.
15. Webber, S. E. and Swenberg, C. E., *Chem. Phys.*, **49**, 231 (1980).
16. Fredrickson and Frank (11b) have considered a polymer chain to be divided into segments between excimer forming sites in their treatment of excimer formation.
17. (a) Soutar, I., Phillips, D., Roberts, A. J. and Rumbles, G., *J. Polym. Sci.: Polymer Physics Edition*, **20**, 1759 (1982); (b) Anderson, R. A., Reid, R. F. and Soutar, I., *Europ. Polym. J.*, **16**, 945 (1980); (c) Anderson, R. A., Reid, R. F. and Soutar, I., *Europ. Polym. J.*, **15**, 925 (1979).
18. Morawetz, H. and Steinberg, I. Z. (Eds), *Luminescence from Biological and Synthetic Macromolecules*, *Annals of the New York Academy of Science*, 366, 1981. A number of papers involving polymer excited states.
19. Fox, R. B., Price, T. R., Cozzens, R. F. and Echols, W. H., *Macromolecules*, **7**, 937 (1974).
20. Yokoyama, M., Tamamura, T., Nakano, T. and Mikawa, H., *J. chem. Phys.*, **65**, 272 (1976).
21. Pasch, N. F. and Webber, S. E., *Chem. Phys.*, **16**, 361 (1976).
22. (a) Klöpffer, W. and Fischer, D., *J. Polym. Sci.*, *Polymer Symposium*, **40**, 43 (1973); (b) Klöpffer, W., Fischer, D. and Naundorf, G., *Macromolecules*, **10**, 450 (1977).
23. Pratte, J. F. and Webber, S. E., *Macromolecules* (to be published).
24. Hargreaves, J. S. and Webber, S. E., *Macromolecules* (submitted).
25. Hargreaves, J. S. and Webber, S. E., *Canad. J. Chem.* (submitted).
26. (a) Surprisingly, triplet energy transfer to covalently bound traps has been reported only for poly(2-vinylnaphthalene-co-4-vinylpyrene); Webber, S. E. and Hargreaves, J. S., *Abstracts: IUPAC 28th Macromolecular Symposium*, Amhurst, Massachusetts, July 12–15, 1982. Consequently there is no section analogous to 'Singlet States: Energy Transfer to Traps or Excimer Forming Sites'. (b) Rippen, G., Kaufmann, G. and Klöpffer, W., *Chem. Phys.*, **52**, 165 (1980).

27. Webber, S. E., Phosphorescence and other delayed emissions in polymers. In: *Polymer Photophysics* (Ed. D. Phillips), Chapman and Hall, London, to be published (1985). This book is similar in spirit to ref. 3, taken into the polymer field.
28. Pratte, J. F. and Webber, S. E., *J. phys. Chem.*, **87**, 449 (1983).
29. Pratte, J. F. and Webber, S. E., *Macromolecules*, **16**, 1193 (1983).
30. Aspler, J. S, Hoyle, C. F. and Guillet, J. E., *Macromolecules*, **11**, 925 (1978).
31. Hargreaves, J. S. and Webber, S. E., *Macromolecules*, **15**, 424 (1982).
32. Fitzgibbon, P. D. and Frank, C. W., *Macromolecules*, **15**, 733 (1982).
33. (a) Ng, D. and Guillet, J. E., *Macromolecules*, **15**, 728 (1982); (b) *ibid.*, 724.
34. Holden, D. A., Wang, P. Y-K. and Guillet, J. E., *Macromolecules*, **13**, 295 (1980).
35. (a) Masuhara, H., Tamai, N., Inoue, K. and Mataga, N., *Chem. Phys. Lett.*, **91**, 109 (1982); (b) Masuhara, H., Ohwada, S., Mataga, N., Itaya, A., Okamoto, K. and Kusabayashi, S., *J. phys. Chem.*, **84**, 2363 (1980).
36. Kim, N. and Webber, S. E., *Macromolecules*, **13**, 1233 (1980).
37. (a) Cozzens, R. F. and Fox, R. B., *J. chem. Phys.*, **50**, 1532 (1969); (b) Cozzens, R. F. and Fox, R. B., *Macromolecules*, **2**, 181 (1969).
38. (a) Rippen, G., Kaufmann, G. and Klöpffer, W., *Chem. Phys.*, **52**, 165 (1980); (b) Klöpffer, W., *Chem. Phys.*, **57**, 75 (1981).
39. (a) Burkhart, R. D. and Aviles, R. G., *Macromolecules*, **12**, 1073 (1979); *ibid.*, 1078; (b) Burkhart, R. D. and Aviles, R. G., *J. phys. Chem.*, **83**, 1897 (1979); (c) Turro, N. J., Chow, M-F. and Burkhart, R. D., *Chem. Phys. Lett.*, **80**, 146 (1981).
40. (a) Somesall, A. C. and Guillet, J. E., *Macromolecules*, **6**, 218 (1973); (b) Pasch, N. F. and Webber, S. E., *Macromolecules*, **11**, 727 (1978).
41. Schnabel, W., *Macromolec. Chem.*, **180**, 1487 (1979).

2

Studies of the Antenna Effect in Polymer Systems

J. E. GUILLET

Department of Chemistry, University of Toronto, Canada

1. INTRODUCTION

The primary photochemical step in photosynthesis is now generally recognized to be a one-electron transfer from the singlet excited state of a chlorophyll species (Chl) to an electron acceptor. This reaction takes place within a reaction center protein that spans the thylacoid membrane of the chloroplast organelle of green leaves and algae. In the simplest photosynthetic systems the electron acceptor contains the quinone moiety, such as an ubiquinone, menaquinone or plastoquinone. An essential feature of this process is that the donation of an electron must lead to a separation of the charged species $Chl^{\dot{+}}$ and $Q^{\dot{-}}$ so that they may undergo further reactive steps in the photosynthetic sequence.

In green plants the primary charge-separation process occurs in reaction sites which contain only a small fraction of the total pigment material. The bulk of the chlorophyll in the chloroplast is photochemically inert, functioning as an 'antenna pigment' by transferring light through non-radiant interactions to the reaction centers.[1] In this way the turnover rate for reactive sites is increased, the occurrence of this energy-transfer process having the same effect as if the extinction coefficient of the reactive center was increased by a factor of over 100. In biological photochemistry, this effect is known as the 'antenna effect'.

It has been shown that macromolecules containing aromatic chromophores can also display efficient electronic energy transfer to low-energy traps.[2] By analogy with the biological process of photosynthesis, we have termed such molecules 'antenna molecules'. Macromolecules containing chromophores attached to a polymeric backbone often display very high efficiency of singlet energy transfer. The function of the connecting macromolecular chain and the plant thylacoid membrane is similar in that both serve as anchors supporting high local concentrations of the chromophores. Synthetic polymers can thus mimic the function of the light-harvesting pigment layers without reproducing their exact structure.

There are a number of reasons for the present interest in antenna polymers. Besides the challenge of pursuing this analogy between natural chloroplasts and synthetic polymer systems, there is considerable relevance to the stabilization of synthetic polymers against photodegradation. It is also possible to use energy transfer to reactive groups to create polymers with enhanced degradation rates. Furthermore, these principles can be utilized to develop polymeric photocatalysts which are effective at extremely small concentrations in solution.

2. THE ANTENNA EFFECT IN POLYMERS

Early studies by Schneider and Springer[3] and by Fox and Cozzens[4] show that efficient fluorescence emission occurred from small amounts of copolymerized chemically-bound fluorescence traps in polymer chains. Schneider and Springer used styrene acenaphthylene copolymers, and Fox and Cozzens used styrene–vinyl naphthalene copolymers. In both studies it was shown that the emission of naphthalene fluorescence in such copolymers was much higher than in a solution of equivalent amounts of the two homopolymers. Both groups proposed that this phenomenon was due to energy migration between the styrene sequence to the naphthalene moieties in the polymer chain.

More recently, Guillet and coworkers have carried out extensive studies of a variety of polymers containing naphthalene and phenanthrene donors with traps containing the anthracene moiety.[5–16] Although such polymers will also demonstrate triplet energy migration, the important process in natural photosynthesis appears to involve only the migration of singlet excitons; thus this article will restrict attention to those processes in which the photon-harvesting process involves singlet excitation energy.

A typical antenna molecule consists of a polymer with a sequence of aromatic groups (such as naphthalene or phenanthrene) and a small number (>1 per molecule) of traps located at the end or at various other positions within the chain.

Guillet *et al.*[5,6,17] have shown that the photon-collecting efficiency is not entirely due to energy migration among the chromophores making up the antenna but to a combination of migration and direct energy transfer to the trap. It seems very likely that energy migration occurs because of dipole–dipole interactions between the excited dipole in the donor group and a similar dipole in the ground state of an acceptor. The theory for such processes has been worked out in detail by Förster[18] who derived the rate constant for transfer from the donor to the acceptor.

$$k_{D^* \to A} = \frac{8{\cdot}8 \times 10^{-25} \kappa^2 \phi_D}{n^4 \tau_D R^6} \int_0^\infty F_D(\nu) \varepsilon_A(\nu) \frac{d\nu}{\nu^4} \quad (1)$$

where $k_{D^* \to A}$ is the rate constant for energy transfer, R is the donor–acceptor separation, κ^2 is an orientation factor, usually assumed to be two-thirds for a random distribution of D and A, ϕ_D is the quantum yield for emission from the donor, n is the refractive index of the medium, τ_D is the mean lifetime of the donor excited state, $F_D(\nu)$ is the normalized spectral distribution of the donor emission, and $\varepsilon_A(\nu)$ is the molar extinction coefficient of the acceptor as a function of ν. The critical distance R_0 is defined as the distance between the donor and the acceptor at which energy transfer occurs with 50% probability. It follows that

$$R_0^6 = \frac{8{\cdot}8 \times 10^{-25} \kappa^2 \phi_D}{n^4} \int_0^\infty F_D(\nu) \varepsilon_A(\nu) \frac{d\nu}{\nu^4} \quad (2)$$

Efficient transfer by this mechanism occurs only when there is a significant overlap between the donor emission and the acceptor absorption spectra. Combining (1) with (2) one obtains, for the rate of exchange between donor and acceptor in a chain

$$k_{DA} = (1/\tau_D)(R_0/R)^6 \quad (3)$$

where R is now the distance between the donor and the acceptor chromophore in the polymer.

A theory for energy migration in polymer chains has been developed by Guillet.[19] If the migration process is visualized as occurring between donor units on the chain (not necessarily adjacent units) the rate constant k_{DD} for the migration step is given by

$$k_{DD} = (1/\tau_D)(R_0'/R_D)^6 \quad (4)$$

where R_0' is the Förster radius for the transfer between identical donor chromophores and R_D is the distance between them. The number of hops, n, is given by

$$n = 2k_{DD}\tau_D = 2(R_0'/R_D)^6 \quad (5)$$

where R_D is now the average distance between chromophores along the polymer chain. The factor 2 is introduced because there will be two chromophores approximately equally spaced along either side of each excited chromophore in the chain. In the absence of any trap in the polymer, this hopping will follow a random walk primarily along the backbone of the chain until the energy is deactivated by other processes. The average displacement of the excitation in this one-dimensional random hopping mechanism will be given by

$$\bar{l} = \sqrt{2nR_D} \tag{6}$$

In the presence of a singlet energy trap, the lifetime of the excitation will be reduced and the number of hops as well as the distance diffused by energy migration will be reduced. The probability of an energy hop from donor to donor versus one directly to the acceptor trap is given by the ratio of the two rate constants:

$$k_{DD}/k_{DA} = (1/2)(R_D/R_0')^6(R_0/R)^6 \tag{7}$$

Since R_0 and R_0' can be calculated from spectroscopic data and R_D can be estimated from the polymer dimensions, this equation permits an evaluation of the relative sphere of influence of the two types of energy transfer. The radius R_F defines a sphere around the acceptor chromophore inside of which direct Förster transfer from the absorbing chromophore to the trap is favored. Outside of R_F at least one energy migration step will occur before the energy is trapped by the acceptor.

This situation is shown in Fig. 1 using naphthalene chromophores as the donors and anthracene as the acceptor. Such a polymer is similar to those synthesized by Guillet and coworkers where the anthracene is introduced by copolymerization with a monomer such as anthrylmethyl methacrylate (AMMA). In this system, the best estimates for R_0 and R_0' are 24 and a minimum of 7 Å, respectively. Since most naphthalene polymers exhibit excimer formation, which requires an approach of close to 3 Å between naphthalene groups, a value of approximately 4 Å is plausible for R_D, the average distance between naphthalene chromophores in simple copolymers such as 1- and 2-vinylnaphthalene (1- and 2-VN) and possibly even for naphthyl esters, such as 1- and 2-naphthylmethyl methacrylate (1- and 2-NMMA). With these assumptions, one can calculate the values of R_1, R_{10} and R_{100} as 12·2, 17·9 and 26·2 Å, respectively, using the relation

$$R_N = 0{\cdot}89(R_D/R_0')R_0 n^{1/6} \tag{8}$$

These correspond to the radii of spheres centered on the anthracene trap in

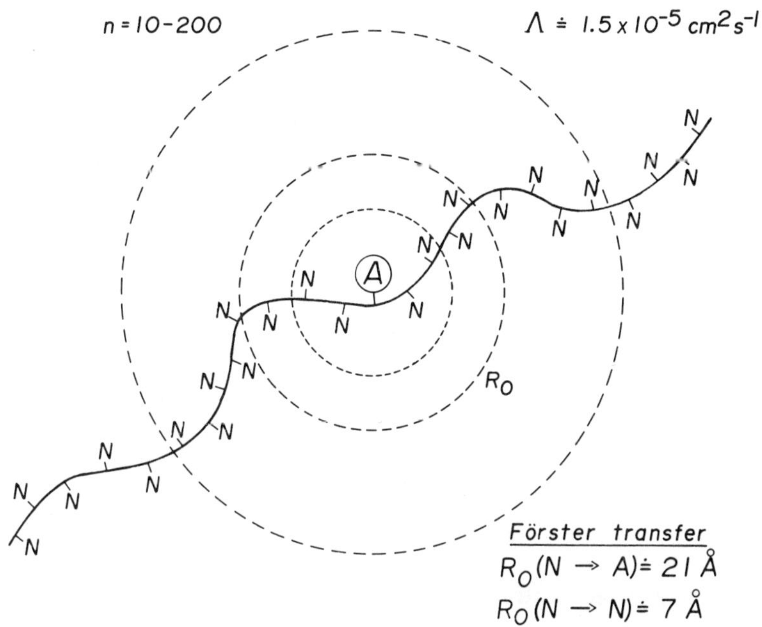

FIG. 1. Mechanism of the antenna effect in a naphthalene-substituted polymer containing an anthracene trap. N = naphthalene; A = anthracene.

which, for example, at R_1 the possibility of a naphthalene to naphthalene transfer is equally probable to the transfer from naphthalene to anthracene. At R_{10} there will be, on the average, 10 naphthalene-to-naphthalene transfers for every naphthalene–anthracene transfer, and at R_{100} there will be 100 naphthalene-to-naphthalene transfers for each transfer from naphthalene to anthracene.

The spheres illustrated in the two-dimensional projection in Fig. 1 also include the standard Förster radius R_0 of 24 Å. From this diagram it can be seen that although a substantial number of naphthalene-to-naphthalene transfers may occur at distances well within the Förster radius, they will not be expected to contribute to efficiency of energy collection, since the energy would be transferred to anthracene in any case as long as the radius was less than the Förster radius. The only increase in efficiency of energy collection due to the energy migration process will occur for naphthalene chromophores outside the Förster radius R_f where migration will occur at a rate corresponding to that in the absence of the anthracene trap. The number of migration steps can be calculated from eqns (4) and (5), substituting the value of the natural lifetime of the naphthalene chromophores and an estimate of their separation distance in the polymer. This gives an average number of hops of 57, and a migration distance $\bar{l}$ of 21·3 Å.

The detection and quantification of the role of energy migration in antenna polymer systems requires very careful experimentation and analysis. Clearly, for systems in which the average distance of migration $\bar{l}$ is comparable with the Förster radius for direct transfer from donor to acceptor, most of the prompt emission from anthracene will come as a result of direct single-step Förster transfer from donor to acceptor, and only at longer times after the initiating pulse will the effects of energy migration be observed.

The distinction between this type of energy transfer and that observed in scintillators and other solid aromatic polymers is that the photon energy is collected within a single polymer molecule, and all transfer processes are intramolecular. The antenna effect thus permits the collection of photon energy from the entire region of space representing the hydrodynamic volume of the polymer, and its transmission to traps located on the polymer chain. Unlike the process in scintillator systems, the efficiency is relatively independent of concentration, and can be efficient even in very dilute solutions.

3. EXPERIMENTAL STUDIES OF THE ANTENNA EFFECT IN POLYMER SOLUTIONS

In early studies of transfer within single polymer molecules in solution, Aspler *et al.*[6] showed that anthracene attached to the ends of poly-(naphthyl methacrylate) (PNMA) chains gave strong fluorescence emission when the naphthalene chromophores were excited. Time-resolved fluorescence studies showed that emission from anthracene increased while that from the naphthalene monomer and excimer decreased with times of the order of 1–20 ns. In more detailed studies, Holden and Guillet[7] showed that the efficiency of energy transfer in the naphthyl methacrylate system was low, but much higher efficiencies could be obtained in polymers of 1-naphthylmethyl methacrylate (1-NMMA) terminated with methyl anthracene or with copolymerized 9-AMMA monomer units. The energy efficiency χ can be calculated for such systems from the relations

$$\phi_F = \chi\phi_A - (1 - \chi)\phi_N \qquad (9)$$

$$\chi/(1 - \chi) = \phi_N I_A/\phi_A I_N \qquad (10)$$

where

$$\chi = \frac{\text{number of photons transferred to anthracene}}{\text{number of photons absorbed by naphthalene}}$$

and ϕ_F, ϕ_N and ϕ_A are the fluorescence quantum yields of the total copolymer, the corresponding naphthalene homopolymer, and the anthracene trap on direct excitation, respectively. Usually the naphthalene excitation was at 280 nm and that for direct excitation of anthracene was at 340 nm. Typical experimental fluorescence ($\lambda_{ex} = 280$ nm) curves for anthracene-terminated poly(naphthylmethyl methacrylate) (PNMMA) in THF at 25 °C and in MTHF glass at 77 K are shown in Fig. 2.

At room temperature there is still a correction for excimer emission (Fig. 2a, dotted line), while at 77 K the emission from N and A units is completely resolved. In each case the concentration of anthracene is less than 1 mol-%, but the emission intensity of A is substantially greater than that of the absorbing N group, illustrating the substantial delocalization of the excitation energy.

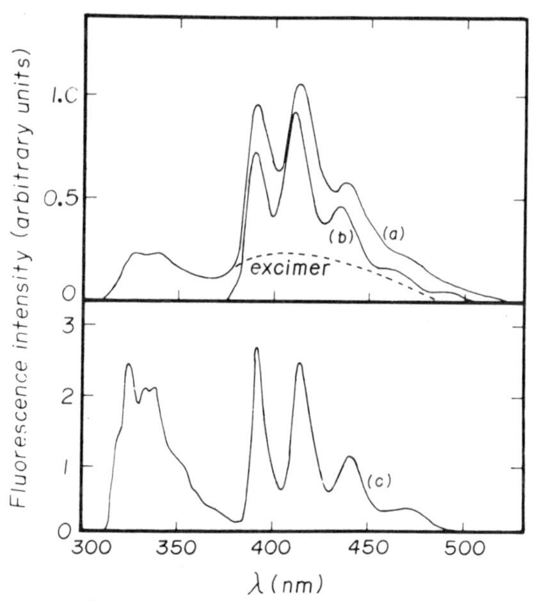

FIG. 2. Steady-state fluorescence spectra in deoxygenated THF at 25 °C of (a) poly(NMMA-co-0·67 % AMMA), $\lambda_{ex} = 280$ nm; (b) 9-anthrylmethyl pivalate (5×10^{-5} M), $\lambda_{ex} = 340$ nm, and (c) poly(NMMA-co-6·7 % AMMA) in 2-MeTHF at 77 K. $\lambda_{ex} = 280$ nm.

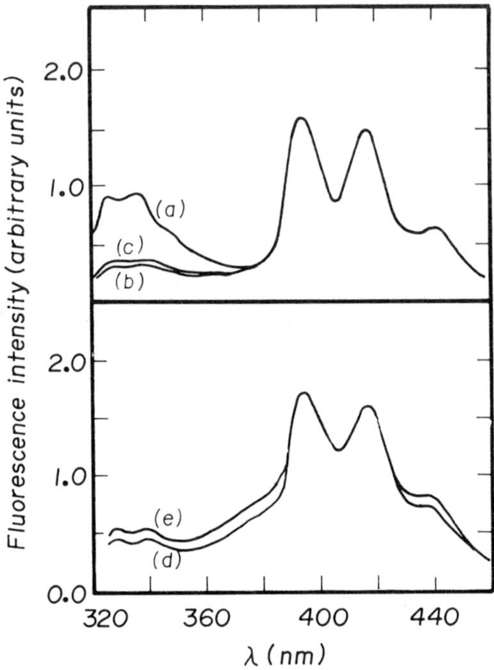

FIG. 3. Time-resolved fluorescence spectra of anthracene-terminated poly-(NMMA) in toluene at 25 °C on excitation at 292 nm. Lower and upper time limits in ns from maximum of the exciting pulse are (a) 0·0–2·2, (b) 11·5–13·8, (c) 21·0–24·0, (d) 44·3–48·6 and (e) 179·2–228·5. The curves are normalized to the same intensity at 390 nm.

TABLE 1
Efficiency χ of Energy Transfer to Anthracene in Copolymers of 1-Naphthylmethyl Methacrylate with 9-Anthrylmethyl Methacrylate

Mol-% anthracene in copolymer	*χ (%) in 2-MeTHF at 77 K*	*χ (%) in THF at 25 °C*
0·174	8	9·9
0·348	15	18
0·665	22	30
1·42	61	69
2·06	70	84

The χ values for a series of copolymers with increasing amounts of anthracene traps are shown in Table 1. The efficiencies in excess of 70% are comparable to those estimated for the antenna chlorophyll in the chloroplasts in green plants. Furthermore, the relatively high efficiency in 2-MeTHF glass at 77 K suggests that collisional processes and molecular mobility are not required for efficient energy migration and transfer.

Time-resolved emission spectroscopy (Fig. 3) and emission decay curves (Fig. 4) show that the anthracene emission is observable even at the shortest time scale, and that in good solvents and expanded coils, energy migration and transfer require 5–20 ns or longer to occur. The rapid quenching of the naphthalene units close to anthracene is demonstrated by the very rapid decay of naphthalene monomer fluorescence in the first few nanoseconds, as shown in Fig. 4B.

Kinetic analysis of the fluorescence decay curves of the donor and acceptor are consistent with a mechanism involving rapid one-step Förster transfer from the N chromophores nearest the trap, followed by a slower contribution from energy migration through more distant N donors.

Ng and Guillet[9,10] have synthesized similar polymers using phenanthrylmethyl methacrylate (PhMMA) as the antenna monomer. This polymer

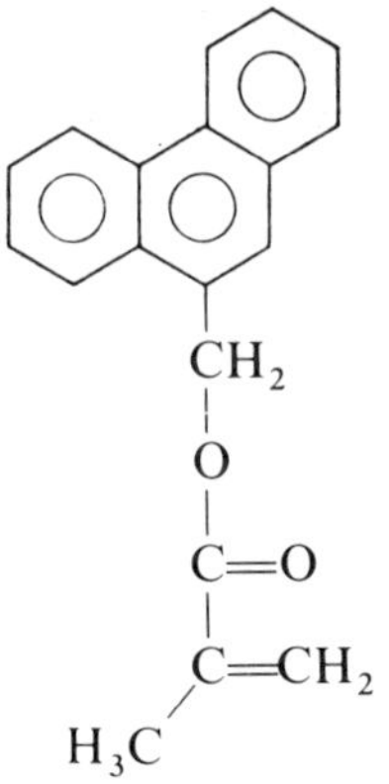

CH_2
O
C=O
$C{=}CH_2$
H_3C

has the advantage that no excimer emission is observed for the phenanthrene moieties, even at room temperature, hence removing a complication of the kinetic analysis. Based on a detailed study of emission kinetics and depolarization studies of copolymers of poly(NMMA) with AMMA and methyl methacrylate spacer groups, Ng and Guillet[10] calculated an average distance for energy migration $\bar{l} = 120$ Å and a value of $\Lambda = 1{\cdot}5 \times 10^{-5}\,cm^2\,s^{-1}$ which is comparable to the bulk diffusion constant of small molecule diffusants.

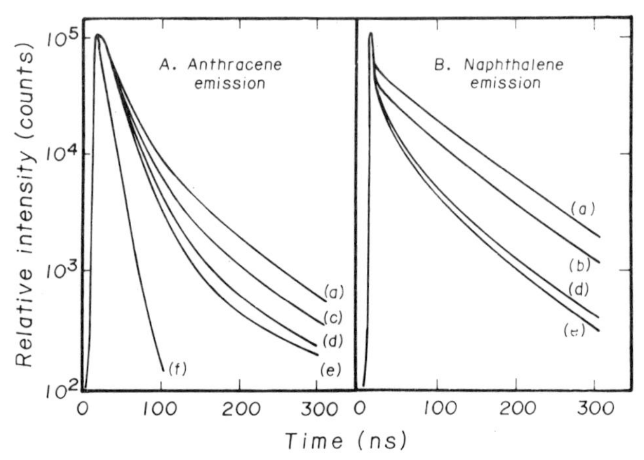

FIG. 4. Fluorescence decays at 280 nm excitation of poly(NMMA-AMMA) samples in nitrogenated THF at 77 K. The mol-% AMMA in the copolymer samples is (a) 0·174, (b) 0·348, (c) 0·665, (d) 1·42 and (e) 2·06. Curve (f) is the decay of the anthracene emission on direct excitation at 366 nm.

In both the NMMA and PhMMA copolymers, the efficiency χ depends on the hydrodynamic volume, which can be estimated by the intrinsic viscosity in solution, being higher in poor solvents where the coil is compressed to smaller dimensions. Typical results for poly(NMMA) are shown in Table 2. As the non-solvent, methanol is added to the solution, the polymer coil contracts, thus reducing the average distance between donor chromophores and the anthracene trap, thereby increasing both the rate and efficiency of energy transfer.

Copolymers of poly(1-) and (2-NMMA) have been prepared with a variety of trap sites, including methyl and phenyl ketones,[15,17,20]

TABLE 2

Effect of Solvent on Efficiency χ of Energy Transfer from Naphthalene to Anthracene in Anthracene-terminated Poly(NMMA) at 25 °C

Solvent	χ (%)
THF	10
Dioxane	11
Dioxane/methanol (60/40 v/v)	21
Dioxane/methanol (40/60 v/v)	47
Dioxane/methanol (20/80 v/v)	66
Poly(methyl methacrylate)	43

TABLE 3
Efficiency χ of Quenching of Naphthalene Fluorescence by Aromatic Ketone in Copolymers of 2-Vinylnaphthalene and Phenyl Vinyl Ketone[a]

Mol-% ketone in copolymer	χ (%)	*No. of donors quenched by each acceptor*
0·17	12	71
0·35	21	60
0·68	35	51
0·995	49	50
1·04	51	49
3·8	71	[b]
8·0	92	[b]
12·0	94	[b]

[a] Solutions in 2-MeTHF at 77 K.
[b] Overlap of quenching domains prevents analysis.

anthraquinone,[21] and ferrocene.[22] All of these groups have been shown to trap migrating singlet excitons in single polymer chains with varying degrees of efficiency. However, the aromatic esters of methacrylic and acrylic acid are not particularly well designed as antenna polymers because the aromatic groups are widely separated, even in adjacent units along the polymer chain. Poly(2-vinylnaphthalene) would seem to represent a better model antenna in that the naphthalene groups are separated by only three carbon atoms and thus obey the so-called Hirayama $n = 3$ rule.

Studies of copolymers of 2-vinylnaphthalene (2-VN) containing phenyl vinyl ketone (PVK) traps were reported by Holden *et al.*[15,17] In this case the traps do not emit fluorescence themselves but the efficiency of energy transfer can be determined by the intramolecular quenching of naphthalene monomer fluorescence from the donor chromophores. The efficiency χ of energy transfer is shown in Table 3 as a function of the mol-% ketone in the polymer along with the number of donors quenched by each acceptor. The experiments were carried out at low temperature (77 K) in MTHF glass in order to reduce complications from the formation of excimers. Even at this low temperature a substantial amount of quenching is observed with relatively low quantities of the phenyl ketone trap. Furthermore, the unquenched fluorescence emission from the naphthalene donors was completely depolarized at all excitation wavelengths. The number of donors, n, quenched by each trap can be calculated from the relation $\chi = F_A n$ where F_A is the mole fraction of PVK in the copolymer. The values listed in Table 2 show that below 1 % of the PVK traps, n is fairly constant at 60 ± 10. Assuming that the excitation migrates in 2-VN sequences on both sides of the acceptor, this value of n corresponds to a mean range of singlet energy migration of 30 ± 5 chromophores.

Further evidence for energy transfer was obtained from fluorescence decay measurements using a single-photon counting apparatus of conventional design. In the absence of the ketone traps, decay of the emission from poly(2-VN) at 77 K is exponential with the decay time $\tau_0 = 80{\cdot}5$ ns. With increasing amounts of PVK traps the donor decay becomes increasingly non-exponential and the decay times shorten considerably. Holden *et al.*[15] showed that it was not possible to fit the decay curves with a model based on a one-step Förster transfer from the absorbing naphthalene to a bound trap, but they could be fitted reasonably well with a model involving a sequence of energy migration steps.

Further evidence for energy migration in an antenna polymer was reported by Holden *et al.*[13] who used copolymers of PhMMA end-trapped with anthracene. These polymers have the advantage that the phenanthrene group does not form excimers, even at room temperature, and hence the kinetics for the migration process can be studied in solutions at room temperature rather than in low-temperature glasses, as in the previous case. In this experiment, studies were made of the efficiency of quenching of polymer fluorescence by oxygen and compared with similar studies on small-molecule models using the Stern–Volmer equation,

$$\phi_0/\phi = \tau_0/\tau = 1 + k_q \tau_0 [Q] \tag{11}$$

In this equation, ϕ_0, τ_0, ϕ and τ represent the fluorescence quantum yields and lifetimes in the absence and presence, respectively, of the quencher. $[Q]$ is the quencher concentration and k_q the bimolecular rate constant of quenching. k_q can be represented in the form

$$k_q = 4\pi N_0 (D_F + D_Q) p R \tag{12}$$

In this modification of the Smoluchowski–Einstein equation, D_F and D_Q are the diffusion coefficients of excited chromophore and quencher, respectively, p represents the reaction probability per collision, R is the sum of the collision diameters for the two species and N_0 is Avogadro's number. In comparison with values of k_q for small-molecule and polymer chromophores it is assumed that p and $[Q]$ are the same for both species and that D_F for the fluorophore on the polymer is essentially 0. If k_q for the polymer is larger than one-half of k_q for the small molecule compound, the excess is attributed to singlet energy migration between the chromophores and the homopolymer.

In these experiments the rate constants for quenching of the anthracene chromophore by oxygen was compared under conditions (1) where the anthracene was excited directly and (2) where it was excited by energy migration from the naphthalene donors. Based on these measurements the authors concluded that the rate of energy migration corresponded to an apparent diffusion constant of $1{\cdot}3 \times 10^{-9}\ m^2/s$ which is very close to that of $1{\cdot}5 \times 10^{-9}\ m^2/s$ derived by Ng and Guillet for the same polymers using an entirely different experimental procedure. This apparent diffusion coefficient is about one-third of the mobility of an oxygen molecule in the same solution.

If the process of energy migration is treated as a random walk along the contour length of the chain, the average displacement $\bar{l}$ of the excitation during the excited state liftime τ of the phenanthrene would be $\bar{l} = (2\Lambda\tau)^{1/2}$. From these results, $\bar{l}$ is estimated to be about 11·5 nm, a distance which represents an appreciable fraction of the contour length of an expanded polymer coil in solution.

In the original studies on naphthalene antenna polymers containing ketone traps[20] it was expected that the transfer of excitation energy from naphthalene absorbing sites to the ketone trap via a migrating exciton would increase the efficiency of photochemical reactions arising from the ketone trap. In particular, it was expected that the Norrish type I and type II processes would lead to rapid chain scission in such polymers. In fact, the opposite was observed experimentally. In addition, in studies of the delayed emission in polymers dissolved in glassy solvents at 77 K it was shown that there was a considerable increase in phosphorescence emission. These results lead to the observation of an effect known as 'cyclic energy transfer' and is illustrated in Fig. 5 for copolymers of 2-NMMA and PVK.[17,20]

The process can be described in the following sequence of steps illustrated in Fig. 5. (1) The absorption of light by the naphthalene chromophore ($\lambda = 280$ nm) to the first excited singlet state with an energy level of approximately 90 kcal, and (2) the energy migration and transfer to the phenyl ketone trap which has a singlet energy level of approximately 61 kcal. However, the phenyl ketone has a very rapid rate of intersystem crossing so that the energy is quickly converted to the triplet of the phenyl ketone by step (3). The triplet phenyl ketone is adjacent to two naphthalene chromophores which quickly quench the triplet, giving rise to triplet naphthalene (4) which may decay by the emission of phosphorescence (5).

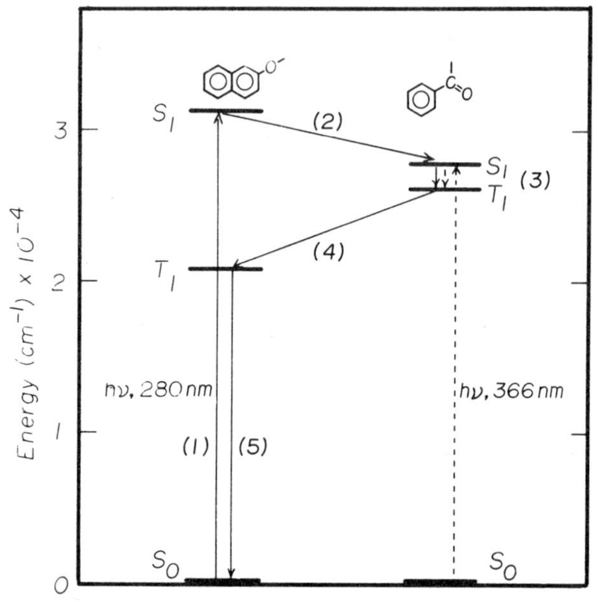

FIG. 5. Energy level diagram for poly(1-naphthyl methacrylate-co-phenyl isopropenyl ketone). Also shown are the principal deactivation pathways on excitation of naphthalene at 313 nm (full line) and on excitation of aromatic ketone at 366 nm (broken line).

Because of the rapidity of the energy migration process, which occurs in 5–10 ns, and the high efficiency of intersystem crossing from singlet ketone to the triplet, plus rapid quenching of the triplet ketone by the naphthalene, this route provides a more efficient route to the population of triplet naphthalene than the direct intersystem crossing in the naphthalene groups themselves. Thus the intensity of phosphorescence goes up, as shown in Table 4.[15]

In addition to the increase in phosphorescence, the antenna effect provides a facile route to delayed fluorescence which also increases, as shown in Table 4. Because of the relatively long lifetime of the naphthalene triplet it is possible to obtain two migrating triplet excitons on the same chain within the lifetime of the first excitation. If these two triplets migrate to adjacent chromophores, they will undergo triplet–triplet annihilation giving rise to delayed emission with the spectral characteristics of fluorescence but with the liftime corresponding to that of phosphorescence. Since the probability of having two excitations on the same chain is proportional to the square of the degree of polymerization of the polymer it is not surprising that a high efficiency of delayed emission is observed in this

TABLE 4
Effect of Copolymer Composition on Relative Phosphorescence Intensity I_p and Efficiency q of Formation of Delayed Singlets in Poly(2-VN-PVK) Samples[a]

Mol-% ketone in copolymer	I_p[b]	q[c]
0·0	1·0	1·0
0·17	1·03	1·2
0·35	1·3	1·1
0·68	1·3	1·4
0·995	1	1·8
1·04	1·7	1·8
3·8	2·0	1·9
8·0		3·0
12·0	3·6	2·3

[a] Solutions in 2-MeTHF at 77 K. Excitation at 280 nm.
[b] From total emission spectrum.
[c] From delayed emission spectrum, corrected for quenching by copolymerized ketone.

case. This represents a further advantage of antenna polymers in that they make it possible to observe multi-photon processes in condensed phases with illumination of relatively low intensities. This may also be an important factor in future work on solar energy conversion.

The first example of a photochemical reaction induced in a polymer via the antenna effect was recently reported by Ren and Guillet.[21] They synthesized copolymers of 1-NMMA containing small amounts of 1-anthraquinonylmethyl methacrylate (1-AQMMA). It was shown that the presence of the anthraquinone group caused a marked reduction in the fluorescence emission from the naphthalene groups adjacent to the anthraquinone traps. Approximately 100 N groups were quenched for each anthraquinone group in the chain. In dry benzene, no photochemical reaction was observed, but in THF containing traces of water, the copolymers were very photosensitive, and on exposure to light the anthraquinone traps were rapidly reduced to the corresponding anthrahydroquinones:

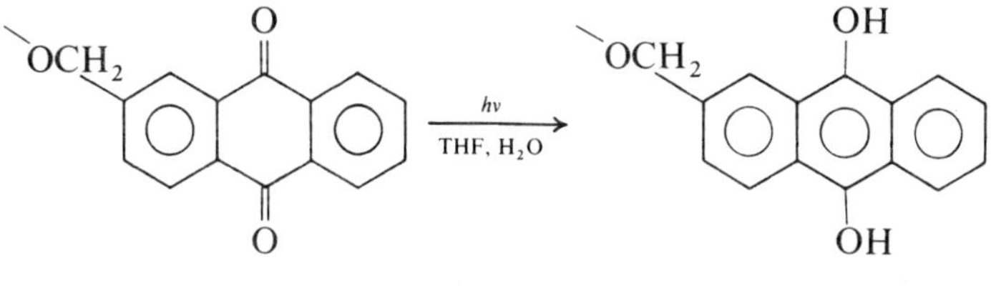

O
OCH$_2$
O
$\xrightarrow[\text{THF, H}_2\text{O}]{h\nu}$
OH
OCH$_2$
OH

This represents the storage of one mole of hydrogen per anthraquinone trap, and because the anthraquinone–anthrahydroquinone system is electrochemically reversible it may suggest a future pathway for solar energy conversion.

4. WATER-SOLUBLE PHOTON-HARVESTING POLYMERS

The successful synthesis of photon-harvesting polymers soluble in organic solvents, such as the copolymers of naphthyl acrylates, methacrylates, and vinylnaphthalene polymers, led to a series of experiments to determine if these principles could be applied to polymers which might provide an effective photon-harvesting system in aqueous media. Holden *et al.*[8] reported the synthesis of copolymers of acrylic acid and 1-NMMA which, when terminated with anthracene, showed highly efficient photon-harvesting in dilute aqueous solution at pH 12. In a typical experiment with an anthracene-terminated copolymer containing 92% acrylic acid and 8% 1-NMMA, the efficiency of energy transfer χ was only 12% in dioxane but increased to 70% when dissolved in dilute aqueous NaOH (pH 12). The surprisingly high efficiency of energy transfer in water solutions was explained by these authors as being due to 'hypercoiling' of the polyelectrolyte chains to form the pseudo-micellar conformations illustrated schematically in Fig. 6.† These unusual conformations bring the hydrophobic aromatic groups closer together in the center of the coil while leaving the hydrophilic carboxylate ions on the exterior. This explains the much higher values of χ observed in aqueous media than in the organic solvent dioxane, where a more random conformation of the polymer would be expected, leading to a greater average distance between the naphthalene donors and the trap.

This work was extended by Gu and Guillet[23] who synthesized water-soluble polymers prepared by partially sulfonating poly(1-) and (2-VN). These polymers also undergo hypercoiling in dilute aqueous solution and when exposed to organic solutions of large aromatic chromophores, will solubilize these molecules, presumably in the central hydrophobic region of the polymeric micelle. These molecules act as efficient traps for migrating singlet excitons created by the absorption of light by the naphthalene moieties. The effect can be quantified by measurements of the efficiency of fluorescence emitted by the trap. In special cases the traps will undergo very rapid photochemical reactions as a result of excitation via the antenna

† This work was first reported at the Eighth Katzir Conference on Luminescence from Biological and Synthetic Macromolecules, June 16–18, 1980, in New York.

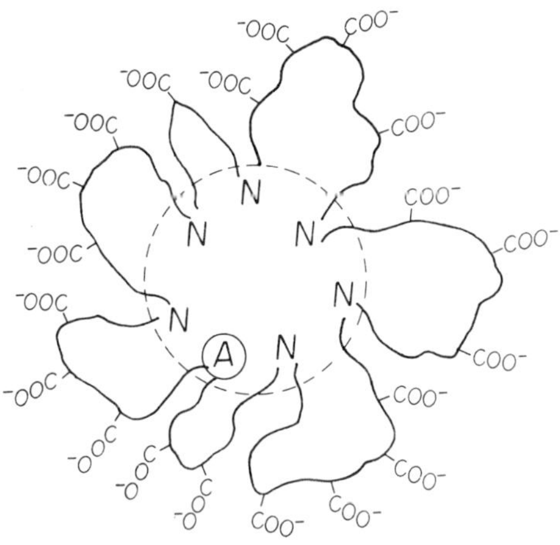

FIG. 6. Proposed antenna structure in water.

effect in these polymeric colloids.[24] After reaction has taken place, extraction of the aqueous solution with an organic solvent removes the products of reaction, and the polymeric catalyst remains unchanged and can be re-used. By analogy with biological systems, these new polymeric photocatalysts, which can be quite specific for certain substrates, have been called 'photozymes'.

5. CONCLUSIONS

Experimental studies of the antenna effect in polymer molecules have demonstrated that efficient photon harvesting takes place in a variety of polymers containing suitable repeating chromophores. Efficiencies approach those observed in natural chloroplasts where the chlorophyll molecules are arranged in a regular manner in the thylakoid membrane. Antenna polymers can be prepared which function well in both organic and aqueous media, and can stimulate both photophysical and photochemical processes. Because of their ability to increase the probability of biphotonic processes in low-level light intensities, these polymers present interesting possibilities as catalysts for solar energy conversion systems.

ACKNOWLEDGEMENT

The financial support of this work by the Natural Sciences and Engineering Research Council of Canada is gratefully acknowledged.

REFERENCES

1. Porter, G. and Archer, M., *Interdiscip. Sci. Rev.*, **1**, 119 (1976).
2. Guillet, J. E., *Pure appl. Chem.*, **49**, 249 (1977).
3. Schneider, F. and Springer, J., *Makromol. Chem.*, **146**, 181 (1971).
4. Fox, R. B. and Cozzens, R. F., *Macromolecules*, **2**, 181 (1969).
5. Hoyle, C. E. and Guillet, J. E., *J. Polym. Sci., Polym. Lett.*, **16**, 185 (1978).
6. Aspler, J. S., Hoyle, C. E. and Guillet, J. E., *Macromolecules*, **11**, 925 (1978).
7. Holden, D. A. and Guillet, J. E., *Macromolecules*, **13**, 289 (1980).
8. Holden, D. A., Rendall, W. A. and Guillet, J. E., *Ann. NY Acad. Sci.*, **366**, 11 (1981).
9. Ng, D. and Guillet, J. E., *Macromolecules*, **15**, 724 (1982).
10. Ng, D. and Guillet, J. E., *Macromolecules*, **15**, 728 (1982).
11. Holden, D. A. and Guillet, J. E., *Macromolecules*, **15**, 1475 (1982).
12. Thomas, J. W., Jr, Frank, C. W., Holden, D. A. and Guillet, J. E., *J. Polym. Sci., Polym. Phys. Ed.*, **20**, 1749 (1982).
13. Holden, D. A., Ng, D. and Guillet, J. E., *Br. Polym. J.*, **14**, 159 (1982).
14. Ng, D., Yoshiki, K. and Guillet, J. E., *Macromolecules*, **16**, 568 (1983).
15. Holden, D. A., Ren, X.-X. and Guillet, J. E., *Macromolecules*, **17**, 1500 (1984).
16. Guillet, J. E., *Polym. Prepr., Am. Chem. Soc., Div. Polym. Chem.*, **20**(1), 395 (1979).
17. Holden, D. A., Shephard, S. E. and Guillet, J. E., *Macromolecules*, **15**, 1481 (1982).
18. Förster, Th., *Ann. Phys.*, **2**, 55 (1948).
19. Guillet, J. E., *Polymer Photophysics and Photochemistry*, Cambridge University Press, Cambridge, 1984, Ch. 9.
20. Merle-Aubry, L., Holden, D. A., Merle, Y. and Guillet, J. E., *Macromolecules*, **13**, 1138 (1980).
21. Ren, X.-X. and Guillet, J. E., manuscript submitted for publication.
22. Buchanan, K. J. and Guillet, J. E., unpublished work.
23. Gu, L.-Y. and Guillet, J. E., manuscript in preparation.
24. Gu, L.-Y. and Guillet, J. E., manuscript in preparation.

3

Fluorescence Studies of Equilibrium and Dynamic Properties of Polymer Systems

HERBERT MORAWETZ

Department of Chemistry, Polytechnic Institute of New York, New York, USA

1. INTRODUCTION

Fluorescence techniques have long been employed for the study of biological macromolecules but intensive studies of their application to the investigation of synthetic polymers have been reported only in recent years. In the present review I shall not be concerned with the many studies of the photophysics of polymers but restrict myself to results in which fluorescence is used as a tool to further our understanding of the equilibrium and dynamic properties of polymeric systems. To this end four phenomena have been employed, i.e. (a) the depolarization of fluorescence, (b) excimer emission, (c) nonradiative energy transfer and (d) the dependence of the emission characteristics of certain fluorophores on the solvent medium.

2. POLYMER MISCIBILITY

It is well known that the mixing of two polymeric species involves a negligible change of entropy so that thermodynamically stable blends are obtained only if the mixing is exothermic. Although one-phase and two-phase systems can be distinguished by a variety of other techniques (DSC, neutron scattering, FT–IR spectroscopy, electron microscopy) two approaches which use fluorescence for the characterization of polymer compatibility have distinct advantages. In the first of these techniques the two polymers are labeled with species A and B, respectively, such that the emission from A (the donor) overlaps the absorption spectrum of B (the acceptor). When A is irradiated, the efficiency of energy transfer to B is highly sensitive to their spacing. Thus the ratio of emission intensities from B and A is much greater in a homogeneous than in a two-phase system.[1-3] A typical result is shown in Fig. 1 in which data are plotted for the blends of poly(methyl methacrylate) with copolymers of styrene and acrylonitrile. The range of copolymer compositions for which miscibility is obtained is clearly indicated by an efficient energy transfer. More recently, the technique has been used to study the mixing of a polystyrene-poly-(t-butylstyrene) block copolymer with the two homopolymers.[4] In that case the label was attached only on the polystyrene portion of the block copolymer and the results showed clearly that this block mixed well with polystyrene but not with poly(t-butylstyrene).

The other technique developed to study polymer compatibility utilizes the phenomenon of excimer fluorescence. In a substance such as poly-(2-vinylnaphthalene) (PVN) excimer emission is much more prominent for the polymer in bulk (where intermolecular excimers can form) than in a dilute system in which only intramolecular interaction of the chromophores leads to excimer formation. Thus, when a trace of PVN was added as a probe to a series of poly(alkyl methacrylates), a minimum in the excimer emission indicated a solution of the PVN probe in the polymethacrylate matrix.[5] Later, this approach was used for a study of the dependence of the PVN probe miscibility on the molecular weight of the polystyrene 'host'[6] and for

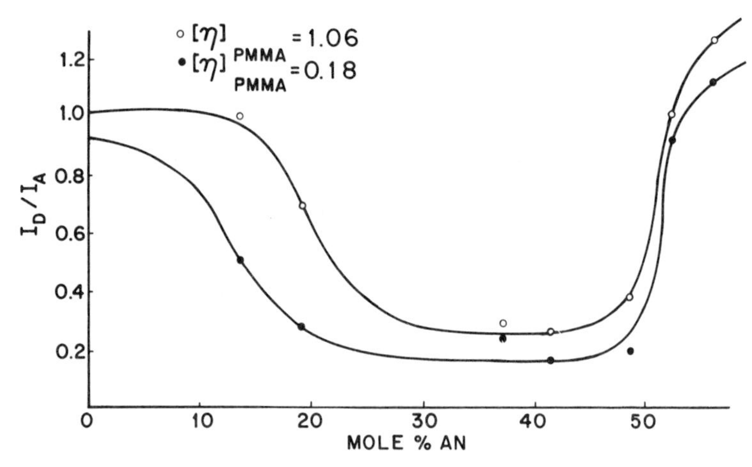

FIG. 1. Ratio of carbazole donor and anthracene acceptor fluorescence intensities in blends of donor-labeled poly(methyl methacrylate) and acceptor-labeled styrene–acrylonitrile copolymers. Intrinsic viscosities of PMMA in benzene at 25°C.

a study of the kinetics of phase separation in the polystyrene–poly(vinyl methyl ether) blends heated above the lower critical solution temperature.[7]

3. CHAIN ENTANGLEMENTS IN POLYMER SOLUTIONS

Dilute polymer solutions in good solvent media are characterized by a strong entropic resistance to the interpenetration of the molecular coils. However, as the polymer concentration increases the swollen coils are eventually forced to form entanglements. This has conventionally been characterized by a 'critical concentration' c^* which is inversely proportional to the intrinsic viscosity or the 3/2 power of the mean square radius of gyration.

Since it is obvious that chain entanglements do not occur suddenly at a well defined concentration but increase in importance over a range of concentrations it seemed desirable to develop a spectroscopic method by which the extent of chain interpenetration could be followed as a function

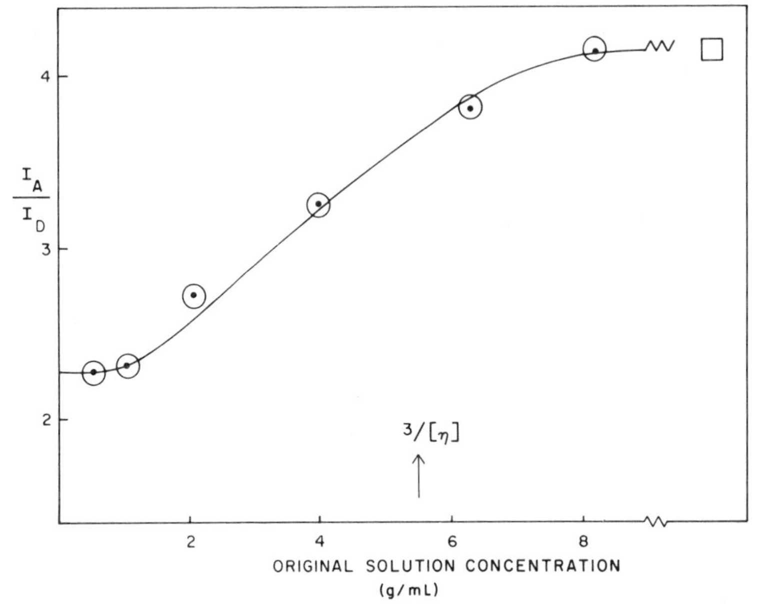

FIG. 2. Ratio of anthracene and carbazole emission intensities from a mixture of donor and acceptor labeled polystyrene ($\bar{M}_\eta = 118\,000$) freeze dried from benzene. Label concentrations $9{\cdot}7 \times 10^{-3}$ moles/kg. The square represents the result obtained with a cast film.

of the solution concentration. To this end we froze rapidly a solution containing a mixture of two similar polymers labeled with a donor and an acceptor fluorophore, respectively, hoping that the extent of chain entanglement would be retained. After subliming the solvent, the powder was pressed into pellets and the reflectance fluorescence spectrum was recorded. As expected, the chemically identical samples exhibited a nonradiative energy transfer efficiency reflecting the concentration of the solution from which they originated.[8] Figure 2 illustrating the results obtained recently by this technique shows that they reveal the increase in chain entanglements over a range of concentrations.

4. DYNAMICS OF CONFORMATIONAL TRANSITIONS IN POLYMER SOLUTIONS

An experimental technique which would allow us to compare the rate of hindered rotations around a bond in the backbone of a long polymer chain and in a small analogous molecule has to be sensitive to conformational transitions with relaxation times in the nanosecond range while being unaffected by rotational motions of the molecule as a whole. These conditions are beautifully met by the phenomenon of intramolecular excimer formation, since the ratio of excimer and 'normal' fluorescence reflects the ratio of the rate constant for the hindered rotation by which the excimer is formed and the rate constant for emission from the excited 'monomer'.[9]

We prepared[10] a polymer with an excimer forming site incorporated into the middle of the chain and an analogous small molecule:

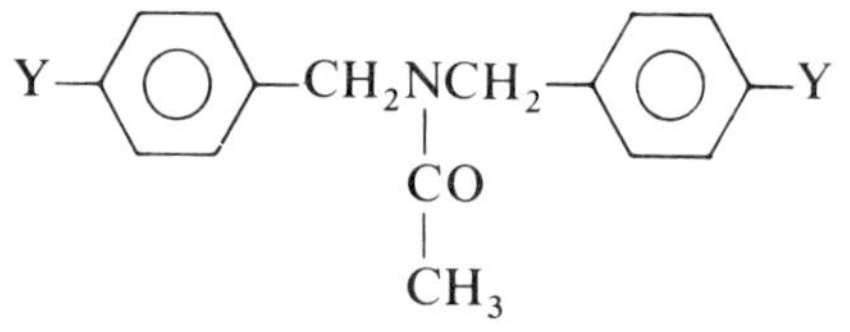

$$\text{Y}-\text{C}_6\text{H}_4-\text{CH}_2\text{N}(\text{COCH}_3)\text{CH}_2-\text{C}_6\text{H}_4-\text{Y}$$

Polymer: $\text{Y} = \text{HO}-(-\text{CH}_2\text{CH}_2\text{O}-)_{68}-\underset{\substack{\|\\ \text{O}}}{\text{C}}\text{NH}(\text{CH}_2)_6\text{NH}\underset{\substack{\|\\ \text{O}}}{\text{C}}\text{OCH}_2-$

Analog: $\text{Y} = \text{CH}_3-$

It was found that the ratio of excimer and monomer emission was only 27 % smaller in the polymer than in the analog and that the temperature dependence of this ratio was identical in the two cases. This seemed to rule out conformational transitions in backbones of polymer chains requiring two simultaneous hindered rotations ('crankshaft-like motions').

Conformational mobility of long polymer chains may also be studied by the rate at which intramolecular diffusion-controlled interactions take place between groups attached to the two ends of a polymer chain. This principle has been used in a study of the behavior of 'monodisperse' polystyrenes carrying pyrene residues at the two chain ends.[11,12] If the pyrene is excited in highly dilute solutions of the polymer, its 'monomer' emission will be observed only if no intramolecular excimer is formed during the lifetime of the excited state. Thus, data on the monomer emission from pyrene-terminated polystyrenes yield information about the rate at which the two chain ends collide for polystyrenes of varying molecular weight. The results show also that the greater chain expansion in a good solvent medium will reduce sharply the frequency with which the chain ends collide with each other.

More recently, studies have been reported on chain molecules carrying many pyrene groups separated by a constant number of styrene residues. Excimer formation in solutions of these substances was found again to be sensitive to the solvent power of the medium; plots of the ratio of excimer and monomer emission intensity (corrected for variations in solvent viscosity) against the Hildebrand solubility parameter of the solvent exhibited a sharp minimum.[13] This was interpreted as reflecting changes in the polymer expansion; this was taken as a maximum (minimizing excimer formation) when the cohesive energy density of the polymer matched that of the solvent. Addition of unlabeled polymer had a surprisingly small effect on excimer emission[14] showing that the conformational mobility of chain molecules is less sensitive to polymer concentration than one might have expected.

Another approach to the study of the dynamics of relatively short chains utilizes energy transfer between a donor and an acceptor group attached to the two chain ends.[15] In this case it could be demonstrated that energy transfer is reduced by a high viscosity of the medium which inhibits the diffusion of the donor and acceptor fluorophores towards each other. In fact, the fluorescence decay curve can be interpreted as reflecting an apparent intramolecular diffusion coefficient of these end-group labels.

5. SELF DIFFUSION IN BULK POLYMERS

Since the mean square diffusion distance $\bar{x}^2$ in a specified direction is related to the time t and the diffusion coefficient D by $\bar{x}^2 - 2Dt$, the smallest diffusion coefficient which can be measured depends critically on the spatial resolution of the experimental method. According to generally accepted theories of deGennes[16] and of Doi and Edwards,[17] the self-diffusion coefficient of chain molecules should be inversely proportional to the square of the chain length, leading to extremely low values for polymers with molecular weights above 10^5. Thus, determinations of D would require very high spatial resolutions for any reasonable experimental time.

The problem was brilliantly solved by the use of a procedure in which the polymer was tagged with a fluorescent label and the label bleached so as to produce a periodic pattern with a spacing of the order of micrometers. The disappearance of the pattern could then be interpreted in terms of diffusion coefficients down to $10^{-13}\,cm^2\,s^{-1}$.[18] An alternative method employed a mixture of donor and acceptor labeled polymers freeze-dried from highly dilute solution, so that each polymer coil collapses to a small globule. If now a pellet pressed from this material is heated to a temperature at which the diffusion coefficient is to be measured, the gradual interdiffusion of the polymer molecules is reflected in an increase of the nonradiative energy transfer.[19] It should be understood that the parameter measured in this experiment is not, strictly speaking, a diffusion coefficient since the chain molecules have to 'unfold' from a tightly packed structure to their equilibrium conformational distribution. On the other hand, this technique has the virtue of experimental simplicity and a very high spatial resolution, since the polymer globules at the outset of the experiment have diameters less than 10^{-6} cm.

6. COOPERATIVE POLYMER ASSOCIATION

It has been known for a long time that polycarboxylic acids form in aqueous solution complexes with polymers acting as hydrogen bond acceptors, polyoxyethylene or poly(vinyl pyrrolidone), and this phenomenon has been studied by a drop in the solution viscosity, a rise of pH, a drop in conductivity, calorimetry and ultracentrifuge sedimentation. In an alternative experimental technique poly(acrylic acid) was tagged with the dansyl group which fluoresces much more intensely in organic media than in water. Thus, when a complex is formed with PVE or PVP, the displacement of the water in the neighborhood of the dansyl label leads to a sharp increase in its emission intensity.[20] Figure 3 illustrates this effect for the formation of the complex with PVP. The results imply the formation of complexes containing equal numbers of acrylic acid and vinyl pyrrolidone monomer residues. As the pH is raised, leading to an ionization of the

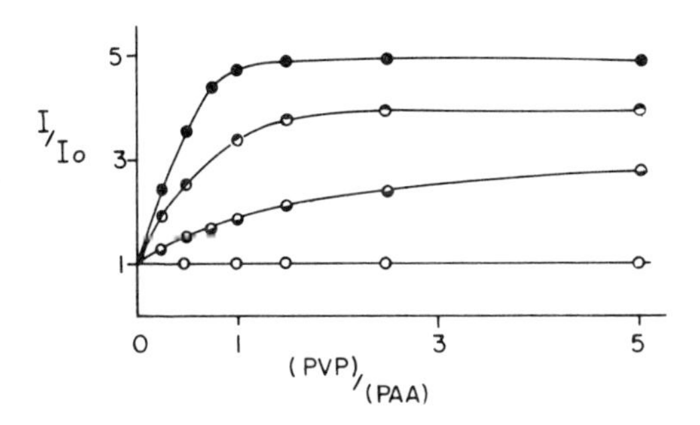

FIG. 3. Increase in fluorescence intensity of $2{\cdot}5 \times 10^{-4}$ N dansyl-labeled poly(acrylic acid) on addition of poly(vinyl pyrrolidone). pH 4·1 (●); pH 4·5 (◓); pH 4·8 (◒); pH 5·7 (○).

carboxyl groups and a consequent destruction of hydrogen bonds, the solvation of the complex increases leading to a smaller change in the emission intensity of the label.

The sensitivity of the dansyl label to the local medium may also be used to follow kinetically the displacement of labeled poly(acrylic acid) from its PVE or PVP complex by unlabeled poly(acrylic acid).[20] This 'complex interchange' may also be followed using the depolarization of fluorescence of labeled poly(acrylic acid) to distinguish between the complexed and the free polymer chains.[21]

Using a stopped flow apparatus with fluorescence detection, we have recently succeeded in measuring the rate of complex formation from labeled poly(acrylic acid) and polyoxyethylene, as well as the rate at which the complex is destroyed when its solution is mixed with a buffer at high pH.[22] These processes are characterized by very low activation energies but unusually high negative entropies of activation (36 e.u. for the formation and 43 e.u. for the decomposition of the complex) suggesting that both the formation of a stable complex and its destruction require extensive conformational rearrangements of the interacting chain molecules.

ACKNOWLEDGEMENT

The investigations reported in this review which were carried out in my laboratory were made possible by the financial support of the National Science Foundation, Polymer Program, through Grant No. DMR-77-05210.

REFERENCES

1. Amrani, F., Hung, J.-M. and Morawetz, H., *Macromolecules*, **13**, 649 (1980).
2. Mikes, F., Morawetz, H. and Dennis, K. S., *Macromolecules*, **13**, 969 (1980).
3. Morawetz, H., *Ann. NY Acad. Sci.*, **366**, 404 (1981).
4. Mikes, F., Morawetz, H. and Dennis, K. S., *Macromolecules*, **17**, 60 (1984).
5. Frank, C. W. and Gashgari, M. A., *Macromolecules*, **12**, 163 (1979).
6. Semerak, S. N. and Frank, C. W., *Macromolecules*, **14**, 443 (1981).
7. Gelles, R. and Frank, C. W., *Macromolecules*, **16**, 1448 (1983).
8. Jachowicz, J. and Morawetz, H., *Macromolecules*, **15**, 828 (1982).
9. Goldenberg, M., Emert, J. and Morawetz, H., *J. Am. chem. Soc.*, **100**, 7171 (1978).
10. Liao, T.-P. and Morawetz, H., *Macromolecules*, **13**, 1228 (1980).
11. Winnik, M. A., Redpath, A. E. C. and Richards, D. H., *Macromolecules*, **13**, 328 (1980).
12. Redpath, A. E. C. and Winnik, M. A., *Ann. NY Acad. Sci.*, **366**, 75 (1981).
13. Li, X., Winnik, M. A. and Guillet, J. E., *Macromolecules*, **16**, 992 (1983).
14. Winnik, M. A., Li, X. and Guillet, J. E., *Macromolecules*, **17**, 699 (1984).
15. Katchalski-Katzir, E., Haas, E. and Steinberg, I. Z., *Ann. NY Acad. Sci.*, **366**, 44 (1981).
16. deGennes, P. G., *J. chem. Phys.*, **55**, 572 (1971).
17. Doi, M. and Edwards, S. F., *J. chem. Soc., Faraday Trans.*, 2, **74**, 1789 (1978).
18. Antonietti, M., Coutandin, J., Grütter, R. and Sillescu, H., *Macromolecules*, **17**, 798 (1984).
19. Shiah, T. Y.-J. and Morawetz, H., *Macromolecules*, **17**, 792 (1984).
20. Chen, H.-L. and Morawetz, H., *Eur. Polym. J.*, **19**, 923 (1983).
21. Anufrieva, E. V., Pantov, V. O., Papisov, I. M. and Kabanov, V. A., *Dokl. Akad. Nauk SSSR*, **232**, 1096 (1977).
22. Bednar, B., Morawetz, H. and Shafer, J. A., *Macromolecules*, in press.

4

Initiation of Free Radical Polymerisation by Acylphosphine Oxides and Derivatives

WOLFRAM SCHNABEL and TAKASHI SUMIYOSHI

Hahn-Meitner-Institut für Kernforschung, Berlin, Federal Republic of Germany

1. INTRODUCTION

Recently, acylphosphine oxides (I) and acylphosphonates (II) having the general structures

$$R^1{-}\underset{\underset{O}{\parallel}}{C}{-}\overset{\overset{O}{\parallel}}{P}{<}^{R^2}_{R^3} \qquad\qquad R^1{-}\underset{\underset{O}{\parallel}}{C}{-}\overset{\overset{O}{\parallel}}{P}{<}^{OR^2}_{OR^3}$$

(I) (II)

were introduced as a new class of UV-photo-initiators.[1] As can be seen from Fig. 1, some of these compounds absorb light relatively strongly in the wavelength range between 350 and 400 nm, which makes them particularly suitable for the initiation of the free radical polymerisation of various monomers in technical processes such as photocuring of TiO_2-pigmented coatings[2] and of thick-walled glass fibre-reinforced polyester laminates.[3] Another advantage of using initiators of type I and II consists in the fact that after curing yellowing only occurs to a negligible extent.

This paper is concerned with kinetic and mechanistic studies of the photolysis of acyl phosphine oxides and acyl phosphonates, which recently were carried out in our laboratory. These investigations revealed that many type I and type II compounds readily undergo α-scission after photo-excitation:

$$R^1{-}\underset{\underset{O}{\parallel}}{C}{-}\overset{\overset{O}{\parallel}}{P}{<}^{R^2}_{R^3} \xrightarrow{h\nu} R^1{-}\underset{\underset{O}{\parallel}}{C}^{\cdot} + {}^{\cdot}\overset{\overset{O}{\parallel}}{P}{<}^{R^2}_{R^3} \qquad (1)$$

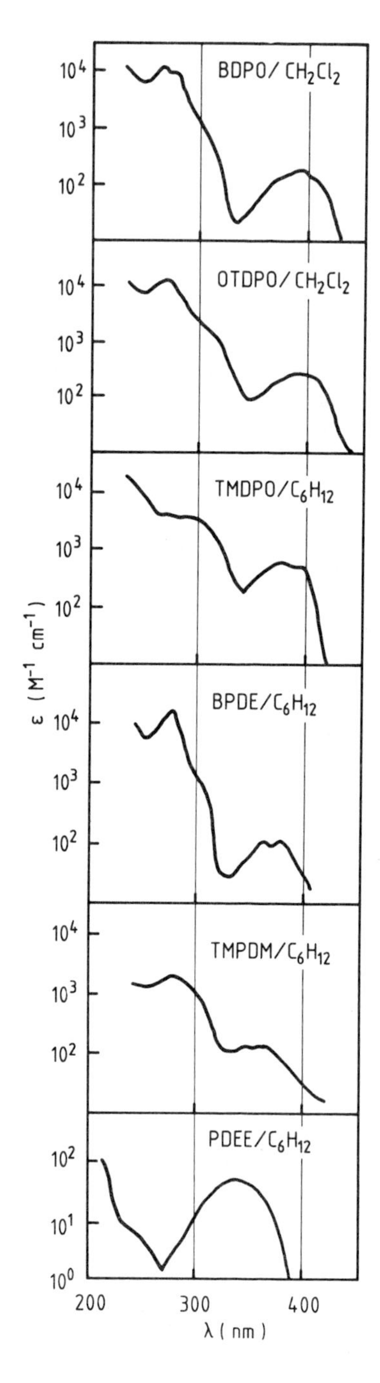

FIG. 1. Optical absorption spectra of acyl phosphine oxides and acylphosphonates recorded in cyclohexane- or CH_2Cl_2-solution.

$$R^1{-}\overset{\displaystyle O}{\underset{}{\overset{\|}{C}}}{-}\overset{O}{\overset{\|}{P}}(OR^2)(OR^3) \xrightarrow{h\nu} R^1{-}\overset{O}{\overset{\|}{C}}{}^{\cdot} + {}^{\cdot}\overset{O}{\overset{\|}{P}}(OR^2)(OR^3) \qquad (2)$$

Quite frequently, reactions (1) and (2) proceed with quantum yields between 0·3 and 1·0 and with lifetimes of a few ns or less.[4] Moreover, phosphonyl radicals generated in these processes were found to be very reactive towards compounds having olefinic unsaturations[5] and quite inert towards aliphatic amines and epoxides.[6] On the basis of these properties it is feasible why these compounds exhibit an excellent performance as free radical photo-initiators.

In the following some interesting features concerning the photolysis of three groups of initiators, namely, benzoyl diphenylphosphine oxides, benzoylphosphonic acid esters and pivaloyl phosphonic acid esters will be described. Moreover, data concerning the reactivity of phosphonyl radicals will be discussed.

2. PHOTOLYSIS OF BENZOYLDIPHENYLPHOSPHINE OXIDES

Benzoyldiphenylphosphine oxide (BDPO) readily undergoes α-scission upon irradiation at $\lambda > 300$ nm:

$$\underset{\text{BDPO}}{C_6H_5{-}C({=}O){-}P({=}O)(C_6H_5)_2} \xrightarrow{h\nu} \left(C_6H_5{-}C({=}O){-}P({=}O)(C_6H_5)_2\right)^* \xrightarrow[\phi = 0{\cdot}5]{} C_6H_5{-}\dot{C}({=}O) + {}^{\cdot}P({=}O)(C_6H_5)_2 \qquad (3)$$

The formation of diphenylphosphonyl radicals can be easily followed in flash photolysis experiments, because these radicals absorb light characteristically around 330 nm.[4] Figure 2(a) shows the transient absorption

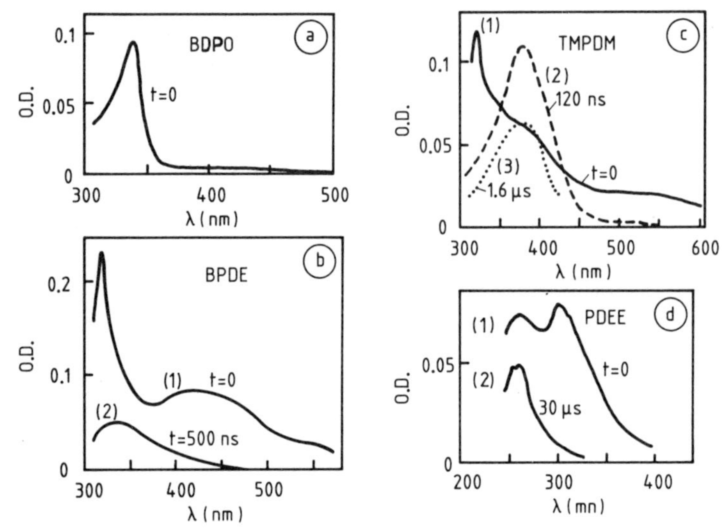

FIG. 2. Transient absorption spectra recorded in flash photolysis experiments. (a) Spectrum of diphenylphosphonyl radical generated by the photolysis of BDPO in Ar-saturated CH_2Cl_2 solution ($1{\cdot}7 \times 10^{-3}$ M). (b) Spectra recorded at the end of the flash (1) and 500 ns later (2). BDPE in Ar-saturated CCl_4-solution ($4{\cdot}8 \times 10^{-3}$ M). (c) Spectra recorded at the end of the flash (1) and 120 ns (2) and 1·6 μs (3) later. TMPDM in Ar-saturated CCl_4 solution ($2{\cdot}5 \times 10^{-3}$ M). (d) Spectra recorded at the end of the flash (1) and 30 μs later (2). PDEE in Ar-saturated cyclohexane solution ($3{\cdot}6 \times 10^{-3}$ M).

spectrum of the diphenylphosphonyl radical. The extinction coefficient at 335 nm in benzene solution was estimated as $\varepsilon = 1{\cdot}9 \times 10^4$ (l/mol cm). This value is about two orders of magnitude higher than the extinction coefficient of benzoyl radicals.

p-Methylbenzoylphosphine oxide (PTDPO) and 2,4,6-trimethyl benzoylphosphine oxide (TMDPO)

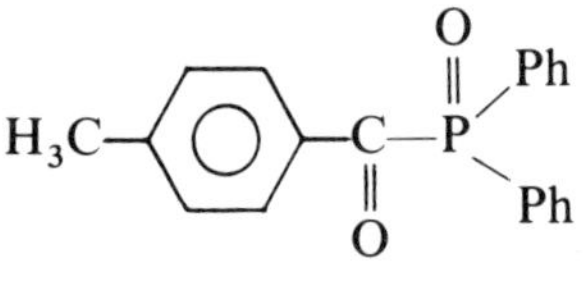

H_3C–C_6H_4–C(=O)–P(=O)Ph₂

PTDPO

H_3C–$C_6H_2(CH_3)_2$–C(=O)–P(=O)Ph₂

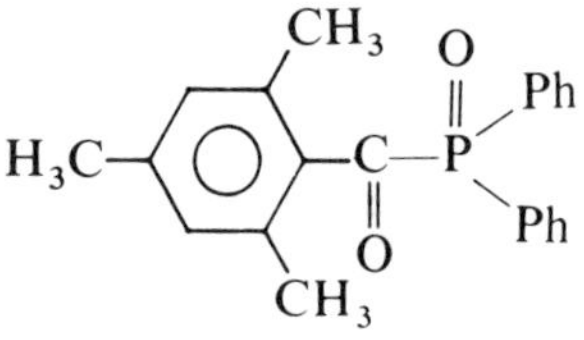

TMDPO

exhibited a behaviour analogous to that of BDPO. Extensive studies were performed with TMDPO. In this case it was inferred from quenching

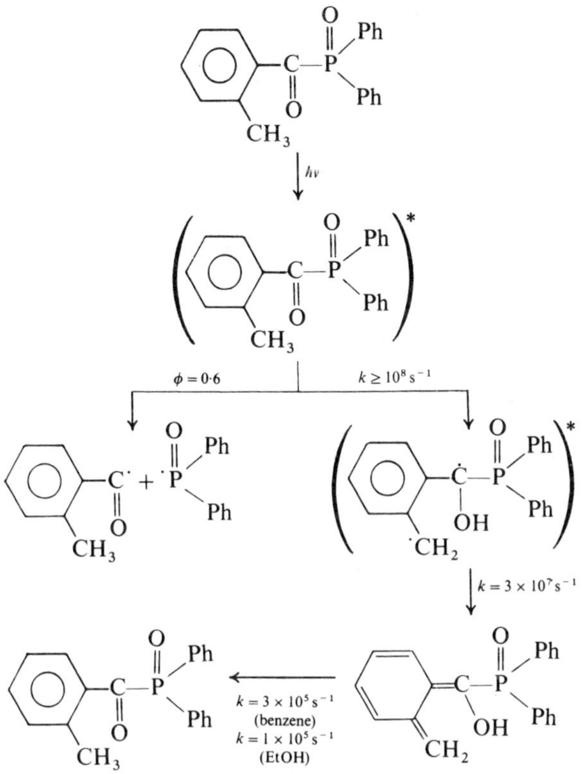

$\phi = 0{\cdot}6$ $k \geq 10^8\,s^{-1}$ $k = 3 \times 10^7\,s^{-1}$ $k = 3 \times 10^5\,s^{-1}$ (benzene) $k = 1 \times 10^5\,s^{-1}$ (EtOH)

SCHEME 1. Photolysis of o-methylbenzoyldiphenylphosphine oxide.

experiments with naphthalene, that triplets are very rapidly formed ($\tau <$ 1 ns). However, it could not be decided whether bond breakage according to reaction (2) involves both singlet- and triplet-excited or just triplet-excited molecules.

In the case of o-methylbenzoyldiphenylphosphine oxide (OTDPO) flash photolysis studies indicated that α-scission competes with enolisation, as illustrated in Scheme 1. Enol triplets relax to the ground state with $k = 3 \times 10^7\,s^{-1}$ and conversion from the (ground state) enol to the keto form occurs with $k \approx 1$ to $3 \times 10^5\,s^{-1}$, depending on the solvent. The quantum yields for the generation of diphenylphosphonyl radicals, $\phi(Ph_2\dot{P}O)$, are rather high: ca. 1·0 (PTDPO), ca. 0·6 (OTDPO and TMDPO) and ca. 0·5 (BDPO). For practical applications TMDPO has obtained prominence because its solvolysis stability is very much higher than that of the other compounds.[2]

3. PHOTOLYSIS OF BENZOYLPHOSPHONIC ACID ESTERS

Benzoylphosphonic acid diethyl ester (BDPE) does not undergo α-scission to a detectable extent upon irradiation at $\lambda > 300$ nm in dilute CH_2Cl_2-solution. Singlets are short-lived ($\tau \leq 1$ ns). Energy transfer to naphthalene occurs with $k_q = 4 \times 10^9 \, M^{-1} s^{-1}$. From quenching studies the triplet quantum yield $\phi(T) = 0{\cdot}9$ was determined. After irradiation with a 20 ns flash of 347 nm light the transient spectrum 1 in Fig. 2(b) with maxima at 310 and 450 nm was detected, which was ascribed to a biradical formed by intramolecular hydrogen abstraction. This species decays in a 1st order process to a not yet identified product which possesses an absorption maximum at 340 nm (spectrum 2 in Fig. 2(b)). According to our present knowledge the mechanism presented in Scheme 2 was postulated.[7]

The dimethyl and diethyl esters of 2,4,6-trimethylbenzoylphosphonic acid, TMPDM and TMPDE respectively, exhibited a behaviour quite different from that observed with BPDE, upon irradiation in dilute

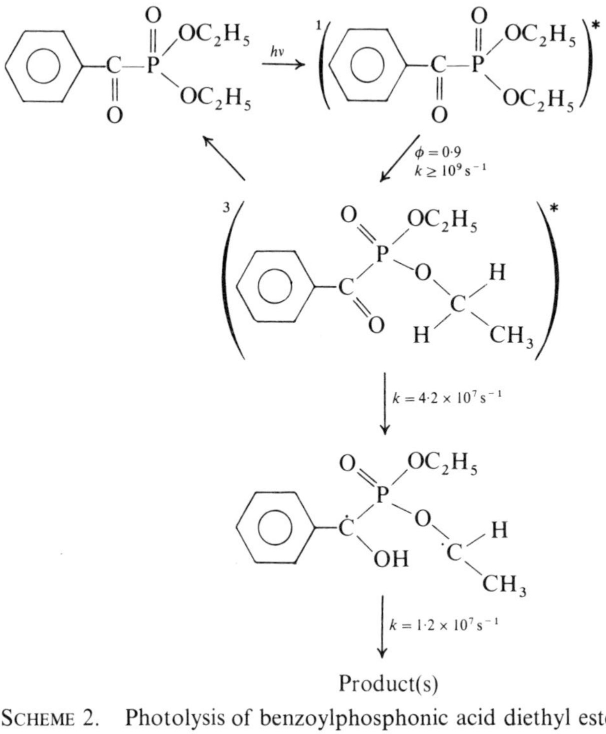

Product(s)

SCHEME 2. Photolysis of benzoylphosphonic acid diethyl ester.

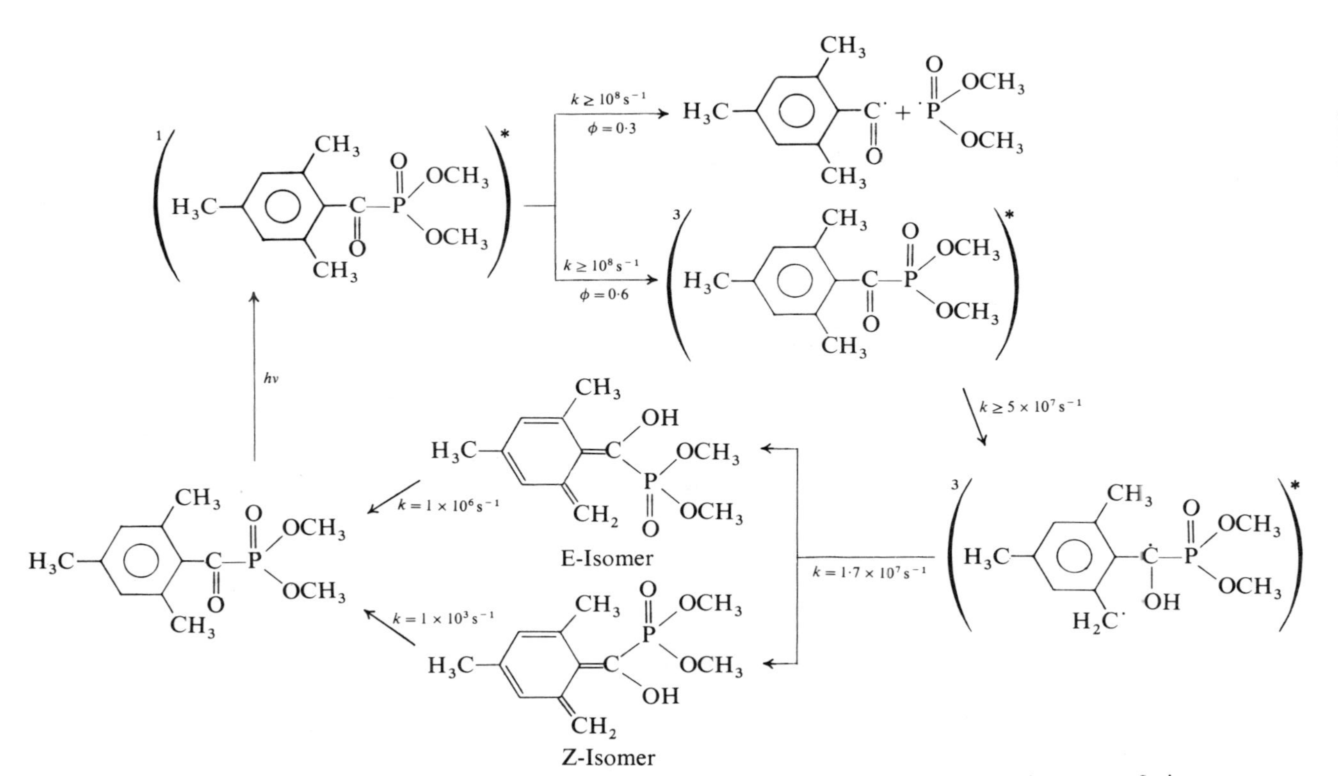

SCHEME 3. Photolysis of 2,4,6-trimethylbenzoylphosphonic acid dimethyl ester in benzene solution.

solution. Triplets were formed with $\phi(T) \approx 0{\cdot}6$. Moreover, both compounds underwent α-scission ($\phi \approx 0{\cdot}3$), mainly, if not exclusively from singlet states, as detected by the formation of adduct radicals generated by reaction of dialkoxyphosphonyl radicals with benzene or styrene. A major route of deactivation of the triplets involves enolisation. Ketone triplets, formed initially, are converted with $k_T^K \geq 5 \times 10^7\,s^{-1}$ to enol triplets. The latter relax with $k_T^E = 1{\cdot}7 \times 10^7\,s^{-1}$ to Z- and E-isomers. Difference optical absorption spectra of the enol triplet and the relaxed enol isomers are shown in Fig. 2(c). The reketonisation rate depends on the nature of the solvent. The routes of deactivation of electronically excited TMPDM are depicted in Scheme 3.

4. PHOTOLYSIS OF PIVALOYLPHOSPHONIC ACID ESTERS

The dimethyl and diethylesters of pivaloylphosphonic acid, PDME and PDEE, respectively, exhibited a very similar behaviour. The singlet lifetime

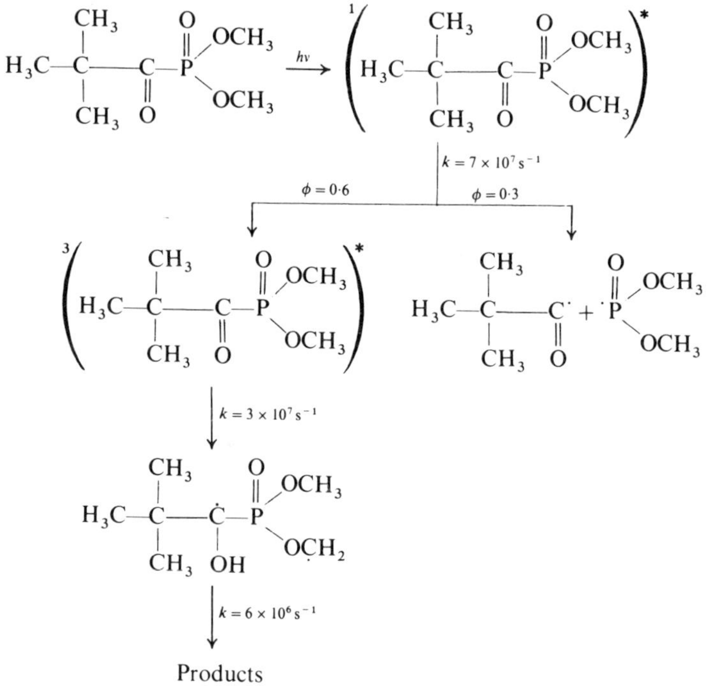

$(CH_3)_3C{-}C(=O){-}P(=O)(OCH_3)_2 \xrightarrow{h\nu} {}^1\big((CH_3)_3C{-}C(=O){-}P(=O)(OCH_3)_2\big)^*$

$k = 7 \times 10^7\,s^{-1}$

$\phi = 0{\cdot}6$ $\phi = 0{\cdot}3$

${}^3\big((CH_3)_3C{-}C(=O){-}P(=O)(OCH_3)_2\big)^*$ $(CH_3)_3C{-}\dot{C}(=O) + {}^{\cdot}P(=O)(OCH_3)_2$

$k = 3 \times 10^7\,s^{-1}$

$(CH_3)_3C{-}\dot{C}(OH){-}P(=O)(OCH_3)(O\dot{C}H_2)$

$k = 6 \times 10^6\,s^{-1}$

Products

SCHEME 4. Photolysis of pivaloyl phosphonic acid dimethyl ester in cyclohexane.

is ca. 14 ns and the triplet lifetime ca. 33 ns as determined in cyclohexane solution. Triplets are formed with $\phi(T) \approx 0{\cdot}6$. Energy transfer to naphthalene occurs with $k_q = 2 \times 10^9\ M^{-1}\ s^{-1}$. Transient-absorption spectroscopy revealed the existence of several radicals as can be seen from Fig. 2(d). The spectrum observed at the end of the flash ($\lambda_{inc} = 347$ nm) contained a relatively rapidly decaying portion in the range 280–360 nm ($k = 6 \times 10^6\ s^{-1}$) which was attributed to biradicals formed upon the decay of triplets as indicated in Scheme 4. As inferred from the detection of adduct styryl radicals formed in the presence of styrene, at $\lambda = 320$ nm ($\varepsilon = 5{\cdot}8 \times 10^3\ M^{-1}\ cm^{-1}$),[8] α-scission occurs with $\phi \approx 0{\cdot}3$. A rather long-lived absorption with a maximum at about 260 μm was ascribed to pivaloyl radicals (see spectrum (2) in Fig. 2(d)). This absorption decayed according to 2nd order kinetics with first lifetimes in the order of magnitude of 10 μs.

5. ON THE REACTIVITY OF PHOSPHONYL RADICALS TOWARDS OLEFINIC COMPOUNDS

In Scheme 5 it is illustrated how phosphonyl radicals were generated for the determination of their reactivity towards monomers frequently used in free radical polymerisation. In the cases of diphenylphosphonyl and phenylpropoxyphosphonyl radicals reaction rates were determined by measuring the rate of decay of the optical absorption of the radicals at $\lambda = 330$ nm as a function of monomer concentration. Because the other three radicals do not absorb characteristically above 300 nm the competition method illustrated in Scheme 6 was applied. This method is based on the fact that phosphonyl radicals add to styrene forming a substituted styryl radical which absorbs strongly at $\lambda = 322$ nm, whereas the addition to other monomers leads to non-absorbing adduct radicals.

Rate constants $k_{R\cdot+St}$ were obtained by measuring the rate of formation of the optical density (OD) at 322 nm and rate constants $k_{R\cdot+M}$ were evaluated from plots of OD_0/OD vs the monomer concentration $[M]$ according to eqn (4)

$$\frac{OD_0}{OD} = 1 + \frac{k_{R\cdot+M}[M]}{k_{R\cdot+St}[St]} \tag{4}$$

OD and OD_0 denote the maximum optical density in the presence and absence of monomer, respectively. Data obtained this way are compiled in Table 1. It is noteworthy that the phosphonyl radicals under investigation are very reactive towards methacrylonitrile (MAN), styrene (St) and

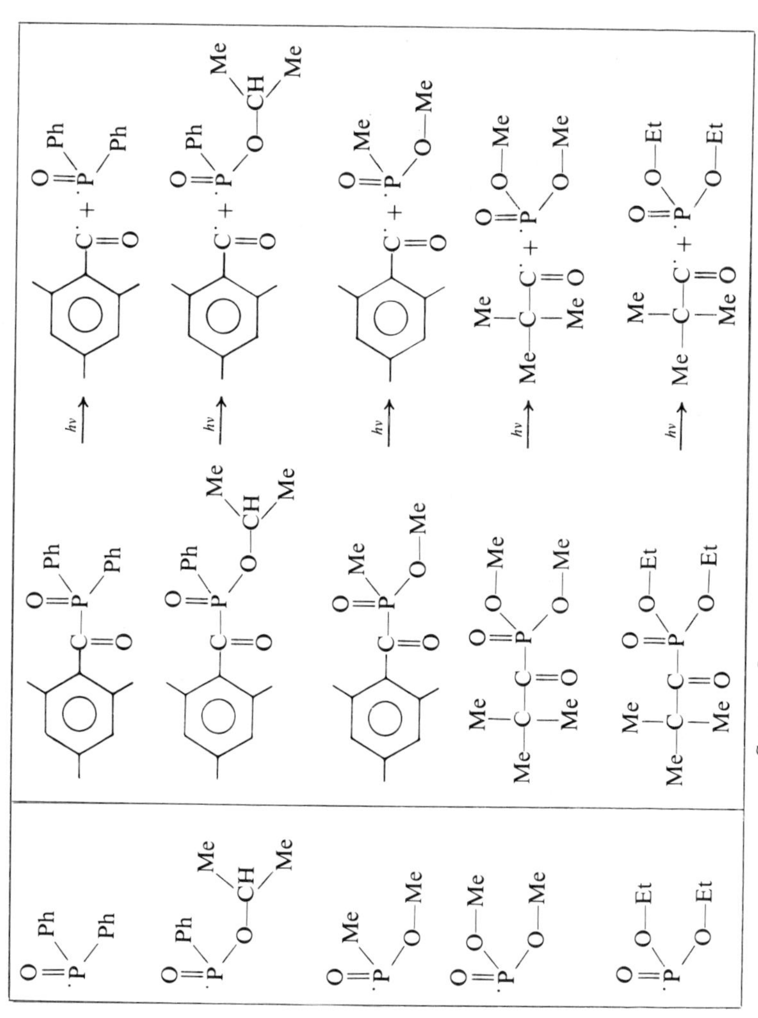

SCHEME 5. Generation of phosphonyl radicals.

TABLE 1

Bimolecular Rate Constants in l/mol s of the Reaction of Phosphonyl Radicals with Various Monomers in Cyclohexane at Room Temperature

Monomer	Radical / Q/e	$\cdot P(=O)(\phi)(\phi)$	$\cdot P(=O)(\phi)(OCH(CH_3)_2)$	$\cdot P(=O)(CH_3)(OCH_3)$	$\cdot P(=O)(OCH_3)(OCH_3)$	$\cdot P(=O)(OC_2H_5)(OC_2H_5)$
MAN	1·12 0·81	$(5{\cdot}0 \pm 0{\cdot}2) \times 10^7$	$(4{\cdot}6 \pm 0{\cdot}2) \times 10^7$	$(4{\cdot}5 \pm 0{\cdot}2) \times 10^7$	$(9{\cdot}2 \pm 1{\cdot}3) \times 10^7$	$(1{\cdot}1 \pm 0{\cdot}1) \times 10^8$
ST	1·00 −0·80	$(6{\cdot}0 \pm 0{\cdot}2) \times 10^7$	$(4{\cdot}5 \pm 0{\cdot}2) \times 10^7$	$(8{\cdot}0 \pm 0{\cdot}2) \times 10^7$	$(2{\cdot}2 \pm 0{\cdot}2) \times 10^8$	$(2{\cdot}5 \pm 0{\cdot}2) \times 10^8$
MMA	0·74 0·40	$(8{\cdot}0 \pm 0{\cdot}5) \times 10^7$	$(5{\cdot}0 \pm 0{\cdot}2) \times 10^7$	$(5{\cdot}8 \pm 0{\cdot}5) \times 10^7$	$(5{\cdot}8 \pm 0{\cdot}2) \times 10^7$	$(5\ 3 \pm 0{\cdot}2) \times 10^7$
AN	0·60 1·20	$(2{\cdot}0 \pm 0{\cdot}1) \times 10^7$	$(2{\cdot}0 \pm 0{\cdot}1) \times 10^7$	$(1{\cdot}8 \pm 0{\cdot}3) \times 10^6$	$(5{\cdot}8 \pm 0{\cdot}2) \times 10^6$	$(2{\cdot}6 \pm 0{\cdot}2) \times 10^6$
MA	0·42 0·60	$(3{\cdot}5 \pm 0{\cdot}2) \times 10^7$	$(2{\cdot}1 \pm 0{\cdot}2) \times 10^7$	$(1{\cdot}3 \pm 0{\cdot}1) \times 10^7$	$(1{\cdot}7 \pm 0{\cdot}2) \times 10^7$	$(1{\cdot}6 \pm 0{\cdot}2) \times 10^7$
BVE	0·087 −1·20	$(4{\cdot}0 \pm 0{\cdot}2) \times 10^6$	$(3{\cdot}0 \pm 0{\cdot}3) \times 10^6$	$(2{\cdot}3 \pm 0{\cdot}2) \times 10^6$	$(2{\cdot}1 \pm 0{\cdot}2) \times 10^7$	$(1{\cdot}4 \pm 0{\cdot}2) \times 10^7$
VA	0·026 −0·22	$(1{\cdot}6 \pm 0{\cdot}1) \times 10^6$	$(1{\cdot}3 \pm 0{\cdot}1) \times 10^6$	$(8{\cdot}2 \pm 0{\cdot}4) \times 10^5$	$(2{\cdot}9 \pm 0{\cdot}2) \times 10^6$	$(1{\cdot}3 \pm 0{\cdot}2) \times 10^6$

MAN: Methacrylonitrile; ST: Styrene; MMA: Methylmethacrylate; AN: Acrylonitrile; MA: Methyl acrylate; BVE: n-Butyl vinyl ether; VA: Vinyl acetate.

$$R^1R^2P^{\cdot}(=O) + H_2C{=}CH(C_6H_5) \xrightarrow{k_{R\cdot+St}} R^1R^2P(=O){-}CH_2{-}C^{\cdot}H(C_6H_5)$$

$\lambda_{max} = 322$ nm
$\varepsilon = 5{\cdot}8 \times 10^3\ M^{-1}\ cm^{-1}$

$$R^1R^2P^{\cdot}(=O) + H_2C{=}CH{-}X \xrightarrow{k_{R\cdot+M}} R^1R^2P(=O){-}CH_2{-}C^{\cdot}H{-}X$$

non-absorbing

SCHEME 6. Competition of styrene and another monomer for phosphonyl radicals.

methylmethacrylate (MMA) and that dialkoxyphosphonyl radicals exhibit the highest reactivity. Actually rate constants of 1–$2 \times 10^8\ M^{-1}\ s^{-1}$ were measured for the reaction of $(H_3CO)_2\dot{P}{=}O$ and $(H_5C_2O)_2\dot{P}{=}O$ with MAN and St. In the cases of the other monomers, i.e. acrylonitrile (AN), methylacrylate (MA), n-butyl vinyl ether (BVE) and vinyl acetate (VA) lower rate constants ranging from 10^6 to $10^7\ M^{-1}\ s^{-1}$ were measured. Interestingly, certain trends in the dependence of rate constants $k_{R\cdot+M}$ on the chemical nature of both the radical and the monomer were recognised with the aid of the Q–e scheme proposed by Alfrey and Price[9] for the copolymerisation of olefinic monomers. This scheme relates the rate constant k, of the reaction of the growing chain radical with a monomer to two reactivity parameters, P_R and Q, and to the two polarity parameters, e_R and e_M, taking into account the influence of substituents on the electron density at the site of the unpaired electron and of the double bond, respectively:

$$k = P_R Q\, e^{-e_R e_M} \tag{5}$$

With the aid of tabulated values of Q and e_M[10] the ratio k/P_R was calculated for various values of e_R. As can be seen from Fig. 3 proportionality between the experimental $k_{R\cdot+M}$ values and k/P_R was found in most cases.

For the radicals $(Ph)_2\dot{P}{=}O$ and i-PrO(Ph)$\dot{P}{=}O$ best fits were obtained with $e_R = 0$, whereas for the radicals $(H_3CO)_2\dot{P}{=}O$, $(H_5C_2O)_2\dot{P}{=}O$ and $(H_3CO)(CH_3)\dot{P}{=}O$ a linear relationship between $k_{R\cdot+M}$ and k/P_R could be approximated with $e_R = 0{\cdot}8$, $0{\cdot}7$ and $0{\cdot}4$, respectively. With respect to those values which obviously do not fit in, it cannot be decided, at present,

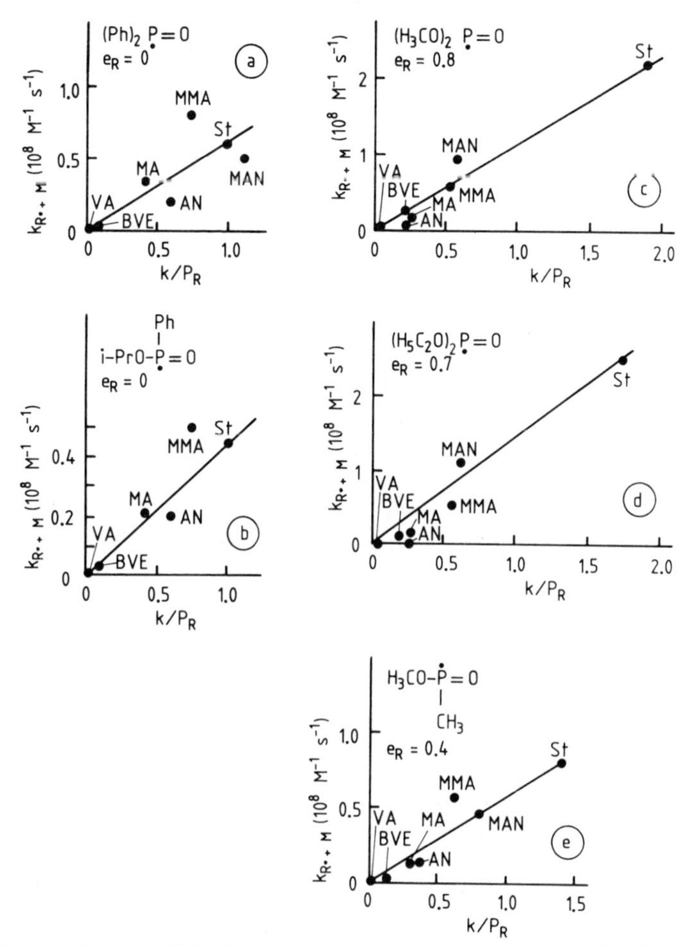

FIG. 3. Application of the Q–e scheme to the rate constants compiled in Table 1. Plots of experimental rate constants vs values of k/P_R calculated according to eqn (5) with the aid of data taken from *Polymer Handbook*.[10] (a) $(Ph)_2\dot{P}{=}O$, $e_R = 0$; (b) i-PrO(Ph)$\dot{P}{=}O$, $e_R = 0$; (c) $(H_3CO)_2\dot{P}{=}O$, $e_R = 0{\cdot}8$; (d) $(H_5C_2O)_2\dot{P}{=}O$, $e_R = 0{\cdot}7$; (e) $H_3CO(CH_3)\dot{P}{=}O$, $e_R = 0{\cdot}4$.

whether this is due to uncertainties in the tabulated Q- and e-values or due to intrinsic factors such as steric effects, etc.

The high reactivity of the phosphonyl radicals can be understood in terms of their pyramidal structure which permits the site of the unpaired electron to be approached quite freely by reactants. In ESR studies it was disclosed that the ratio of spin densities in 3s and 3p orbitals of phosphorus is $\rho_{3p}/\rho_{3s} \simeq 3$, in the case of dialkoxyphosphonyl radicals[11] and $\rho_{3p}/\rho_{3s} \approx 5{\cdot}5$, in the case of diphenylphosphonyl radicals.[12] The increase from 3 to 5·5 was thought to be indicative for a rather flattened tetrahedral structure of $(Ph)_2\underset{\cdot}{P}{=}O$ relative to that of $(H_3CO)_2\underset{\cdot}{P}{=}O$.[11] The higher reactivity of dialkoxyphosphonyl radicals relative to that of diphenylphosphonyl radicals can be explained on the basis of these structural differences.

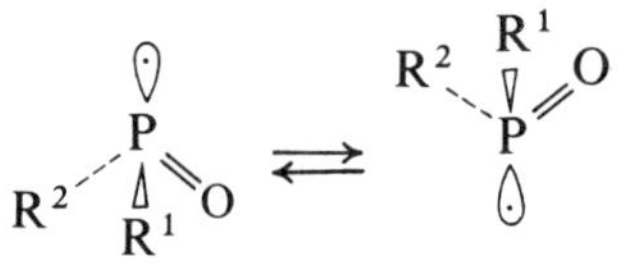

In this connection it is interesting to draw attention to the differences in e_R-values of the different phosphonyl radicals.

The rather low value of the 3p/3s ratio of orbital spin density close to 3 found for dialkoxyphosphonyl radicals indicates a pronounced delocalisation of the unpaired electron in sp^3 orbitals in contrast to the case of diphenyl phosphonyl radicals where the higher value of the 3p/3s ratio indicates little or negligible delocalisation. These properties are in accordance with differences in the e_R-values found during the data fitting process (vide ante: $e_R \approx 0{\cdot}8$ for $(H_3CO)_2\underset{\cdot}{P}{=}O$ and $e_R \approx 0{\cdot}0$ for $(Ph)_2\underset{\cdot}{P}{=}O$). If e_R-values can be taken as a measure of the electron density (at the radical site), the latter should be the higher the lower or the more negative e_R. This is actually what was found in this work: in the case of dialkoxy radicals with strong electron delocalisation e_R is much higher than in the case of diphenylphosphonyl radicals with negligible electron delocalisation.

6. POLYMERISATION EXPERIMENTS

A number of photopolymerisation experiments were carried out, which demonstrated the high initiator efficiency of those compounds that undergo α-scission. Only negligible amounts of polymer were formed, on the other hand, when compounds such as BPDE were used that do not undergo α-scission.

It is noteworthy that phosphorus was incorporated into the polymer, whenever compounds capable of initiating the polymerisation were used. This was inferred from neutron activation analyses using the nuclear reaction $P^{31}(n, \gamma)P^{32}$.

ACKNOWLEDGEMENTS

This work was carried out in cooperation with Dr A. Henne, BASF AG, Ludwigshafen. The initiators were synthesised in the laboratories of BASF AG. Some of the flash photolysis experiments were carried out by Dipl.-Chem. W. K. Wong. Mrs U. Fehrmann performed the polymerisation experiments and assisted in the characterisation of the initiators and purification of solvents. Dr G. Beck maintained the flash photolysis set-up and Dr Lilie assisted in the computer evaluation of data.

REFERENCES

1. (a) DOS 2830927 (1980) (BASF AG), P. Lechtken, I. Buethe and A. Hesse;
   (b) DOS 2909994 (1980) (BASF AG), P. Lechtken, I. Buethe, M. Jacobi and W. Trimborn;
   (c) DOS 3023486 (1980) (Bayer AG), H.-G. Heine, H. J. Rosenkranz and H. Rudolph.
2. Jacobi, M. and Henne, A., *Radiat. Curing*, **10**, 16 (1983).
3. Nicolaus, W., Hesse, A. and Scholz, D., *Plastverarbeiter*, **31**, 723 (1980).
4. Sumiyoshi, T., Henne, A., Lechtken, P. and Schnabel, W., *Z. Naturforsch.*, **39a**, 434 (1984).
5. Sumiyoshi, T., Henne, A., Lechtken, P. and Schnabel, W., *Polymer*, in press.
6. Wong, W. K. and Schnabel, W., unpublished results (1984); Wong, W. K., Diploma Thesis, Technical University, Berlin (1984).
7. Sumiyoshi, T., Henne, A. and Schnabel, W., *J. Photochem.*, in press.
8. Brede, O., Helmstreit, W. and Mehnert, R., *J. prakt. Chem.*, **316**, 402 (1974).
9. Alfrey, T. and Price, C. C., *J. Polym. Sci.*, **2**, 101 (1947).
10. Brandrup, J. and Immergut, E. H. (Eds), *Polymer Handbook*, 2nd edn, Wiley-Interscience, New York, 1975, p. 387.
11. Kerr, C. M., Webster, K. and Williams, Ff., *J. Phys. Chem.*, **79**, 2650 (1975).
12. Geoffroy, M. and Lucken, E. A., *Molec. Phys.*, **22**, 257 (1971).

5

The Photochemical Behaviour of Some Monomers which Contain Electron Donor and Electron Acceptor Pairs

WU SHIH-KANG[1] and LI FU-MIAN[2]

[1] *Institute of Photographic Chemistry, Academia Sinica, China*
[2] *Department of Chemistry, Peking University, China*

1. INTRODUCTION

Photopolymerisation of vinyl monomers initiated by charge transfer complexes has attracted attention in recent years. Tsuda and coworkers[1] have reported the polymerisation of methyl methacrylate initiated by N,N,dimethylaniline. They discovered that in the following general formulation of aromatic amine

$$A-C_6H_4-N(CH_3)_2$$

the stronger the electron repulsion of the A group, the faster initiation rate it has. The order of the polymerisation rate of methyl methacrylate by different aromatic amines is as follows:

$$CH_3O-C_6H_4-N(CH_3)_2 > CH_3-C_6H_4-N(CH_3)_2$$

$$> C_6H_5-N(CH_3)_2 > F-C_6H_4-N(CH_3)_2 > Br-C_6H_4-N(CH_3)_2$$

$$> Cl-C_6H_4-N(CH_3)_2 > O_2N-C_6H_4-N(CH_3)_2$$

Some work has been done on the kinetics and mechanism of this kind of polymerisation. For instance, Wei[2] reported that during photo-initiation, the formation of an exciplex between excited aromatic amine and the vinyl monomers such as MMA and AN may be observed in the fluorescence spectrum. The mechanisms of monomer initiation by exciplex formation have also been proposed in different ways. The experimental evidence of the formation of ionic radicals in the initiation step has been obtained in our laboratory. Cao and Feng[3] proposed that the initiation of methyl methacrylate in this case depends on the presence of a small amount of oxygen. Recently, we have reported the photopolymerisation of one kind of interesting monomer which contains both vinyl group as an acceptor and aromatic amine group as an electron donor in a single molecule. From the results of this study, the initiation mechanism is further clarified. In this paper, we would like to report some new results in this interesting field.

2. PHOTODISSOCIATION OF CHARGE TRANSFER COMPLEXES

A study on the fluorescence spectra, absorption spectra and photoconductivity of the systems which were composed of dimethyl toluidine (DMT) as an electron donor and a series of cyano-containing vinyl compounds as electron acceptors has been carried out in our laboratory.[4] The fluorescence and absorption spectra were measured by the commercially available equipment such as Hitachi MPF-4 and Hitachi 340 respectively. But, the photoconductivity was measured with a home-made apparatus. It is shown in Fig. 1.

The main body of the photoconduction cell was constructed from teflon block. It possesses two quartz windows and stainless steel electrodes. The distance between the two electrodes is 6 mm and the area of the electrodes is $1{\cdot}39\,cm^2$. The voltage applied in the experiment was 5300 V (DC) and the resistance used was 10^{11} ohm. For ensuring the reliability of the micro current measurement, an electrostatic meter (a vibrating-reed electrometer, Takeda Riken TR-84B) was used. The light source was a globe high pressure mercury lamp (200 W). A quartz lens focuses the light beam which

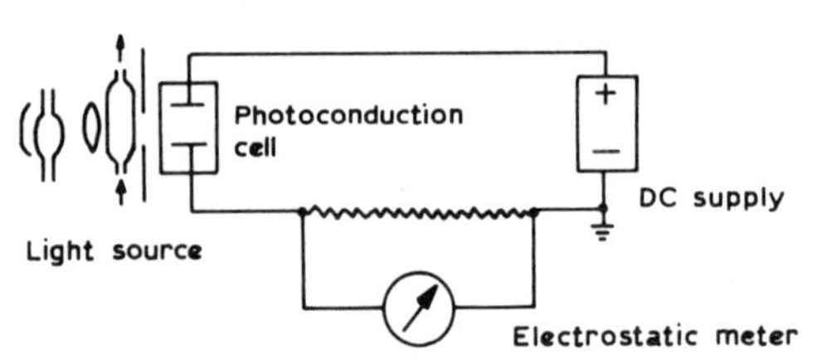

FIG. 1. The schema of the photoconduction apparatus.

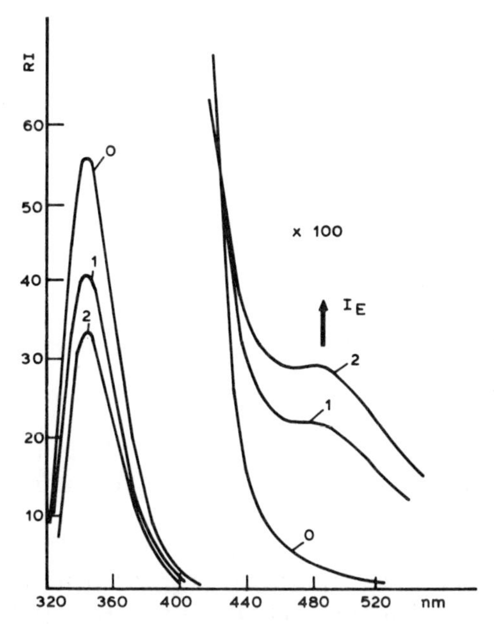

FIG. 2. Fluorescence spectra of AN–DMT in cyclohexane. Excitation wavelength: 303 nm; DMT: $3{\cdot}4 \times 10^{-4}$ M; AN: O—O; 1–$1{\cdot}4 \times 10^{-2}$; 2–$2{\cdot}8 \times 10^{-2}$ M.

penetrates a quartz filter cooled with water, a quartz window, and goes into the photoconduction cell.

The cyano-containing compounds used in this experiment are

1. Methyl acrylonitrile (MAN)
2. Acrylonitrile (AN)
3. Fuma-dinitrile (FDN)
4. Tetracyano ethylene (TCE)

According to what is known about the chemical structure of these compounds, the order of the electron deficiency or the electron acceptivity of the double bond in these compounds is as follows:

Tetracyano ethylene > Fuma-dinitrile > Acrylonitrile
> Methyl acrylonitrile

The experimental results showed that, in the fluorescence spectra, the exciplex emission could just be observed for the MAN–DMT and AN–DMT systems (Fig. 2). Due to greater availability of the charge transfer interaction in the ground state of FDN–DMT and TCE–DMT systems and because it can be easily decomposed by irradiation, we can not observe the exciplex emission in the fluorescence spectra of these systems. But all of these systems evidently exhibit photoconduction. It was shown that these systems could dissociate to ionic radicals after irradiation through the formation of exciplex or the charge transfer complex in the ground state. So, it is evident that the formation of ionic radicals may be considered to be a very important step in the process of monomer initiation.

These processes may be illustrated as follows.

$$D \xrightarrow{h\nu} D^* \xrightarrow{A} [D^*A] \qquad \text{Exciplex formation}$$

$$[D^*A] \longrightarrow [D^* \cdots\cdot A] \longrightarrow D^{\cdot +} + A^{\cdot -}$$

$$D + A \longrightarrow [D \cdots\cdot A]$$

$$[D \cdots\cdot A] \xrightarrow{h\nu} D^{\cdot +} + A^{\cdot -} \qquad \text{Charge transfer complex formation in ground state}$$

3. THE PHOTOCHEMICAL BEHAVIOUR OF SOME MONOMERS WHICH CONTAIN ELECTRON DONOR AND ELECTRON ACCEPTOR PAIRS

Recently, some monomers[5] which contain both electron donor and electron acceptor in a single molecule were synthesised in our laboratory.[6] The structures of these compounds are as follows:

1. Dimethylaminobenzyl methacrylate (DMABMA)

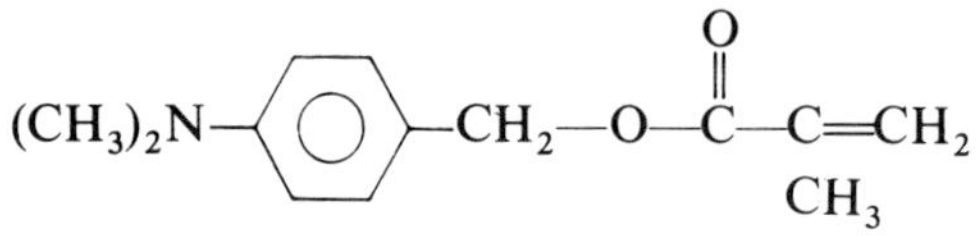

$$(CH_3)_2N-C_6H_4-CH_2-O-\overset{O}{\overset{\|}{C}}-\underset{CH_3}{C}=CH_2$$

2. Dimethylaminophenyl acryloamide (DMAPAA)

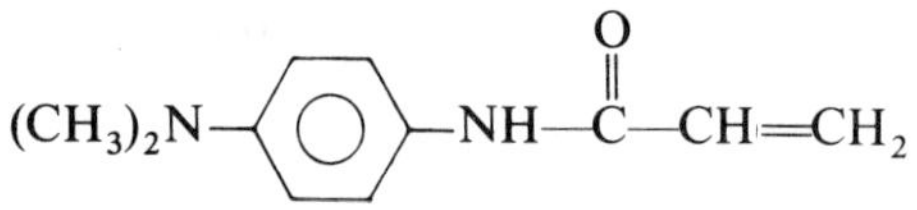

$$(CH_3)_2N-C_6H_4-NH-\overset{O}{\overset{\|}{C}}-CH=CH_2$$

3. Dimethylaminophenyl methacryloamide (DMAPMA)

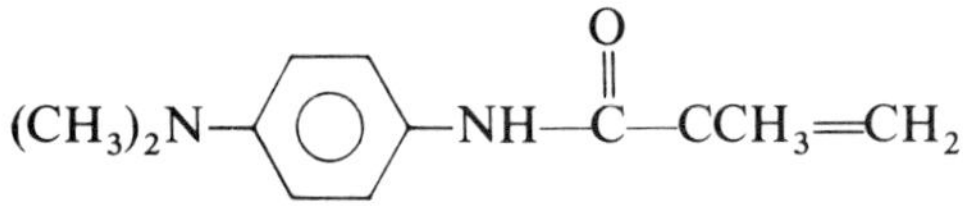

$$(CH_3)_2N-C_6H_4-NH-\overset{O}{\overset{\|}{C}}-CCH_3=CH_2$$

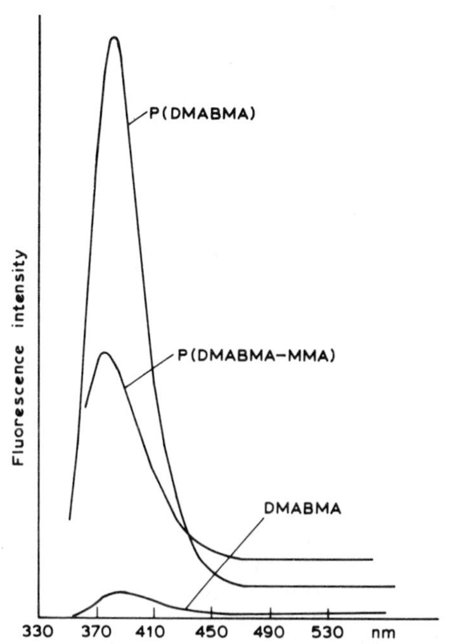

FIG. 3. Fluorescence emission spectra of DMABMA, P(DMABMA), and P(DMABMA–MMA) in dichloroethane. $\lambda_{ex} = 329$ nm.

The details of synthesis and purification as well as the polymerisation kinetics of these monomers by usual initiator will be published elsewhere.

It could be shown that in these compounds the aromatic amine is used as an electron donor and the double bond is used as an electron acceptor. For the sake of convenience, the experimental results[7] of the compound DMABMA will first be discussed in detail.

The fluorescence emission spectra of DMABMA, its polymer and copolymer with methyl methacrylate measured with excitation by 329 nm in the solution of 1,2-dichloroethane are shown in Fig. 3. It is obvious that the emission intensities of different species at 354 nm are very different. In the same molar concentration of DMABMA and P(DMABMA), the emission intensity of the latter is much higher than that of the former. Besides, the quantity of dimethylaminobenzyl group in the copolymer is much less than that in DMABMA, but the emission intensity of the copolymer is much higher. This difference in the emission intensity can be attributed to the double bond of the vinyl group in the monomer molecule. Because the double bond, just described above, is used as an electron acceptor, so it can accept an electron from the excited aromatic amino

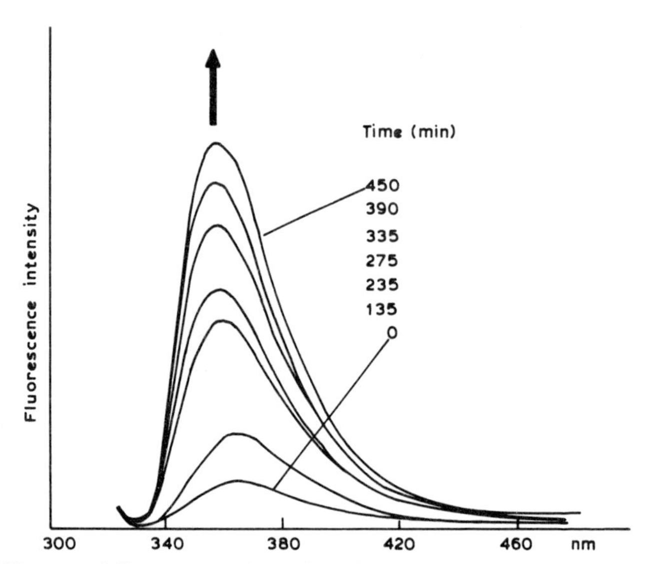

FIG. 4. Change of fluorescence intensity of DMABMA solution at different irradiation intervals during photopolymerisation.

group, and the fluorescence of the aromatic amine is quenched by the exciplex formation. Therefore the fluorescence emission of DMABMA is very low. On the contrary, after polymerisation of this monomer, the double bond in the monomer has disappeared; in this case, the fluorescence emission of the aromatic amine can not be quenched by itself and keep its intensity at a high degree.

From Fig. 4 it is interesting to find that in the process of irradiation of DMABMA solution, the emission intensity of this solution increases gradually at different irradiation intervals. It indicated that after irradiation the photopolymerisation had occurred. The loss of double bond in the system leads to an increase in the emission intensity. In order to examine whether this double bond can quench the fluorescence emission of the amino group or not, a small amount of MMA monomer was added to the solution of P(DMABMA). Evidently, the fluorescence emission of the dimethylaminobenzyl group was quenched dramatically. However, addition of compounds without a double bond such as methyl propionate will not provide a similar quenching effect. It is obvious that the fluorescence emission of aromatic amine is quenched by the double bond, not by the ester group. The fluorescence emission of the aromatic amine group is quenched by the olefinic monomer, but it is well known that the energy of the excited singlet state of olefinic monomer is higher than that of

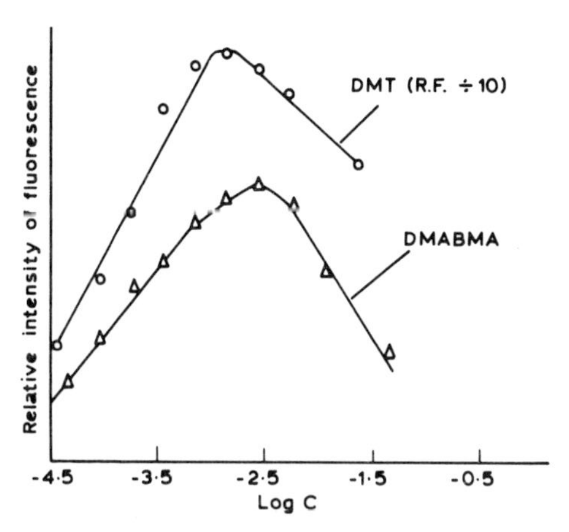

FIG. 5. Change of fluorescence intensity of DMT and DMABMA solution in cyclohexane and different concentrations.

the excited amino group. Therefore, quenching of the fluorescence emission by energy transfer is impossible. In this case, the fluorescence quenching is obviously related to the formation of exciplex. However, we could not observe the fluorescence emission of exciplex, although we have tried several techniques, such as changing the solvents and changing the concentration of the solution. In these experiments, we have observed that as the concentration of DMABMA solution in cyclohexane increased to a certain degree, the fluorescence emission intensity at 354 nm decreased gradually. In order to eliminate the influence of self-quenching due to high concentration, at the same time we measured the fluorescence spectra of dimethyl toluidine, which does not contain a double bond, in different concentrations, for comparison. It was observed that, under similar conditions, the fluorescence intensity of DMT was about 20 times higher than the fluorescence intensity of DMABMA. As the concentration of both solutions gradually increased from 5×10^{-5} to 1×10^{-2} M, the fluorescence intensity underwent an enhancing and then a quenching stage. Figure 5 shows that at low concentration the rate of the enhancement of DMT fluorescence is obviously higher than that of DMABMA. However, as the concentration increases to a certain degree, the fluorescence changes to a quenching stage. The trend of fluorescence quenching of the DMABMA is faster than that of DMT. Although the fluorescence of the exciplex was not found in the spectra, the phenomena of fluorescence quenching described above may have demonstrated the existence of intra- or intermolecular exciplex formation.

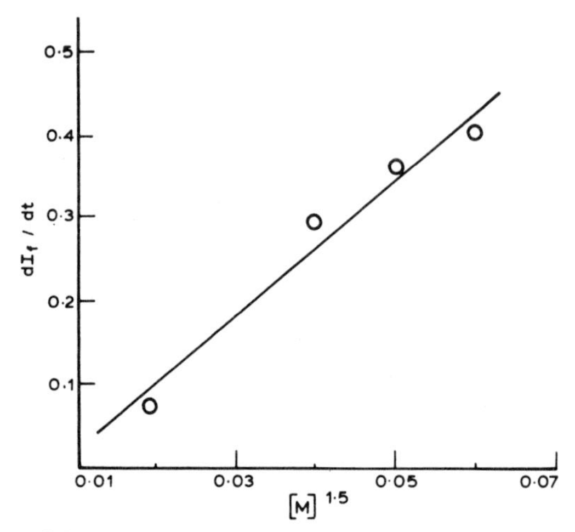

FIG. 6. A plot of dI_f/dt versus the original concentration of DMABMA solution.

With respect to the kinetics of photopolymerisation of DMABMA, it was studied according to the phenomena of fluorescence quenching of DMABMA during irradiation. The 1,2-dichloroethane solutions of DMABMA with different original concentrations had nitrogen bubbled through for a few minutes. Then, the fluorescence emission spectra which were excited by 310 nm were measured after irradiating the solutions at different intervals with a 80 W high pressure mercury lamp. The rate of photopolymerisation should be connected to the rate of change of fluorescence. So, dI_f/dt could be used as a parameter for studying the kinetics of photopolymerisation of these monomers. We measured the dI_f/dt at different original concentrations. The plot of dI_f/dt versus the original concentration [M] of solution is shown in Fig. 6. It shows that the rate of photopolymerisation dI_f/dt is in fact a linear function of $[M]^{1 \cdot 5}$, i.e., the kinetic process of this photopolymerisation accords with the following formula:

$$R_p = dI_f/dt = K[M]^{1 \cdot 5}$$

It indicated that the compound DMABMA is used not only as a monomer but also as an initiator in the polymerisation process. So, we could obtain such results in a free radical initiation polymerisation.[6]

The other two monomers[8] listed above are similar in structure to DMABMA. They contain both electron donors and electron acceptors in a single molecule. The fluorescence spectra of monomer DMAPAA and its polymer are shown in Fig. 7. It is evident that the large difference of

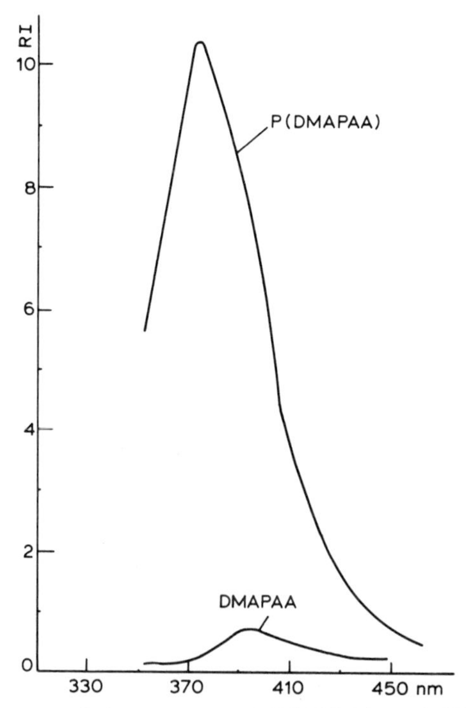

FIG. 7. Fluorescence emission spectra of DMAPAA and P(DMAPAA) in dichloroethane. Excitation wavelength = 329 nm.

emission intensity excited by the light with 338 nm between monomer and polymer which contain the same concentration of the dimethyl aniline group (1×10^{-3} M) is still present. The emission intensity of the monomer is about one-tenth as strong as the polymer. It is further indicated that in the presence of the double bond used as an electron acceptor in a molecule which contains an electron donor group, charge transfer or exciplex formation could occur easily. In this case, the fluorescence emission of the fluorescent group contained in the molecule should be quenched by itself. However, due to the loss of double bond after polymerisation the emission intensity of polymer remains strong enough.

To clarify the charge transfer properties of the fluorescence quenching of the monomer solution, we have measured the fluorescence spectra of the monomer in different polarity solvents. It is shown in Fig. 8 that the fluorescence emission intensities of DMAPAA dissolved in different polarity solvents at the same concentration are different. In acetonitrile, which has the strongest polarity of the solvents used, the emission intensity

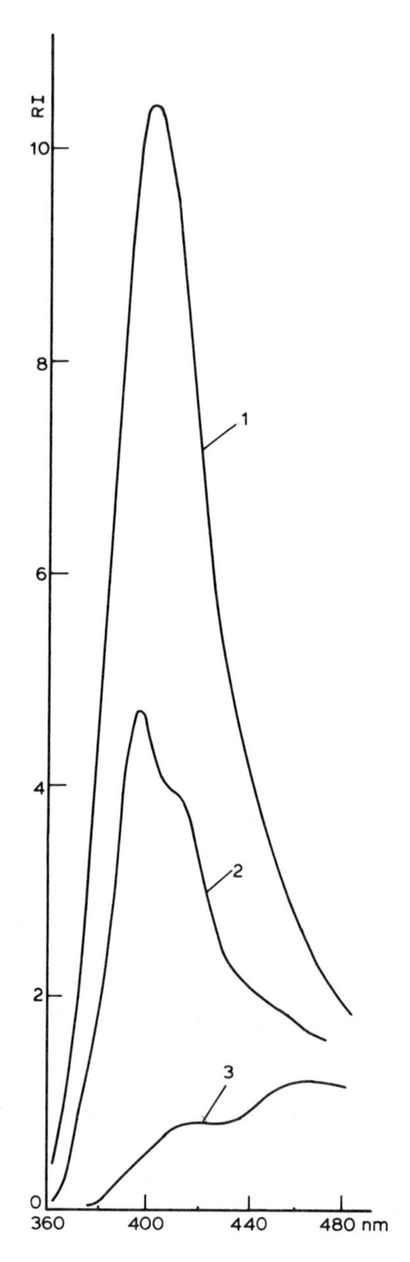

FIG. 8. Fluorescence intensity of DMAPAA in different solvents, with same concentration. Excitation wavelength 340 nm: 1, in toluene; 2, in dichloroethane; 3, in acetonitrile.

is the lowest. On the contrary, when the monomer was dissolved in toluene, which possesses the weakest polarity in the solvents used, its fluorescence intensity increased to a high degree. These results indicated that the polar environment is favourable for charge transfer, so, the exciplex formation is also favoured. Therefore it can promote the fluorescence quenching of monomers. In a non-polar solvent such as toluene, the charge transfer occurs with difficulty, so the exciplex is also difficult to generate. In this case, the fluorescence of monomer cannot be quenched easily, therefore it has stronger fluorescence emission. All of these have shown that the phenomena of fluorescence quenching described above are increased by the charge transfer process.

Another novel similar compound shows identical behaviour regarding its fluorescence emission spectra, so, we would not introduce its spectral behaviour in detail here. But it is interesting to find that the irradiation effect of the monomer DMAPAA solution is similar to that of monomer DMABMA. It means that the emission itensity of the DMAPAA solution increases gradually at different irradiation intervals. However, the fluorescence emission intensity of the other monomer DMAPMA is not changed after irradiation for a long time. It is indicated that the monomer DMAPMA cannot be polymerised by photo-irradiation. We compared the molecular structure of the three kinds of monomers studied and found that monomers DMABMA and DMAPMA both contain methyl groups in the alpha position which is able to repel the electron of the double bond. In this case, the methyl group makes the double bond to be an electron-rich one. The electron acceptability of this double bond must be decreased. On the other hand, monomers DMAPAA and DMAPMA both contain amide groups in the molecule. It can be considered to be a negative factor for electron donor ability of the nitrogen in the molecule, because it is possible to form an intermolecular hydrogen bond between the amide group and the nitrogen atom. It will decrease the electron donor ability of the nitrogen atoms and will influence the effect of the charge transfer. For the monomers DMABMA and DMAPAA, the former contains a methyl group but without an amide group, and the latter has an amide group, however, without a methyl group. Both monomers contain one negative factor described above, the former DMABMA has a weak acceptor and the latter DMAPAA has a weak electron donor. It is believed that in this case, a certain degree of charge transfer is presented in these systems. Therefore, the photopolymerisation can occur.

The other monomer DMAPMA, which cannot be polymerised by irradiation, possesses a different molecular structure to the others. It contains a methyl group and an amide group both in a single molecule. It means that it has two negative factors to encourage the charge transfer to occur. Therefore the photopolymerisation of this monomer would not be realised.

According to the results obtained in this work, it is evident that the degree of charge transfer between the electron donor and the electron acceptor is a very important parameter for photopolymerisation.

From the results described above and the results[9] of photopolymerisation obtained in Peking University—the aromatic amine fragment was attached to the end of poly(acrylonitrile) and poly(methyl methacrylate) macromolecules which were obtained by using dimethyl toluidine as an initiator of photopolymerisation—a possible mechanism of photoinitiation of these monomers was proposed as follows:

$$(CH_3)_2N-C_6H_4-CH_2-O-\overset{O}{\overset{\|}{C}}-\underset{CH_3}{\underset{|}{C}}=CH_2 \xrightarrow{h\nu}$$

$$(CH_3)_2N-C_6H_4-CH_2O\overset{O}{\overset{\|}{C}}\underset{CH_3}{\underset{|}{C}}=CH_2^* \xrightarrow{(CH_3)_2N-C_6H_4-CH_2O\overset{OCH_3}{\overset{|}{C}}C=CH_2}$$

$$\left[\begin{array}{c} (CH_3)_2N-C_6H_4-CH_2O\overset{OCH_3^*}{\overset{|}{C}}C=CH_2 \\ \downarrow \qquad\qquad\qquad \uparrow \\ CH_2=\underset{CH_3}{\underset{|}{C}}\overset{O}{\overset{\|}{C}}OCH_2-C_6H_4-N(CH_3)_2 \end{array}\right] \text{Exciplex formation}$$

Photodissociation ↙ ↓ (H) abstraction

$$\left[(CH_3)_2N-C_6H_4-CH_2O\overset{O}{\overset{\|}{C}}\underset{CH_3}{\underset{|}{C}}=CH_2\right]^{\dot{+}} + \left[(CH_3)_2N-C_6H_4-CH_2O\overset{O}{\overset{\|}{C}}\underset{CH_3}{\underset{|}{C}}=CH_2\right]^{\dot{-}}$$

$$(CH_3)(\dot{C}H_2)N-C_6H_4-CH_2O\overset{O}{\overset{\|}{C}}\underset{CH_3}{\underset{|}{C}}=CH_2 + (CH_3)_2N-C_6H_4-CH_2O\overset{OH}{\overset{|}{C}}\underset{CH_3}{\underset{|}{C}}-\dot{C}H_2$$

↓

Polymerisation

4. CONCLUSIONS

1. The low fluorescence emission of the monomers studied here which contain both electron donor and acceptor in a single molecule is attributed to the presence of a charge transfer process and the exciplex formation.

2. The photoconductive behaviour of similar compound pairs indicated that after irradiation of these monomers the ionic radicals could be generated.

3. According to the fact that some monomers can be polymerised after irradiation, but some monomers cannot, it shows that although the charge transfer phenomena can be observed in all of the systems studied in this work, the degree of charge transfer is very important for photopolymerisation.

4. According to the kinetics of polymerisation in this work and some previous results, the polymerisation can be considered to be a free radical process.

5. A possible initiation process of these monomers has been proposed.

REFERENCES

1. Ishida, I., Sondo, S. and Tsuda, K., *Makromol. Chem.*, **178**, 3221 (1977).
2. Wei, Y., Master dissertation in Dept. of Chemistry, Peking University, 1980.
3. Cao Wei-Xiao and Feng Xin-De, *Polymer Communication* (*Chinese*), No. 6, 357 (1980); No. 2, 96 (1982).
4. Wu Shih-Kang, Yao Shao-Ming and Jiang Yong-Cai, *Scientia Sinica* (*B*), **27**, 341 (1984).
5. Hrabak, F. *et al.*, *Makromol. Chem.*, **197**, 2593 (1978).
6. Li Fu-Mian, Ye Wei-Peng and Feng Xin-De, *Polymer Communication* (*Chinese*), No. 5, 398 (1982).
7. Wu Shih-Kang, Jiang Yong-Cai, Li Fu-Mian and Feng Xin-De, *Polymer Bulletin*, **8**, 275 (1982).
8. Wu Shih-Kang, Zhu Qing-Qing, Li Fu-Mian and Feng Xin-De, *Photographic Science and Photochemistry* (in press).
9. Li Tong, Cao Wei-Xiao and Feng Xin-De, *Polymer Communication* (*Chinese*), No. 4, 260 (1983).

6

Photoinitiation of Radical, Cationic and Concurrent Radical-Cationic Polymerization

S. Peter Pappas

Polymers and Coatings Department, North Dakota State University, Fargo, USA

1. INTRODUCTION

Photoinitiation of radical and cationic polymerization with multifunctional monomers results in crosslinked polymer networks and constitutes the basis of important commercial processes with broad applicability, including microlithography[1] and curing.[2] This report deals with UV curing, which offers unique advantages over thermal curing, including rapid polymer network formation on heat-sensitive substrates, as well as reduced energy consumption, low emissions and minimal space requirements.

2. LIGHT INTENSITY AND PHOTOINITIATOR CONCENTRATION

The rate of formation of initiator species (R_i) equals the intensity of light absorbed by the photoinitiator (I_a) times the fraction (ϕ) of excited-state photoinitiator (PI*) which produces initiator species (eqn (1)).

$$R_i = I_a \cdot \phi \tag{1}$$

The absorbed light intensity is related to the incident light intensity (I_0) by eqn (2), where A, the photoinitiator absorbance, is the product of the molar absorptivity (ε), path length of light (or film thickness) (d), and PI concentration (c), as shown in eqn (3).

$$I_a = I_0(1 - 10^{-A})/d \tag{2}$$

$$A = \varepsilon dc \tag{3}$$

These relationships show that I_a (and R_i) increase with both I_0 and PI concentration (c). However, as c increases, A, I_a and R_i increase within shorter path lengths. For radical polymerization, where high concentrations of radicals favor termination, the average rate of polymerization (R_p) has been shown to decrease for $A > 2{\cdot}5$.[3] Furthermore, even with substantially lower A values ($A \gtrsim 0{\cdot}3$), the available light intensity at successively longer path lengths decreases significantly with increasing path length (or film thickness).

These considerations predict optimum concentrations of photoinitiator in UV curing which decrease with increasing film thickness. The predictions have been experimentally verified utilizing a photocalorimetric device as a measure of polymerization rate.[4] Calculations have also been made in a pigmented system where the effect of decreasing available light intensity with film thickness is more pronounced due to higher absorptivities.[5]

This 'light screening' by photoinitiator and other absorbing species, such as pigments, retards through-cure of films and is expected to contribute substantially towards poor adhesion. PI concentration (c) corresponding to a particular absorbance (A) may be calculated as a function of film thickness (d) utilizing eqn (3). The importance of utilizing photoinitiators with both strongly and weakly absorbing bands to achieve both surface and through-cure in air by radical polymerization has been discussed.[6]

Surface cure by radical polymerization in air requires high rates of radical formation (R_i) not only to consume surface oxygen but also to compete with the rate of diffusion of new oxygen to the surface. This may be achieved by utilizing high intensity lamps which provide high photon flux (I_0), the rate of photons delivered to the surface per unit area. Thus, I_0 is determined by lamp intensity and, importantly, lamp focus, but is unaffected by line speed or number of lamps (which may influence the extent of cure).

High photon flux (I_0) together with relatively low PI concentration (c) are also expected to enhance depth of cure by radical and cationic polymerization. Furthermore, high I_0 allows utilization of lower concentrations of relatively expensive PI. In this regard, it should be apparent that the optimum concentration of PI is a function of I_0, which should be taken into account particularly when comparing systems with different UV lamps.

3. REACTIVE FUNCTIONALITY

Reactive functionality (monomers and oligomers) in UV curing by radical polymerization includes acrylated resins, in general,[7] and for flexible coatings,[8] and thiol-ene compositions.[9] Properties of reactive diluents and their performance in UV curable coatings, inks and adhesives (radical polymerization) have been presented.[10]

Design of epoxy-functional coatings for UV curing by cationic polymerization has been systematically treated.[11] Furthermore, useful procedures for the synthesis of multifunctional vinyl ethers,[12] including aromatic derivatives,[13] have been reported. In general, vinyl ethers are substantially more reactive than epoxies in cationic polymerization.

Important factors and concepts for controlling viscosity of UV curable coatings, including pigmentation effects, have been discussed.[14]

4. PHOTOINITIATORS FOR RADICAL POLYMERIZATION

Photoinitiators for radical polymerization may be placed in two classes: (a) those which undergo intramolecular bond cleavage, notably acetophenone derivatives, and (b) those which undergo intermolecular H-abstraction from a H-donor, notably benzophenone, benzil and quinone derivatives. A relatively current overview of photoinitiators for radical polymerization is available.[15]

It is suggested that photoinitiators which undergo intramolecular bond-cleavage be classified as type-PI_1, since initiator radicals are produced by a unimolecular process. Photoinitiators which undergo intermolecular H-abstraction may correspondingly be classified as type-PI_2, since a bimolecular reaction is involved.[16]

4.1. Intramolecular Bond Cleavage (PI_1)

An important criterion for photoinitiators in this class (PI_1) is the presence of a bond with a dissociation energy lower than the excitation energy of the reactive excited state, on the one hand, and sufficiently high, on the other hand, to provide thermal stability.

Benzoin ethers represent the first class of photoinitiators utilized in UV curing, predominantly in the area of particle board finishing. Benzoin ethers (BE) undergo photocleavage to produce benzoyl (B) and ether (E) radicals, as shown in eqn (4).[6]

$$\underset{\text{BE}}{\mathrm{Ph{-}\overset{\overset{\displaystyle O}{\|}}{C}{-}\overset{\overset{\displaystyle OR}{|}}{C}H{-}Ph}} \xrightarrow{h\nu} \underset{\text{B}}{\mathrm{Ph\overset{\overset{\displaystyle O}{\|}}{C}^{\cdot}}} + \underset{\text{E}}{\mathrm{Ph\overset{\overset{\displaystyle OR}{|}}{C}H^{\cdot}}} \quad (4)$$

Under conditions of relatively low radical and high monomer

concentrations (acrylate and methacrylate), the B and E radicals are reported to be comparably efficient as initiators, based on results with ^{14}C-labeled benzoin ether derivatives,[17] and independent generation of the E radical.[18] However, under conditions of relatively high radical and low monomer concentrations (styrene), the B radical is primarily responsible for initiation; and the E radical participates, predominantly, as a chain terminating agent.[19]

Other aspects of benzoin ethers as photoinitiators, as well as the photoactivity of α,α-dimethoxy-α-phenylacetophenone (DMPA) and α,α-diethoxyacetophenone (DEAP) have been discussed.[6] DMPA is commercially available as Irgacure 651. (Note that Irgacure is a trademark of Ciba-Geigy.)

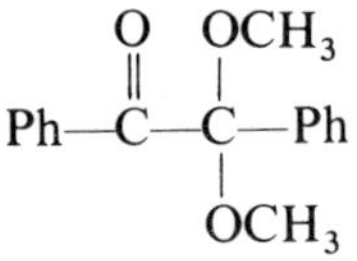

$$\underset{\text{DMPA}}{\mathrm{Ph{-}C({=}O){-}C(OCH_3)_2{-}Ph}} \qquad \underset{\text{DEAP}}{\mathrm{Ph{-}C({=}O){-}CH(OEt)_2}}$$

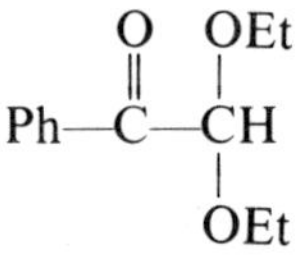

A newer class of PI_1 type photoinitiators includes α-hydroxy-α,α-dimethyl-acetophenone (**1**) and 1-benzoylcyclohexanol (**2**). Laser flash photolysis studies on **1** have provided evidence for photocleavage, as shown in eqn (5).[20] Photolysis of **2** is expected to proceed analogously.

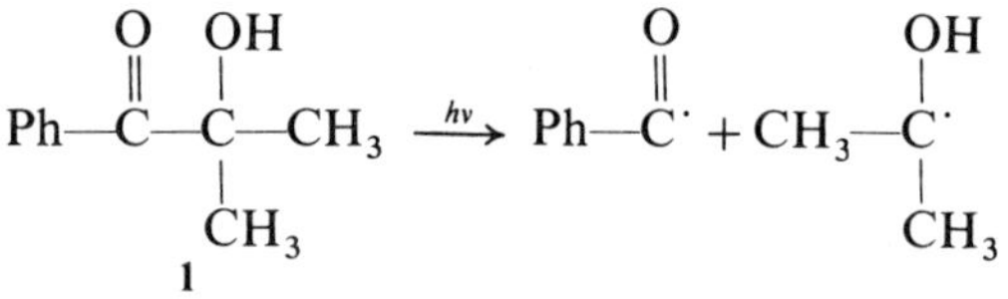

$$\underset{\mathbf{1}}{\mathrm{Ph{-}C({=}O){-}C(OH)(CH_3){-}CH_3}} \xrightarrow{h\nu} \mathrm{Ph{-}\dot{C}{=}O + CH_3{-}\dot{C}(OH){-}CH_3} \qquad (5)$$

$$\underset{\mathbf{2}}{\mathrm{Ph{-}C({=}O){-}C(OH)\langle CH_2{-}CH_2{-}CH_2{-}CH_2{-}CH_2\rangle}}$$

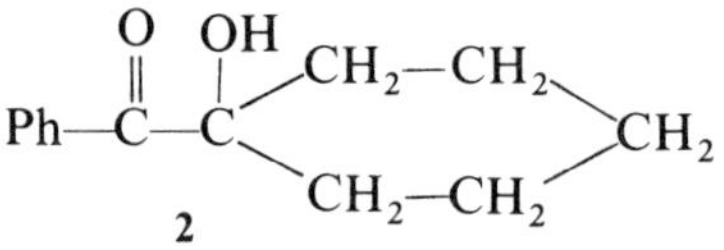

Photoinitiators **1** and **2** are commercially available as Darocure 1173 and Irgacure 184, respectively. The α-hydroxyacetophenone photoinitiators are promoted (in the trade literature) for their nonyellowing characteristics, which is consistent with the absence of benzylic radicals from photolysis.[6] (Note that Darocure is a trademark of EM Chemicals.)

A surface-active analog **3** of photoinitiator **1** has been shown, recently, to enhance surface hardness of films cured in air, when utilized in relatively low concentration (about 0·1 %) together with **1**.[21]

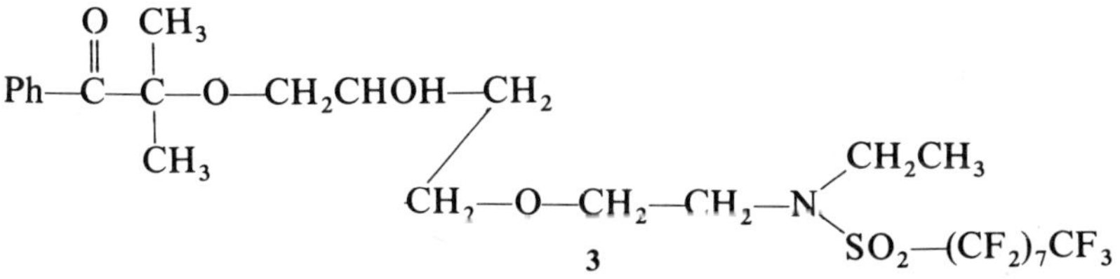

$$Ph-C(=O)-C(CH_3)_2-O-CH_2CHOH-CH_2-O-CH_2-CH_2-N(CH_2CH_3)-SO_2-(CF_2)_7CF_3$$

3

All of the aforementioned photoinitiators possess the acetophenone chromophore and exhibit relatively low absorptivity in the near UV spectral region (300–400 nm). This characteristic is generally desirable for clear coatings since low absorptivity by the photoinitiator in this region increases the depth of light penetration.[6] However, for pigmented coatings, it is necessary that the photoinitiator exhibit relatively high absorptivity in the near UV region in order to compete effectively with the pigment for the light energy.

Auxochromic substituents, such as oxygen and sulfur, enhance absorptivity of aromatic ketones in the near UV spectral region. A new PI_1-type photoinitiator **4** embodies this principle and exhibits high absorptivity in the near UV region (resulting primarily from the p-thiomethyl group in conjugation with the ketone group). This photoinitiator, commercially available as Irgacure 907, is recommended for pigmented coatings and printing inks.

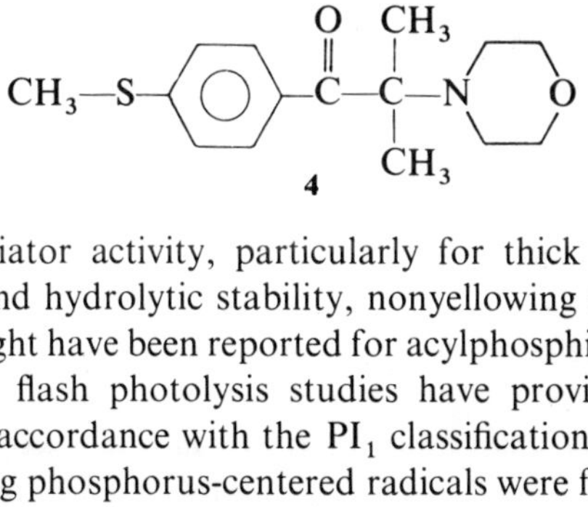

$$CH_3-S-C_6H_4-C(=O)-C(CH_3)_2-N(CH_2CH_2)_2O$$

4

High photoinitiator activity, particularly for thick films (in the cm range), thermal and hydrolytic stability, nonyellowing and activation by near UV-visible light have been reported for acylphosphine oxides, such as **5**.[22] Recent laser flash photolysis studies have provided evidence for photocleavage in accordance with the PI_1 classification, as shown in eqn (6).[23] The resulting phosphorus-centered radicals were found to be highly reactive with unsaturated monomers.[24]

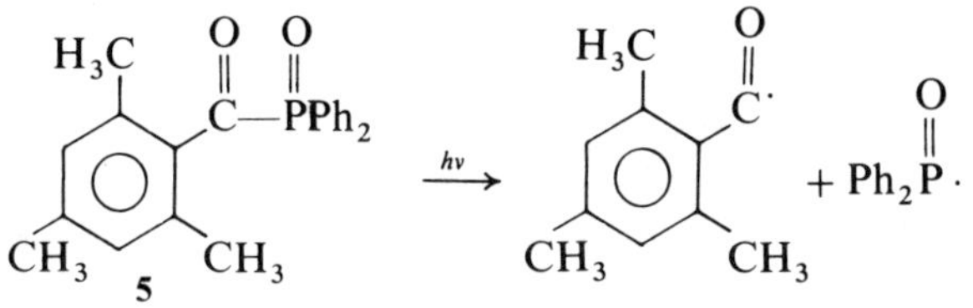

$$2,4,6\text{-}(CH_3)_3C_6H_2-C(=O)-P(=O)Ph_2 \;(\mathbf{5}) \xrightarrow{h\nu} 2,4,6\text{-}(CH_3)_3C_6H_2-\dot{C}=O + Ph_2\dot{P}=O \qquad (6)$$

4.2. Intermolecular H-Abstraction (PI_2)

Photoinitiators of this type include benzophenone, Michler's ketone, thioxanthones, benzil and quinones. In contrast to PI_1-type photoinitiators which are capable of generating initiator radicals independently, PI_2-type photoinitiators must undergo a bimolecular reaction with H-donors. For the efficient production of radicals, this bimolecular H-abstraction reaction must compete with nonproductive processes, including quenching of the photoinitiator excited state by oxygen as well as by monomers. Thus, a high reactivity of excited state photoinitiator with the H-donor is particularly important when UV curing is conducted in air. Furthermore, the reactivity of the radical derived from the H-donor is also a critical factor. Tertiary amines with abstractable α-H atoms are particularly effective H-donors for UV curing of acrylate monomers (and oligomers) in air, which may be attributed to: (a) rapid formation of an excited state complex (exciplex) between excited state photoinitiator and tertiary amines, and (b) initiation of acrylate polymerization by the resulting α-amino radical, as previously discussed.[6,16]

UV curing of an acrylate–urethane system with benzophenone/methyldiethanolamine is reported to provide faster surface cure and softer films than obtained with PI_1-type photoinitiators.[25] Both characteristics may be attributed to the high chain transfer activity of amines with peroxide and acrylate radicals.

Benzophenone/Michler's ketone combinations represent an effective photoinitiator system for UV curing of printing inks, based on reactive acrylate functionality.[26] However, the yellow color of Michler's ketone prevents its utilization as a photoinitiator in white pigmented coatings, in which case thioxanthone derivatives in combination with tertiary amines are reported to be effective.[27]

Benzil and quinones, such as 9,10-phenanthrene quinone and camphor quinone, also function as photoinitiators by intermolecular H-abstraction from tertiary amines. Since these α-dicarbonyl compounds absorb light at wavelengths longer than 380 nm, they may be used with visible light sources. Applications for such systems include UV curable orthopedic devices and dental materials.[28] A potentially important market for visible-light active photoinitiator systems is architectural (interior and exterior) paints.

A new class of PI_2-type photoinitiators, based on the 3-ketocoumarin ring system **6**, has been disclosed.[29] The absorptivity of this system can be varied throughout the UV-visible light region by proper selection of substituents R_1–R_4. A variety of coinitiators, including tertiary amines, are also claimed, which, together with appropriate analogs of **6**, compare favorably to the benzophenone/Michler's ketone photoinitiator system.

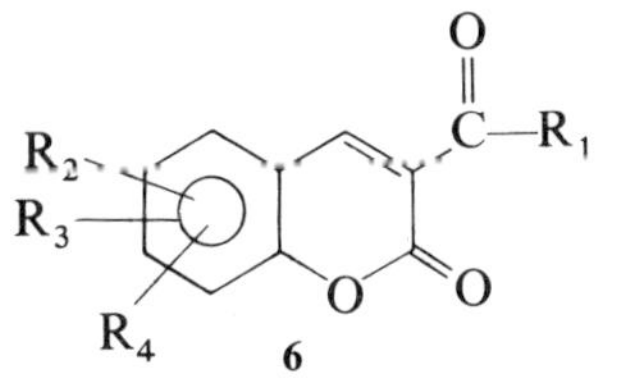

6

5. PHOTOINITIATOR COMBINATIONS (PI_1 AND PI_2)

The utilization of photoinitiator mixtures of type PI_1 and PI_2 for reducing air-inhibition in the absence of amines has been reported.[30] Furthermore, a eutectic mixture (liquid) of benzophenone (PI_2) and Irgacure 184 (PI_1) is commercially available as Irgacure 500. A possible explanation for the synergistic effect has been presented.[16]

6. PHOTOINITIATORS FOR CATIONIC POLYMERIZATION

Photoinitiators for cationic polymerization, of commercial significance, are onium salts with complex metal halide anions.[31,32] A key feature of this technology is the low nucleophilicity of the anions which reduces termination processes and allows ambient temperature cationic polymerization to proceed. The absence of air inhibition represents a distinguishing feature of cationic as compared to radical polymerization.

6.1. Triarylsulfonium and Diaryliodonium Salts

Thermally-stable photoinitiators for cationic polymerization include triarylsulfonium (**7**) and diaryliodonium (**8**) salts with metal halide anions.

$$Ar_3S^+ \; X^- \qquad\qquad Ar_2I^+ \; X^-$$

7 **8**

$$X^- = SbF_6^-, \; AsF_6^- \text{ and } PF_6^-$$

Irradiation of these salts results in cleavage of Ar–S (or Ar–I) bonds to yield reactive radical cations, as shown in eqns (7) and (8), respectively, corresponding to the PI_1 classification.

$$Ar_3S^+ \; X^- \xrightarrow{h\nu} Ar_2S^{+\cdot} \; X^- + Ar\cdot \qquad (7)$$

$$Ar_2I^+ \; X^- \xrightarrow{h\nu} ArI^{+\cdot} \; X^- + Ar\cdot \qquad (8)$$

Recently, direct evidence for this photocleavage has been obtained from laser flash photolysis studies.[33,34] Furthermore, the radical cations were found to be highly reactive with nucleophiles, including cyclohexene oxide. Thus, the radical cation may directly initiate polymerization of epoxies, which is supported by polymer end-group analysis.[35] Protons (H^+), formed from reaction of the radical cations with nucleophiles, also function as initiators for cationic polymerization, which continues after light exposure due to the long life-time of protons. Continued polymerization after light exposure further distinguishes cationic from radical polymerization.

Absorptivity of triarylsulfonium salts is enhanced in the 300–400 nm region by the introduction of thiophenoxy substituents. Examples are 4-diphenyl-sulfoniodiphenylsulfide (**9**)[36] and 4,4′-bis(diphenylsulfonio)-diphenylsulfide (**10**)[37] salts.

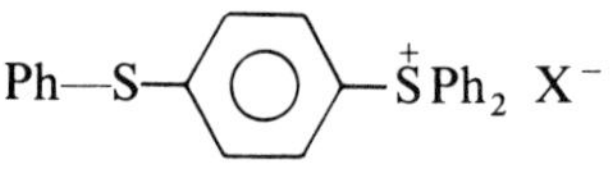

$$Ph{-}S{-}C_6H_4{-}\overset{+}{S}Ph_2\ X^-$$

9

$$Ph_2S^+{-}C_6H_4{-}S{-}C_6H_4{-}\overset{+}{S}Ph_2\ 2\ X^-$$

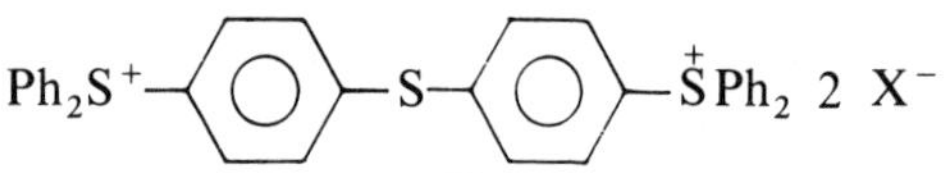

10

7. REDOX PHOTOSENSITIZATION OF ONIUM SALTS

The spectral response of onium salts may be extended into the 300–400 nm region (as well as into the visible) by the utilization of photosensitizers (PS). Evidence for two distinct mechanisms of redox photosensitization has been obtained.

7.1. Direct Electron Transfer to Onium Salts

One mechanism of redox photosensitization involves direct electron transfer from excited-state photosensitizer (PS*) to the onium salt to produce a reactive photosensitizer radical cation (eqn (9)). This species ($PS^{+\cdot}$) may initiate cationic polymerization of monomers (M), or produce protonic initiators by reaction with nucleophiles. The process is outlined in Scheme 1 with triarylsulfonium salts. An analogous scheme applies to diaryliodonium salts.

$$PS \xrightarrow{h\nu} PS^*$$

$$PS^* + Ar_3S^+\ X^- \longrightarrow PS^{+\cdot}\ X^- + Ar_3S^\cdot \qquad (9)$$

$$PS^{+\cdot}\ X^- \xrightarrow{M} \text{Polymer}$$

SCHEME 1.

Evidence in support of this mechanism has been obtained for phenothiazine, anthracene and perylene in combination with either triarylsulfonium or diaryliodonium salts.[34,38] Thioxanthone and xanthone also function as photosensitizers by this mechanism with diaryliodonium, but not triarylsulfonium, salts due to the greater ease of reduction of iodonium salts.

7.2. Indirect Electron Transfer to Onium Salts

A second mechanism of redox photosensitization involves photogeneration of radicals followed by electron transfer to the onium salts.[38,39] Evidence in support of this mechanism has been obtained for benzophenone, as photosensitizer, in combination with diphenyliodonium

$$Ph_2C{=}O \xrightarrow{h\nu} (Ph_2C{=}O)^*$$

$$(Ph_2C{=}O)^* + (CH_3)_2CH{-}OH \longrightarrow Ph_2\dot{C}{-}OH + (CH_3)_2\dot{C}{-}OH \quad (10)$$

$$Ph_2\dot{C}{-}OH + Ph_2I^+ \longrightarrow Ph_2\overset{+}{C}{-}OH + Ph_2I^{\cdot} \quad (11)$$

$$(CH_3)_2\dot{C}{-}OH + Ph_2I^+ \longrightarrow (CH_3)_2\overset{+}{C}{-}OH + Ph_2I^{\cdot} \quad (12)$$

$$Ph_2I^{\cdot} \longrightarrow PhI + Ph^{\cdot} \quad (13)$$

$$Ph^{\cdot} + (CH_3)_2CH{-}OH \longrightarrow (CH_3)_2\dot{C}{-}OH + Ph{-}H \quad (14)$$

$$Ph_2\overset{+}{C}{-}OH \longrightarrow Ph_2C{=}O + H^+$$

$$(CH_3)_2\overset{+}{C}{-}OH \longrightarrow (CH_3)_2C{=}O + H^+$$

SCHEME 2.

hexafluoroarsenate, as outlined in Scheme 2.[38] Photogeneration of radicals with benzophenone requires the presence of a H-donor, which undergoes H-transfer to excited state benzophenone, as shown in eqn (10) of Scheme 2 with 2-propanol as the H-donor. Both of the resulting ketyl radicals may reduce the iodonium salt (X^- is omitted for simplicity) to yield protonated benzophenone and acetone (eqns (11) and (12), respectively) together with diphenyliodide radical. The diphenyliodide radical readily dissociates into iodobenzene and a phenyl radical (eqn (13)), as determined from electrochemical studies.[40] H-abstraction from 2-propanol by the phenyl radical regenerates the acetone ketyl radical (eqn (14)), which may again participate in iodonium salt reduction (eqn (12)).

This mechanism predicts that quantum yields of formation of iodobenzene and protons may exceed one, which has been observed with both 2-propanol and tetrahydrofuran as H-donors.[38] This amplification of photons in the generation of protons results from the chain propagation sequence of eqns (12)–(14).

Evidence in support of the indirect redox photosensitization process (Scheme 2) has also been obtained with acetophenone, thioxanthone and xanthone in combination with iodonium salts. On the other hand, photosensitization of sulfonium salts by this mechanism is relatively inefficient, which may be attributed to the more favorable reduction potential of diphenyliodonium salts ($-0{\cdot}2$ V)[40] as compared to triphenylsulfonium salts ($-1{\cdot}2$ V).[41] The broader range of potential photosensitizers for iodonium salts, including dyes with low-lying excited states,[42] may also be ascribed to a more favorable reduction potential, as compared with sulfonium salts. The utilization of type-PI_1 photoinitiators, such as DMPA, as the radical source in indirect redox photosensitization of diaryliodonium salts has also been reported.[39]

8. HYBRID SYSTEMS—CONCURRENT RADICAL-CATIONIC POLYMERIZATION

A further extension of the concept of redox sensitization of cationic polymerization by photogeneration of radicals (Scheme 2) is the design of photocurable compositions which contain photoinitiators and functionality for both radical and cationic polymerization. Enhanced cure rates and cure depths[43] as well as desirable application and film properties[44] have been reported for such systems. Suppression of air-inhibition has been reported in photoinitiated polymerization of mixtures of acrylate/epoxy and acrylate/vinyl ether functional monomers, utilizing complex sulfonium salt **9** as the photoinitiator.[12]

A properly designed photoinitiator system for redox sensitization may provide enhanced light utilization and more efficient utilization of reactive intermediates. For example, ketyl radicals, generated from radical photoinitiators which undergo H-abstraction, are inefficient initiators for radical polymerization. On the other hand, ketyl radicals efficiently reduce diaryliodonium salts en route to protons for cationic polymerization (Scheme 2, eqn (11)).

In general, hybrid systems may provide greater flexibility in monomer and oligomer utilization toward the goal of achieving desirable application and film properties, such as viscosity control, hardness, flexibility, gloss and adhesion.

For example, cyclic monomers which polymerize cationically by ring-opening reactions tend to undergo little or no volume change and may even expand on polymerization.[45] In contrast, acrylates undergo substantial shrinkage during polymerization which introduces internal stresses in the resulting films. These stresses may retard adhesion as well as impart poor mechanical properties. On the other hand, induction periods are generally observed in photoinduced cationic polymerization of cyclic monomers, such as epoxides, which may result, at least in part, from the presence of basic impurities in commercial materials. Thus, hybrid systems may provide rapid initial radical polymerization together with continuing cationic polymerization after light exposure.

ACKNOWLEDGEMENTS

The author wishes to acknowledge his collaboration with Bob Asmus, Drs Lee Carlblom, Ashok Chattopadhyay, Les Gatechair, Joe Jilek, Betty Pappas, and Professor Wolfram Schnabel. We thank NATO (research grant 080.80) and the Graphic Arts Technical Foundation for financial assistance.

REFERENCES

1. Thompson, L. F., Willson, C. G. and Bowden, M. J. (Eds.), *Introduction to Microlithography*, ACS Symp. Ser. No. 219, Am. Chem. Soc., Washington, D.C., 1983.
2. Pappas, S. P. (Ed.), *UV Curing: Science and Technology*, Vol. 2, Technology Marketing Corp., Norwalk, CT, 1984.
3. Lissi, E. A. and Zanocco, A., *J. Polym. Sci., Polym. Chem. Ed.*, **21**, 2197 (1983).
4. Bush, R. W., Ketley, A. D., Morgan, C. R. and Whitt, D. G., *J. Radiation Curing*, **7**(2), 20 (1980).
5. Wicks, Z. W. and Pappas, S. P., In: *UV Curing: Science and Technology* (Ed. S. P. Pappas), Technology Marketing Corp., Norwalk, CT, 1978, pp. 79–95.
6. Pappas, S. P. and McGinniss, V. D., in ref. 5, pp. 1–22.
7. Blanding, J. M., Osborn, C. L. and Watson, S. L., *J. Radiation Curing*, **5**(2), 13 (1978).
8. Martin, B. M., in ref. 2, pp. 107–42.
9. Morgan, C. R. and Ketley, A. D., *J. Radiation Curing*, **7**(2), 10 (1980); see also, Morgan, C. R., Magnotta, F. and Ketley, A. D., *J. Polym. Sci., Polym. Chem. Ed.*, **15**, 627 (1977).

10. Tu, R. S., in ref. 2, pp. 143–246.
11. Watt, W. R., in ref. 2, pp. 247–82.
12. Crivello, J. V., Lee, J. L. and Conlon, D. A., *J. Radiation Curing*, **10**(1), 6 (1983).
13. Crivello, J. V. and Conlon, D. A., *J. Polym. Sci., Polym. Chem. Ed.*, **21**, 1785 (1983).
14. Wicks, Z. W., Jr and Hill, L. W., in ref. 2, pp. 77–105.
15. Gatechair, L. R. and Wostratzky, D., *J. Radiation Curing*, **10**(3), 4 (1983).
16. Pappas, S. P., *Radiation Curing*, **8**(3), 28 (1981).
17. (a) Carlblom, L. H. and Pappas, S. P., *J. Polym. Sci., Polym. Chem. Ed.*, **15**, 1381 (1977); (b) Pappas, S. P. and Chattopadhyay, A. K., *J. Polym. Sci., Polym. Letters Ed.*, **13**, 483 (1975).
18. Pappas, S. P. and Asmus, R. A., *J. Polym. Sci., Polym. Chem. Ed.*, **20**, 2643 (1982).
19. Hageman, H. J., van der Maeden, F. P. B. and Janssen, P. C. G. M., *Makromol. Chem.*, **180**, 2531 (1979).
20. Eichler, J., Herz, C. P., Naito, I. and Schnabel, W., *J. Photochem.*, **12**, 225 (1980).
21. Hult, A. and Rånby, B., *Am. Chem. Soc., Polym. Prepr.*, **25**(1), 329 (1984).
22. Jacobi, M. and Henne, A., *J. Radiation Curing*, **10**(4), 16 (1983).
23. Sumiyoshi, T., Henne, A., Lechtken, P. and Schnabel, W., *Z. Naturforsch.*, **39a**, 434 (1984).
24. Henne, A., personal communication.
25. Berner, G., Kirchmayr, R. and Rist, G., *J. Oil Col. Chem. Assoc.*, **61**, 105 (1978).
26. Hencken, G., *Farbe und Lack*, **81**, 916 (1975).
27. Davis, M. J., Doherty, J., Godfrey, A. A., Green, P. N., Young, J. R. A. and Parrish, M. A., *J. Oil Col. Chem. Assoc.*, **61**, 256 (1978).
28. Dart, E. C., *et al.*, U.S. Patents 3,874,376 (1975), 4,071,424 (1978), and 4,110,184 (1978) (assigned to ICI).
29. Specht, D. P., Houle, C. G. and Farid, S. Y., U.S. Patent 4,289,844 (1981) (assigned to Eastman Kodak).
30. Gruber, G. W., U.S. Patents 4,017,652 and 4,024,296 (1977) (assigned to PPG).
31. Crivello, J. V., in ref. 5, pp. 13–77.
32. Pappas, S. P., in ref. 2, pp. 1–25 and *Progr. Org. Coatings*, **13**(1), **35** (1985).
33. Pappas, S. P., Pappas, B. C., Gatechair, L. R. and Schnabel, W., *J. Polym. Sci., Polym. Chem. Ed.*, **22**, 69 (1984).
34. Pappas, S. P., Pappas, B. C., Gatechair, L. R., Jilek, J. H. and Schnabel, W., *Polym. Photochem.*, **5**, 1 (1984); also in: *Photochemistry and Photophysics in Polymers* (Ed. N. S. Allen and W. Schnabel), Elsevier Applied Science Publishers, London, 1984, pp. 1–22.
35. Ledwith, A., Preprints, *22nd Fall Symp., Soc. Photogr. Sci. Eng.*, Arlington, VA (1982), pp. 44–5.
36. Crivello, J. V. and Lam, J. H. W., *J. Polym. Sci., Polym. Chem. Ed.*, **18**, 2677 and 2697 (1980).
37. Watt, W. R., Hoffman, H. T., Pobiner, H., Schkolnick, L. J. and Yang, L. S., *J. Polym. Sci., Polym. Chem. Ed.*, **22**, 1789 (1984).

38. Pappas, S. P., Gatechair, L. R. and Jilek, J. H., *J. Polym. Sci., Polym. Chem. Ed.*, **22**, 77 (1984).
39. Ledwith, A., *Makromol. Chem., Suppl.*, **3**, 348 (1979).
40. Bachofner, H. E., Beringer, F. M. and Meites, L., *J. Am. chem. Soc.*, **80**, 4269 (1958).
41. McKinney, P. S. and Rosenthal, S., *J. Electroanal. Chem.*, **16**, 261 (1968)
42. Crivello, J. V. and Lam, J. H. W., *J. Polym. Sci., Polym. Chem. Ed.*, 2441 (1978).
43. Tsao, J.-H. and Ketley, A. D., U.S. Patent 4,156,035 (1979) (assigned to W. R. Grace).
44. Perkins, W. C., *J. Radiation Curing*, **8**(1), 16 (1981).
45. Bailey, W. J., Iwama, H. and Tsushima, R., *J. Polym. Sci.*, Symposium No. 56, 117 (1976); see also, Endo, T., Saigo, K. and Bailey, W. J., *J. Polym. Sci., Polym. Lett. Ed.*, **18**, 457 (1980).

7

Photochemical Synthesis of Block Polymers

GEORGES J. SMETS and TAKAO DOI
Laboratory of Macromolecular and Organic Chemistry,
Katholieke Universiteit Leuven, Belgium

1. INTRODUCTION

Photochemical block polymerisation consists of a photochemical reaction by which active sites are produced at the end of a polymeric chain, which themselves initiate the polymerisation of a second monomer. The method is thus a two-stage procedure: first the synthesis of a $(M_1)_x$ prepolymer containing one or several photosensitive groups, secondly photolysis of these groups in the presence of monomer M_2. These active sites are usually free radicals, i.e. macroradicals, and their production results from the photodissociation of the photolabile groups present in the prepolymer.

The method presents several attractive aspects, namely

(a) selectivity of initiation by adequate choice of photolabile groups;
(b) absence of degradation by using specific irradiation wavelength;
(c) applicability to most vinyl monomers;
(d) low temperature of reactions, which prevents side reactions and decreases the importance of chain transfer reactions;
(e) high yield of block polymer.

On the other hand, the monomer sequences will have a molecular weight distribution characteristic of free radical polymerisations and lack the homogeneity of the classical living polymers.

From the organic point of view the synthesis of the photolabile prepolymer is different according to whether $(M_1)_x$ is a condensation or a polyvinyl polymer. In the first case the light sensitive groups have to be incorporated within the polymer chain, in the second case they are usually present as end groups of the prepolymer, except when a multifunctional initiator is used for the polymerisation of M_1. This statement is the basis of this review.

The further use of these prepolymers for the subsequent block polymerisation of M_2 is similar, but depends evidently on the nature of the photosensitive groups.

2. BLOCK POLYMERIZATION WITH INCORPORATED PHOTOLABILE GROUPS

Block polymers resulting from the incorporation of light sensitive groups in a condensation polymer, followed by their photolysis in the presence of vinylmonomer M_2 have an overall structure which depends on the mode of termination by disproportionation and/or by addition of the second monomer. As far as the polymer contains only one photolabile group per macromolecule, the structure will be PC–PV and/or PC–PV–PC

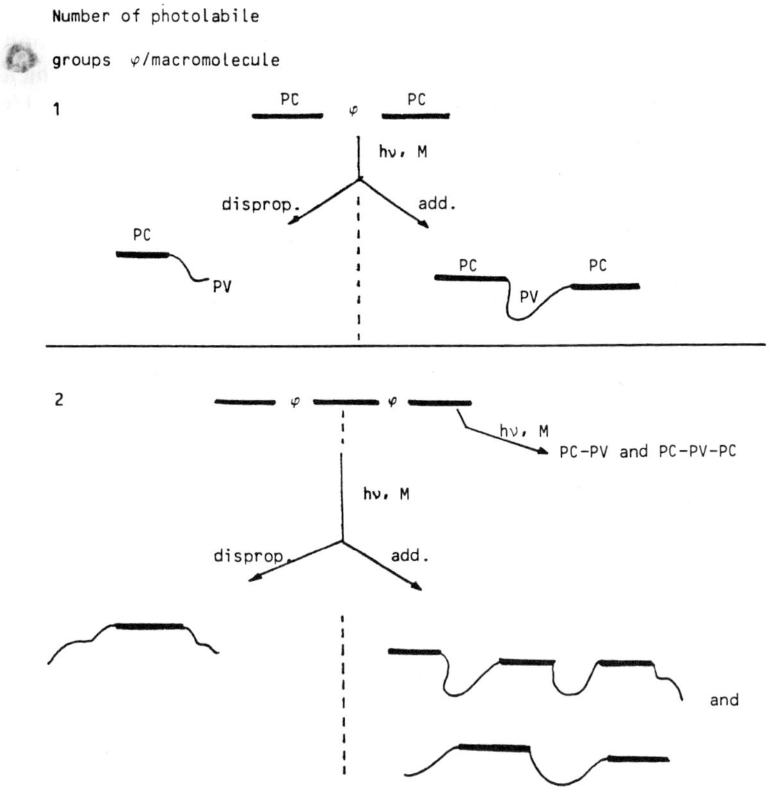

FIG. 1. Block polymerisation scheme as a function of the number of photolabile groups φ per macromolecule.

respectively (PC indicates a condensation polymer sequence, and PV a polyvinyl one).

If the number of photolabile groups exceeds unity, the block polymers will also contain PV–PC–PV triblocks and multiblocks $(PC\text{–}PV)_x$ resulting from the middle sequences. These different possibilities are represented in Fig. 1, where φ indicates a photolabile group.

A high efficiency of block polymerisation requires evidently a high quantum yield of photolysis ϕ_s of the photolabile group into free radicals.

The first example of this method is the incorporation of keto-oxime ester groups in a polytetrachlorobisphenol-A-adipate prepolymer. The photolysis of keto-oxime ester is well known; it results from the n-π^* absorption of the C=N group followed by scission of the N—O bond, generating two free radicals.[1] The quantum yield of photolysis is 0·94 in benzene (reaction 1).

$$\mathrm{Ar{-}CO{-}\underset{\underset{CH_3}{|}}{C}{=}N{-}O{-}CO{-}R \xrightarrow{h\nu} ArCO{=}N^{\cdot} + {}^{\cdot}OCO{-}R} \qquad (1)$$

$$\mathrm{ArCO=N^{\cdot} \longrightarrow ArCO^{\cdot} + CH_3CN \qquad {}^{\cdot}OCO{-}R \longrightarrow CO_2 + R^{\cdot}}$$

The synthesis of the prepolymer is represented by reaction (1a). This synthesis assumes an equivalent reactivity of both bisphenol and p-hydroxyphenyl,1,2-propanedione oxime towards adipoyl chloride. Indeed

$$m\ \mathrm{HO{-}C_6H_2Cl_2{-}C(CH_3)_2{-}C_6H_2Cl_2{-}OH} + n\ \mathrm{HO{-}C_6H_4{-}\overset{O}{\overset{\|}{C}}{-}\overset{CH_3}{\overset{|}{C}}{=}N{-}OH} + (m+n)\ \mathrm{Cl{-}\overset{O}{\overset{\|}{C}}{-}(CH_2)_4{-}\overset{O}{\overset{\|}{C}}{-}Cl} \longrightarrow \qquad (1a)$$

$$\mathrm{{-}O{-}C_6H_2Cl_2{-}C(CH_3)_2{-}C_6H_2Cl_2{-}OCO(CH_2)_4COO{-}C_6H_4{-}\underset{\underset{O}{\|}}{C}{-}\overset{CH_3}{\overset{|}{C}}{=}N{-}O\overset{O}{\overset{\|}{C}}(CH_2)_4\overset{O}{\overset{\|}{C}}{-}O{-}}$$

I: Photolabile prepolymer

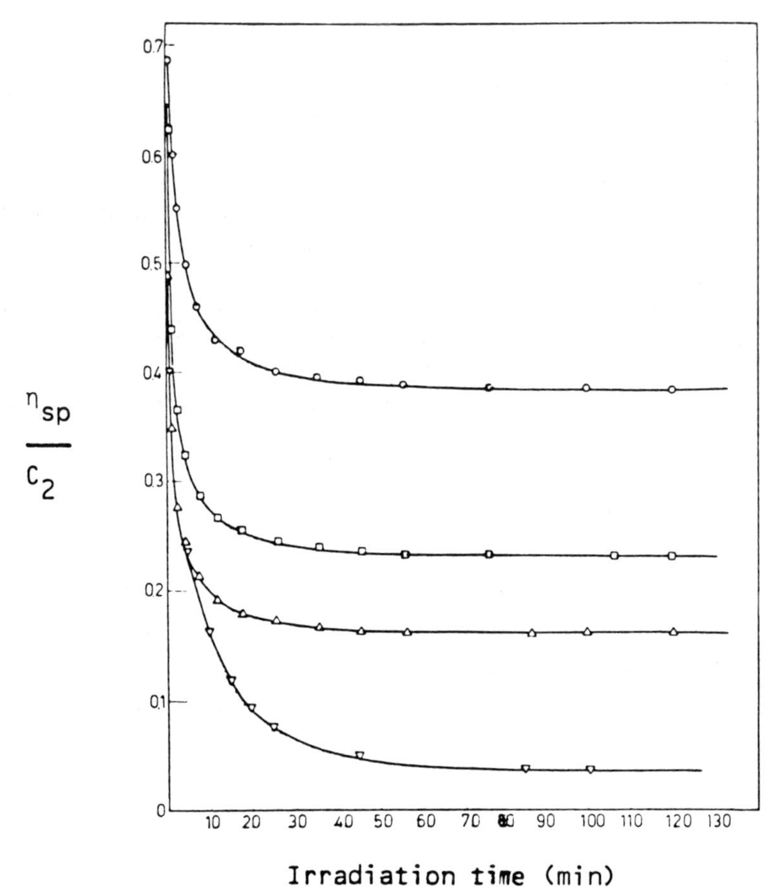

FIG. 2. Decrease of reduced specific viscosity with time of irradiation. $t°$: 20 °C, solvent dioxane; molar fraction α-keto-oxime: ○, 0·05; □, 0·1; △, 0·15; ▽, 1·0.

on irradiation in solution the inherent viscosities of the prepolymers decrease and the final viscosities depend on the mole fraction (5–15 %) of the incorporated oxime (Fig. 2).[2]

Photolysis in the presence of styrene gives high yields of polytetrachlorobisphenol-A-adipate-b-polystyrene block polymers with 38–94 mole % polystyrene, depending on the initial monomer concentrations (reaction 2). Only 5–10 % polystyrene was extractable with boiling cyclohexane.

The same reaction principle was applied recently by incorporating p,p′-dihydroxybenzoin methylether (BME) in polybisphenol-A-carbonate.

Previously BME must be transformed via its bischloroformate into bis(bisphenol-A) dicarbonic ester, of which the reactivity is equal to that of

$$\mathrm{I} \xrightarrow{h\nu} \sim\!\mathrm{O{-}C_6H_2Cl_2{-}C(CH_3)_2{-}C_6H_2Cl_2{-}O{-}CO{-}(CH_2)_4{-}CO{-}O{-}C_6H_4{-}\dot{C}{=}O} + \mathrm{CH_3CN} + \mathrm{CO_2}$$

$$+ \sim\!\mathrm{O{-}CO(CH_2)_4^{\cdot}} \xrightarrow{\mathrm{ST}} \text{block polymers} \qquad (2)$$

bisphenol-A itself (reaction 3). Further condensation with phosgene produces the photolabile BME-polycarbonate prepolymer (reaction 4).

As in the preceding case, the photolysis of the benzoin ether groups was almost complete after one-half hour irradiation ($\phi_s = 0{\cdot}74$)[3] (Fig. 3). The number of chain scissions per macromolecule as measured by the decrease of number average molecular weight agrees remarkably with the BME-content determined by NMR, e.g. 1·27 and 0·77 compared to 1·26 and 0·84, respectively.

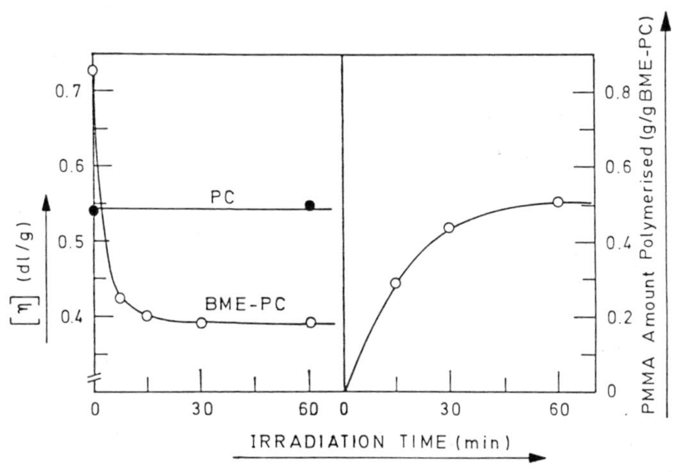

FIG. 3. Photolysis of BME-polycarbonate. (Left) Decrease of intrinsic viscosity under irradiation ($\lambda > 300$ nm) in absence of monomer. Curve PC represents the blank experiment, without the BME-group. (Right) in presence of methyl methacrylate.

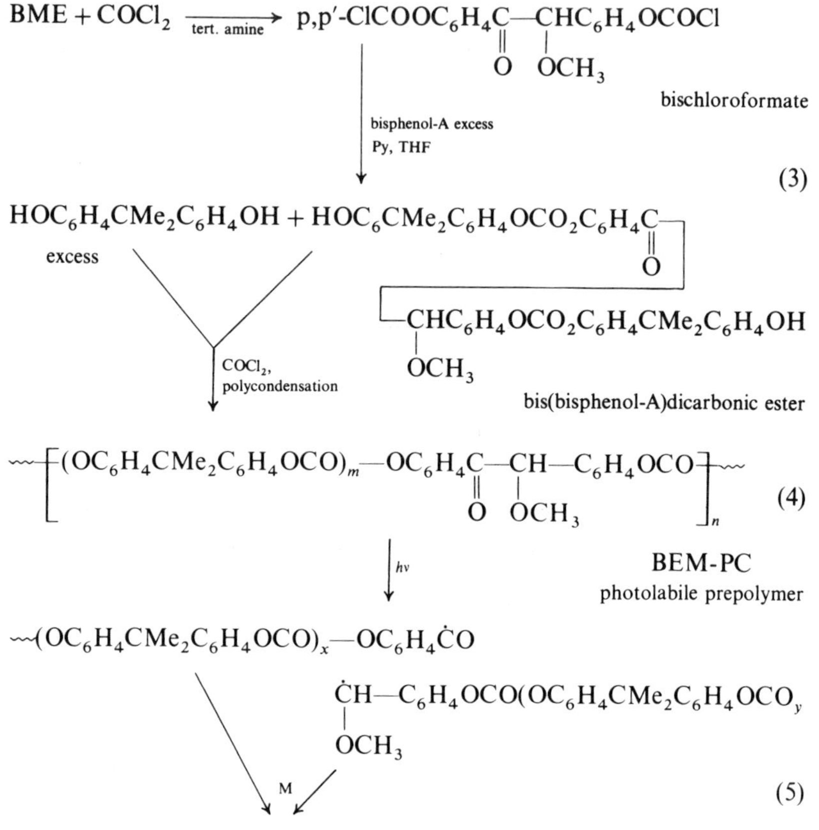

Irradiation of these BME-PC in the presence of vinyl monomers, methyl methacrylate and ethylacrylate (reaction 5) gives rise to the formation of polycarbonate-polyvinyl block polymers, of which the poly(meth)acrylate weight fraction increases proportionally to the monomer concentration. Again the reaction products are almost pure block polymers;[4,5] on differential scanning calorimetry they show two distinct glass transition domains. The method permits a regulation of the lengths of the blocks, the polycarbonate ones by the BME-content, the polyvinyl one by the monomer concentration.

The synthesis of cellulosic three block polymers recently described by Mezger and Cantow[6] is based upon the incorporation of a diaryl disulphide photolabile group in a prepolymer, which was obtained by coupling 1-hydroxy-terminated cellulose esters with bis(4-isocyanato phenyl) disulphide.

$$2\ \text{Cell}\begin{smallmatrix}\diagup \text{OH}\\ \diagdown \text{H}\end{smallmatrix} + OCNC_6H_4S\text{—}SC_6H_4NCO \longrightarrow \text{Cell}\begin{smallmatrix}\diagup \text{O—CONHC}_6\text{H}_4\text{S—SC}_6\text{H}_4\text{NHCO—O}\diagdown\\ \diagdown \text{H} \qquad\qquad\qquad\qquad\qquad\qquad \text{H}\diagup\end{smallmatrix}\text{Cell} \xrightarrow{h\nu} 2\ \text{Cell}\begin{smallmatrix}\diagup \text{O—CONH—C}_6\text{H}_4\text{S}^{\cdot}\\ \diagdown \text{H}\end{smallmatrix} \xrightarrow{M} \text{block polymer} \quad (6)$$

The middle diaryldisulphide group is easily photolysed by UV irradiation[7,8] and produces two macroradicals; on the other hand it has a high transfer constant with respect to a growing polystyryl chain ($C_1 = 1 \pm 0{\cdot}2$). Consequently, in the presence of styrene, irradiation produces a multiphase polymer system of triacylcellulose–block polystyrene–block triacylcellulose and in the presence of chloroprene a thermoreversible elastomer of tributyrylcellulose–block polychloroprene–block tributyrylcellulose.

A similar reaction principle was used by Vlasov *et al.*[9] who prepared symmetric polypeptidic chains containing a photolabile disulphide or azo

$$H_2N\text{—}CH_2\text{—}CONH\text{—}C_6H_4\text{—S—S—}C_6H_4\text{—}NH\text{—}CO\text{—}CH_2\text{—}NH_2$$

$$\text{or } H_2N\text{—}NH\text{—}CO\overset{CH_3}{\underset{CN}{C}}\text{—N=N—}\overset{CH_3}{\underset{CN}{C}}CONHNH_2$$

$$\big\downarrow \text{NCA}$$

$$\text{—[NH—}\underset{\substack{|\\(CH_2)_2\\|\\CO_2CH_2Ph}}{CH}\text{CO]}_x\text{—(NH—CH}_2\text{CONHC}_6\text{H}_4\text{S—S—C}_6\text{H}_4\text{NHCOCH}_2\text{NH)—[CO}\underset{\substack{|\\(CH_2)_2\\|\\CO_2CH_2Ph}}{CH}\text{NH—]}_x$$

or

$$\sim[NH{-}NHCO\underset{CN}{\overset{CH_3}{\underset{|}{\overset{|}{C}}}}{-}N{=}N{-}\underset{CN}{\overset{CH_3}{\underset{|}{\overset{|}{C}}}}CONHNH_2]\sim$$

group by reaction of $\omega . \omega'$ diamino compounds with the N-carboxy-anhydride of γ. benzylglutamate.

Photolysis of these two photolabile polypeptides in the presence of several vinyl monomers affords a great variety of polypeptide-b-polyvinyl block polymers which are able to form supermolecular structures.

Block polymers prepared by nitrosation of polyamides and polyurethane followed by photolysis in the presence of a vinyl monomer have a less well-defined structure than the preceding polymers.[10] The principle of the method is based upon the photoisomerisation of nitrosoamide groups into diazoester, which are unstable and afford two macroradicals with evolution of nitrogen.

$$\sim CH_2CH_2\underset{NO}{\underset{|}{N}}CORNHCH_2\sim \xrightarrow{h\nu}$$

$$\sim CH_2CH_2N{=}N{-}OCORNCH_2\sim \longrightarrow \qquad (7)$$

$$N_2 + \sim CH_2\dot{C}H_2 + \dot{O}CORNHCH_2\sim (\text{or } CO_2 + \dot{R}NHCH_2\sim)$$

The degree of nitrosation of the polymer chain is determined for the structure of the final block polymer. The N-nitrosoacyl groups are, however, relatively unstable, and the quantum yield of production of radicals is unknown. Moreover the characterisation of the soluble block polymers was only based on the viscosimetric Fikentscher's *K*-values while no polymer fractionation was mentioned. Nevertheless, block polymers were obtained with several vinyl monomers such as vinyl acetate, vinyl pyridine and (meth)acrylic derivatives.

3. BLOCK POLYMERISATION WITH PHOTOLABILE END GROUPS

The photodissociation of photosensitive end groups normally produces two free radicals of which only one is a macroradical. Consequently the formation of block- and homopolymers must be expected. Only special synthetic methods may prevent or severely restrict the formation of homopolymers. In this respect Bamford's metal carbonyl initiating method must be first mentioned; indeed it permits a very versatile synthesis of block polymers.[11,12] This method is based on the fact that transition metal carbonyls such as $Mn_2(CO)_{10}$, $Re_2(CO)_{10}$, $Os_3(CO)_{12}$ react easily with reactive organic halides as well as with olefines with strong electron attracting substituents giving rise to free radicals which can initiate the polymerisation.

$$Mn_2(CO)_{10} + RCCl_3 \xrightarrow{h\nu} Mn_2(CO)_{10}Cl + R\dot{C}Cl_2 \xrightarrow{M_1} RCCl_2(M_1)_x \tag{8}$$

Such an initiation reaction is photosensitive up to high irradiation wavelength (460 and 380 nm for Mn and Re carbonyls respectively) and the quantum yields for initiation at 25 °C are close to unity. This photoinitiation method provides thus a convenient route for synthesising vinyl polymers functionalised with different end groups such as CCl_3, CBr_3. These functionalised polymers are themselves reactive photosensitive prepolymers which can start again, in the presence of metal carbonyl, the polymerisation of monomer M_2 (reaction 9).

$$\sim\!\!(M_1)_m\!-\!CX_3 \xrightarrow[Mn_2(CO)_{10}]{h\nu} \sim\!\!(M_1)_m\!-\!\dot{C}X_2 \xrightarrow[M_2]{} \sim\!\!(M_1)_m\!-\!CX_2\!-\!(M_2)_n\!\sim \tag{9}$$

Three-block polymers $(M_1)_mCX_2(M_2)_{2n}CX(M_1)_m$ are formed if termination proceeds by combination. If the prepolymer carries two terminal CX_3, groups alternating multiblocks are obtained.

The iniferter (*ini*tiator-trans*fer*-*ter*minator) method of Otsu and coworkers constitutes a second synthetic method via photosensitive end groups. It is based on the early observations that organic disulphides and especially tetralkyl thiuramdisulphides dissociate into free radicals under UV irradiation[8,13,14] and that in the presence of styrene or methyl methacrylate telechelic polymers were obtained with following structure:

$$Et_2NCS\!-\!S(M_1)_nS\!-\!CSNEt_2$$

These telechelics have a functionality equal to about two as determined by UV spectrometry and $\bar{M}_n$ values.

These telechelics can be used in their turn as photoiniferters for chain extension with other vinyl monomers, giving rise to block polymers, which still carry an appreciable number of thiocarbamate end groups[15] and have a four- to fivefold higher molecular weight.

$$\mathrm{Et_2NCS{-}S(M_1)_nS{-}CSNEt_2 + M_2 \longrightarrow Et_2NCS{-}S(M_2)_x(M_1)_n(M_2)_xS{-}CSNEt_2} \quad (10)$$

In the frame of the present review these telechelics polymers behave as prepolymers with photosensitive end groups. From the kinetic point of view one observes in the homopolymerisation of styrene that the yield and the number average molecular weight of the polymers increase with the time of reaction. Though there is not yet a complete kinetic analysis available, the following reaction mechanism can be presented in the case of tetraalkylthiuram disulphide (TD) on the basis of the experimental data.

3.1. Formation of Telechelic Prepolymer

$$\underset{\mathrm{TD}}{\mathrm{Et_2NCS{-}S{-}S{-}CSNEt_2}} \xrightarrow[(\Delta T)]{h\nu} \mathrm{2Et_2NCS{-}\dot{S}} \quad (11)$$

$$\mathrm{Et_2NCS{-}S^{\cdot} + CH_2{=}\underset{X}{\underset{|}{C}}H} \xrightarrow[k_p]{} \mathrm{Et_2NCS{-}SCH_2{-}\underset{X}{\underset{|}{\dot{C}}}H} \quad (12)$$

$$\mathrm{Et_2NCS{-}S(CH_2{-}\underset{X}{\underset{|}{C}}H)^{\cdot}_m + TD} \xrightarrow[k_{tr}]{} \mathrm{Et_2NCS{-}S(CH_2{-}\underset{X}{\underset{|}{C}}H)_mS{-}CSNEt_2 + Et_2NCS{-}\dot{S}} \quad (13)$$

$$\mathrm{2Et_2NCS{-}S(CH_2{-}\underset{X}{\underset{|}{C}}H)^{\cdot}_m} \xrightarrow[k_{t_a}]{} \mathrm{Et_2NCS{-}S(CH_2{-}\underset{X}{\underset{|}{C}}H)_m(\underset{X}{\underset{|}{C}}H{-}CH_2)_mSCSNEt_2} \quad (14)$$

$$\mathrm{Et_2NCS{-}S(CH_2{-}\underset{X}{\underset{|}{C}}H)^{\cdot}_m + Et_2NCSS^{\cdot}} \xrightarrow[k_t']{} \mathrm{Et_2NCS{-}S(CH_2\underset{X}{\underset{|}{C}}H)_mSCSNEt_2} \quad (15)$$

The telechelics result from reactions (13), (14) and (15), namely by chain transfer reaction (13), chain addition termination (14) and termination by primary dithiocarbamate radicals (15). Considering that the functionality of the telechelics is equal to two for methyl methacrylate as well as for styrene[15] independently of their different chain termination, it must be concluded that reaction (14) is only of secondary importance or even negligible.

The role of organic disulphides is therefore triple: photochemical initiation (reaction 11), chain transfer reaction (13) and *primary* radical termination (reaction 15).

The polymerisation should also involve the photodissociation of the end groups into a reactive macroradical which propagates again by monomer addition, and a small radical of the dithiocarbamate type which is relatively stable (much less reactive) and practically unable to initiate a new chain (reaction 16). The result of such a mechanism is a repeated insertion of monomer molecules within the chain end; this procedure is then responsible for the chain lengthening with time of reaction.[16]

$$\sim CH_2{-}\underset{\displaystyle X}{\underset{|}{CH}}{-}SCSNEt_2 \underset{}{\overset{h\nu}{\rightleftharpoons}} \sim CH_2{-}\underset{\displaystyle X}{\underset{|}{\dot{C}H}} + \dot{S}CSNEt_2 \qquad (16)$$

$$\downarrow CH_2{=}CHX$$

$$\sim(CH_2{-}\underset{\displaystyle X}{\underset{|}{CH}})_n{-}CH_2{-}\underset{\displaystyle X}{\underset{|}{C}}HSCSNEt_2$$

$$\updownarrow h\nu$$

$$\sim(CH_2{-}\underset{\displaystyle X}{\underset{|}{CH}})_n{-}CH_2{-}\underset{\displaystyle X}{\underset{|}{\dot{C}H}} + \dot{S}CSNEt_2 \text{ etc.}$$

3.2. Photochemical Block Polymerisation

The block polymerisation results from the photodissociation of the end groups of the telechelic prepolymers and insertion of a second monomer at the polymer chain ends following a reaction mechanism identical to reaction (16).

By this method prepolymer PSt was transformed in PSt-b-polyvinyl acetate, b.polymethylmethacrylate, b-polyethylacrylate, and prepolymer PMMA into PMMA-b-PSt. Even three and four component blocks were prepared, e.g. with acrylonitrile. The reaction products have usually to be separated into residual prepolymer (0–6 %), M_2 homopolymer (12–34 %) while the block polymers represent about 75 %. The yields depend on the monomer and the time of irradiation.[17]

Recently Otsu and Kuriyama prepared $(M_1)_n(M_2)_m$ and $(M_1)_n(M_2)_m(M_1)_n$ block polymers selectively by using benzyl diethylamino dithiocarbamate, $C_6H_5CH_2$—S—CSNEt$_2$, and p.xylylene bis N.N-diethyldithiocarbamate, Et_2NCS—$SCH_2C_6H_4CH_2S$—CSNEt$_2$, as mono- and bifunctional photoiniferters for the polymerisation of styrene and methylmethacrylate,[18] and their further use as photoiniferter prepolymer.

By the same method, TD-polystyrene prepolymer was used as iniferter for the block polymerisation of 2.acrylamido-2 methylpropane sulphonic acid giving formation of a ABA triblock polymer which exhibits heterogeneous behaviour due to the simultaneous presence of hydrophobic (PSt) and hydrophilic segments —(CH_2—CH—)$_x$.[19]

$$-(CH_2-\underset{\substack{|\\ \text{CONH}\\ |\\ (CH_3)_2C-CH_2-SO_3H}}{CH}-)_x$$

Following Otsu this photoiniferter polymerisation proceeds through a ‘living’ radical mechanism. It is however to be observed that in the present case the chain initiation is slow with respect to the propagation, with a resulting increasing polydispersity with the time of reaction ($\bar{M}_w/\bar{M}_n$ from 2·5 to 5·4 in the case of TD as photoiniferter of styrene).

4. BLOCK POLYMERISATION WITH A HETEROFUNCTIONAL INITIATOR

The use of oligoperoxides with peroxidic linkages of different thermal stability for a two-step synthesis of block polymers was first described by Ivanchev *et al.*[20–23] It is based on the difference of decomposition activation energies E_{a_d} of both peroxides, and can be illustrated by the following example:

$$-(OCO(CH_2)_7CO-O-O-CO\underset{\substack{|\\ Br}}{C}H(CH_2)_6CO-O)_x-$$

$$x = 10\text{–}18$$

where the E_{a_d}'s of the α-bromoperoxide and the unsubstituted compound are respectively 95·7 and 123·3 kJ mole^{-1}. It was found that the difference of 27·6 kJ is sufficient to enable a two-step synthesis of polystyrene-b-polymethylmethacrylate. The higher the degree of oligomerisation x, the higher the yield of block polymer.

An alternative method is based on the selective photochemical decomposition of one of the groups of the heterofunctional initiator, e.g. azonitrile groups incorporated in an azo-oligoperoxide,[24] such as the compound obtained by reaction of hydrogen peroxide with 4,4′azo bis-(4-cyanopentanoic acid chloride):

$$\mathrm{ClCO(CH_2)_2\underset{\displaystyle CN}{\underset{|}{C}}(CH_3)N{=}N{-}\underset{\displaystyle CN}{\underset{|}{C}}(CH_3)(CH_2)_2COCl + H_2O_2}$$

$$\mathrm{-[OCO(CH_2)_2\underset{\displaystyle CN}{\underset{|}{C}}(CH_3){-}N{=}N{-}\underset{\displaystyle CN}{\underset{|}{C}}(CH_3)(CH_2)_2{-}COO{-}]_{x'}} \quad (17)$$

where $x' = 8$–10

The selective photolysis of the azo groups at 350 or 371 nm in the presence of styrene leads to a styrene prepolymer containing acyl peroxidic groups. The number of these groups per molecule was evaluated from the ratio $(Mn_0/Mn_t) - 1$ where Mn_0 and Mn designate the number average molecular weights before and after thermal decomposition in boiling chlorobenzene; it amounts to 0·64–0·87. This peroxidic polystyrene prepolymer was dissolved in tetrahydrofuran or in trichlorobenzene and used for the free radical polymerisation of vinyl chloride by heating at 70°C or by further UV irradiation. After reaction the product mixture was

TABLE 1

Block Polymerisation of Peroxidic Styrene Prepolymer with Vinyl Chloride[a]

	Tetrahydrofuran	*Trichlorobenzene*
Mn_0	102·500	50·000
$(Mn_0/Mn) - 1$	0·64	0·61
VC ml	80	50
solvent ml	20	35
react. product VC %	40	54
polystyrene	45 (66·700)[c]	31 (36·900)
PVC + block polymer[b]	55 (42·200)	69 (39·700)
PVC	18	20
block	37	49
prepolym. in block	25	33

[a] $t°$ 70 °C, 4 h with 3 g prepolymer.

[b] The separation of PVC and block polymer by preparative GPC is incomplete and reflects also differences in molecular weight instead of chemical differences.

[c] The values in parentheses indicate number average molecular weight of the polystyrene and of the PVC + block polymer mixture.

extracted for polystyrene and fractionated by preparative GPC. Some results are summarised in Table 1.

As expected on the basis of the low peroxide content (0·61–0·64) the reaction product contains inert polystyrene. On the other hand homopolyvinyl chloride is always formed mainly by chain transfer with the monomer, while the relative importance of transfer with solvent appears from the comparison THF/TCB. Figure 4 shows the GP-chromatograms with infra-red detector equipment of the PVC block polymer mixture (after extraction of homopolystyrene with cyclohexane). It can be seen that the polystyrene blocks are more concentrated in the high molecular fractions, while the lowest fractions are homopolyvinyl chloride.

Similar photochemical essays have been carried out with di-t.butyl.4.4′ azo bis(4-cyano peroxypentanoate) instead of an oligoazoperoxide initiator. This initiator, prepared by reaction of the corresponding bis acid

$$\text{t.BuOOCOCH}_2\text{CH}_2\underset{\displaystyle\text{CN}}{\underset{|}{\text{C}}}\text{Me—N=N—}\underset{\displaystyle\text{CN}}{\underset{|}{\text{C}}}\text{MeCH}_2\text{CH}_2\text{—COOOBut}$$

chloride with t.butylhydroperoxide, is easily soluble in styrene monomer and can be obtained with a high degree of purity (>99 %). Photolysis at 367 nm in the presence of styrene monomers gives a telechelic polystyrene with two perester end groups; its composition determined by ^{13}C-NMR corresponds to 2 perester for 242 styrene units, which agrees fairly well with the number average molecular weight 25 600 (246 units) as determined by

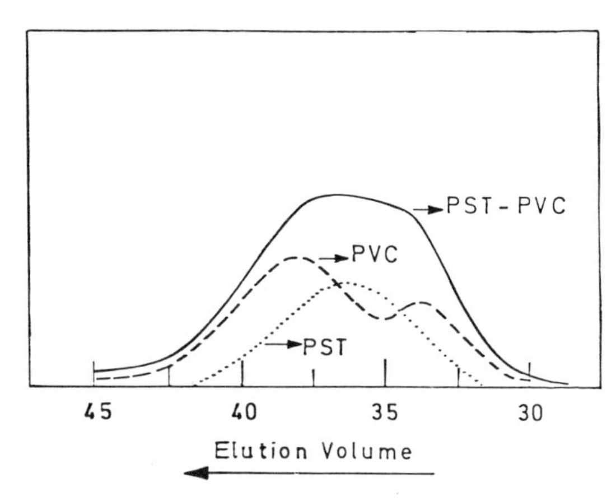

FIG. 4. Gel permeation chromatograms (with infra-red detection) of polystyrene-b-vinyl chloride block polymer mixture.

GPC. This photochemical method for the synthesis of the $\omega.\omega'$ bisperoxy-polystyrene has to be preferred to the thermal and redox methods described by Piirma and Chou[25] because it offers a well-defined and pure polymeric initiator. On heating at 95 °C in the presence of methyl methacrylate, block polymers are obtained in high yield without residual homopolystyrene. In this context mention should be made of similar bifunctional initiators prepared by Simionescu *et al.*[26] carrying two acyl peroxides, e.g. m.chlorobenzoyl peroxidic groups instead of peresters. So far however no block polymers have been described.

In conclusion, photochemical block polymerisation is a versatile and selective two-stage method applicable to most vinyl monomers. The overall structure of the block polymers is determined by a well-defined structure of the prepolymer, which itself requires knowledge of the reactivities of the components and of the reaction kinetics.

ACKNOWLEDGEMENTS

The authors are indebted to the Ministry of Scientific Programmation, Belgium, for financial support and to Asahi Glass Ltd for a post-graduate fellowship of one of them (T.D.).

REFERENCES

1. Delzenne, G., Laridon, U. and Peeters, H., *Europ. Polym. J.*, **6**, 933 (1970).
2. Lanza, E., Berghmans, H. and Smets, G., *J. Polym. Sci., Polym. Phys. Ed.*, **11**, 95 (1973).
3. Ledwith, A., *Pure appl. Chem.*, **49**, 431 (1977).
4. Doi, T., post-graduate fellowship, Asahi Glass, K.U. Leuven 1983–4, private communication.
5. Smets, G., 1st SPSJ International Polymer Conference, Kyoto, Aug. 1984, *Polym. J.*, in press.
6. Mezger, T. and Cantow, H. J., *Makromol. Chem. Rapid Commun.*, **4**, 313 (1983).
7. Otsu, T., *J. Polym. Sci.*, **26**, 236 (1957).
8. Otsu, T., Nayatami, K., *Makromol. Chem.*, **27**, 149 (1958).
9. Vlasov, G. P., Rudkovskaya, G. D. and Ovsyannikova, L. A., *Makromol. Chem.*, **183**, 2635 (1982).
10. Craubner, H., *J. Polym. Sci., Polym. Chem. Ed.*, **18**, 2011 (1980); **20**, 1935 (1982).
11. Bamford, C. H., In: *Reactivity, Mechanism and Structure in Polymer Chemistry* (Ed. A. D. Jenkins and A. Ledwith), J. Wiley, 1974, Chap. 3, pp. 52–113 and references therein.

12. Bamford, C. H. and Mullick, S. U., *J. chem. Soc. Faraday I*, **1977**, 1260; **1978,** 1634, 48.
13. Otsu, T., *J. Polym. Sci.*, **21**, 559 (1956).
14. Otsu, T., Nayatani, K., Muto, I. and Imai, M., *Makromol. Chem.*, **27**, 142 (1958).
15. Otsu, T. and Yoshida, M., *Makromol. Chem. Rapid Commun.*, **3**, 127 (1982).
16. Otsu, T., Yoshida, M. and Kyriyama, A., *Polym. Bull.*, **7**, 45 (1982).
17. Otsu, T. and Yoshida, M., *Polym. Bull.*, **7**, 197 (1982).
18. Otsu, T. and Kyriyama, A., *Polym. Bull.*, **11**, 135 (1984).
19. Konishi, H., Shinagawa, Y., Azuma, A., Okano, T. and Kiji, J., *Makromol. Chem.*, **183**, 2941 (1982).
20. Prisyazhnyuk, A. I. and Ivanchev, S. S., *J. Org. Chem. USSR*, **5**, 1151 (1969); *Polym. Sci. USSR*, **12**, 514 (1970).
21. Ivanchev, S. S. and Zherebin, Y. L., *J. Org. Chem.*, **9**, 1628 (1973); *Polym. Sci. USSR*, **16**, 956 (1974).
22. Zherebin, Y. L., Ivanchev, S. S. and Domareva, N. D., *Polym. Sci. USSR*, **16**, 1033 (1974).
23. Ivanchev, S. S., *Polym. Sci. USSR*, **20**, 2157 (1978) and references therein.
24. Van Broeckhoven, W., Ph.D. thesis, K.U. Leuven, 1982.
25. Piirma, I. and Chou, L. P., *J. appl. Polym. Sci.*, **24**, 2051 (1979).
26. Simionescu, C., Sik, K. G., Comatina, E. and Dumitriu, S., *Europ. Polym. J.*, **20**, 467 (1984).

8

Applications of Transition Metal Derivatives in Free-radical Syntheses and Modification of Polymers by Photochemical Methods

CLEMENT H. BAMFORD

Bioengineering and Medical Physics Unit, University of Liverpool, UK

1. INTRODUCTION

Block and graft copolymers remain of great academic and practical interest, finding a very large number of disparate applications. Polymer modification, which implies changing one or a group of properties to give the polymer 'speciality' characteristics, has also assumed great importance; often it requires preparative reactions similar to those used for block and graft synthesis. Currently, speciality polymer synthesis is a very active and expanding field.

When maximum precision in polymer construction is necessary, for example in some structure-property investigations, it is customary to employ anionic techniques. More recently, group-transfer polymerisation[1] has been shown to offer similar synthetic possibilities in some systems. Well-known limitations are associated with anionic polymerisation, particularly with regard to the types of monomer to which it is applicable. On the other hand, free-radical polymerisation permits choice from a much wider range of components, but suffers from the disadvantage of producing broad molecular weight distributions and therefore less well-defined products. For this reason it has been regarded as the cinderella route to multicomponent polymers (compare ref. 2). It is our belief that, subject to certain conditions, free-radical synthesis can yield polymers sufficiently precise for many purposes and certainly worthy of scientific investigation. In some instances the polymers cannot be conveniently synthesised by other routes. The minimal requirements for such a controlled synthesis of block and graft copolymers are (i) it must not produce significant quantities of homopolymers, (ii) the kinetics must be sufficiently developed to allow calculation of average block lengths and also average crosslink or graft densities.

Free-radical techniques with these characteristics would seem to have a useful place in polymer chemistry. The present paper reviews methods of photo-initiating free-radical polymerisation which are based on transition metal derivatives, and which it is hoped will further our development of the kind of controlled polymerisations to which we have referred.

2. FREE-RADICAL GENERATION INVOLVING THE USE OF TRANSITION METAL DERIVATIVES

We present here only a brief account with the background information necessary for an appreciation of the applications of these processes in polymer synthesis and modification.

Two distinct modes of radical-generation involving transition metal derivatives have been described. Both require the presence of a co-initiator: in Type 1 this is an organic halide while in Type 2 it is an unsaturated species such as a suitable olefine or acetylene (including acetylene itself).

2.1. Type 1 Initiation

An account of early work, including a list of systems studied in detail and relevant kinetic information has been given by Bamford.[3] For further kinetic studies see refs 4 and 5.

Radical formation is basically an electron-transfer process from transition metal to halide, the former assuming a higher oxidation state and the latter generally splitting into an ion and a radical fragment. For example, with molybdenum carbonyl the reaction may be represented by (1)

$$Mo^{\circ} + CCl_4 \longrightarrow Mo^{I}Cl^{-} + \dot{C}Cl_3 \tag{1}$$

It is clearly necessary for the transition metal to be initially in a low (preferably the zeroth) oxidation state.

The basic reaction (1) may occur thermally, or as a result of irradiation. Manganese and rhenium carbonyls ($Mn_2(CO)_{10}$, $Re_2(CO)_{10}$, respectively) are the most convenient derivatives for photo-initiation, and most of this paper is concerned with processes in which one or the other is used. The initiating systems $Mn_2(CO)_{10}/CCl_4$, $Re_2(CO)_{10}/CCl_4$ were first studied by Bamford *et al.*[6,7] who reported quantum yields of initiation under appropriate conditions close to unity. An unexpected feature was the existence (with common vinyl monomers) of a prolonged after-effect with $Re_2(CO)_{10}$. This is only found with $Mn_2(CO)_{10}$ in the presence of certain coordinating species. We shall see later that these after-effects have proved of considerable value in practice.

The early view, that photodissociation into two dissimilar fragments occurs, only one of which reacts rapidly with the co-initiator, has persisted throughout studies on photo-initiation by these carbonyls, including those of Type 2 initiation.[8-11] Indeed, it appears to offer the simplest explanation of the experimental data, especially the overall stoichiometry (e.g. generation of two radicals per $Mn_2(CO)_{10}$ consumed with a quantum yield of unity). However, the proposal receives no support from the results of flash photolysis of $Mn_2(CO)_{10}$ in hydrocarbon solution. According to the most recent report,[12] two primary processes are of about equal importance:

$$Mn_2(CO)_{10} + h\nu \begin{cases} \nearrow 2Mn(CO)_5 \\ \searrow Mn_2(CO)_9 + CO \end{cases} \qquad (2)$$

Dark reactions also occur, involving recombinations to reform $Mn_2(CO)_{10}$. The mechanism may be different in coordinating liquids, a situation encountered in other related systems (cf. refs 13–15). Provisionally, the mechanism in (3) may be advanced; here S represents a coordinating molecule (cf. ref. 2).

$$\begin{aligned} Mn_2(CO)_{10} + h\nu &\longrightarrow Mn_2(CO)^*_{10} \xrightarrow{S} [(CO)_5Mn{-}S{-}Mn(CO)_5] \\ &\longrightarrow (CO)_5Mn{-}S + Mn(CO)_5 \qquad \text{(a)} \\ Mn(CO)_5 + CCl_4 &\longrightarrow Mn(CO)_5Cl + \dot{C}Cl_3 \qquad \text{(fast)} \qquad \text{(b)} \quad (3) \\ S{-}Mn(CO)_5 + CCl_4 &\longrightarrow Mn(CO)_5Cl + \dot{C}Cl_3 + S \qquad \text{(slow)} \qquad \text{(c)} \end{aligned}$$

Recombinations between manganese-containing fragments which do not react with carbon tetrachloride regenerate $Mn_2(CO)_{10}$. A similar mechanism could hold for $Re_2(CO)_{10}$; in this case the vinyl monomer could function as S. Reaction (3c) is responsible for the prolonged after-effect. This mechanism closely resembles that advanced by Bamford and Mullik.[11]

The photolysis of several arene chromium tricarbonyls and photo-initiation by these derivatives have been studied by Bamford and Al-Lamee.[13-15] In the presence of an active halide the carbonyls give Type 1 initiation and some of the systems have been put to practical use (see below).

Two mechanistic aspects of Type 1 initiation are relevant for the purposes of this paper.

(i) In principle, electron transfer to a halide Cl—R may occur in two ways

$$e^- + \text{Cl—R} \nearrow \text{Cl}^- + \dot{\text{R}} \quad \text{(a)}$$
$$\qquad\qquad\qquad\qquad\qquad\qquad\qquad\qquad\qquad\qquad \text{(4)}$$
$$e^- + \text{Cl—R} \searrow \dot{\text{Cl}} + \text{R}^- \quad \text{(b)}$$

the relative importance of which depends *inter alia* on the electron-affinity of R. If $R = CCl_3$, (4a) is predominant, but if R is a trichloroacetate residue, e.g. $Cl_2CCOOEt$, (4b) appears to be of comparable importance with some metal derivatives, e.g. those of nickel, although it is insignificant with Mo, Mn and Pt compounds.[3]

(ii) A second important feature is the form of the dependence of the rate of radical generation on halide concentration. In all Type 1 systems studied (thermal and photochemical) there is effectively no initiation when no halide is present. With increasing halide concentration the rate rises and eventually achieves a plateau value such that the order in [halide] is zero. The initial slope of the 'halide curve' is important since it determines the minimum halide concentration required to obtain the required reaction rate. For synthetic purposes, when the halide concentration is necessarily low, it is desirable to use a system with a 'sharp' halide curve, so that conditions are not too far from the plateau.

The sharpness of the curve is determined by the natures of the transition metal and the solvent, and increases with the activity of the halide. Fortunately, manganese and (especially) rhenium give very sharp curves. Halide activity is governed by factors which determine the ease of accepting electrons according to (1); thus it increases with multiple substitution (i.e. in the series $CH_3Cl < CH_2Cl_2 < CHCl_3 < CCl_4$) and with introduction of electron attracting groups into the molecule. Bromine derivatives are much more active than the corresponding compounds of chlorine and saturated F and I compounds are ineffective. Note that N-chlorinated or N-brominated amides and imides are highly active (see below).

No metal atoms become bound to the polymers in these processes.

2.2. Type 2 Initiation

Recognition of initiating processes of this type started with the observation of Bamford and Mullik[9] that pure liquid tetrafluoroethylene at $-93\,^{\circ}C$

containing a low concentration of $Mn_2(CO)_{10}$ or $Re_2(CO)_{10}$ polymerises rapidly when irradiated. It was concluded that the reaction is a free-radical polymerisation, with C_2F_4 acting as co-initiator to form initiating radicals of type $(CO)_5MnCF_2\dot{C}F_2$. In verification, the systems $Mn_2(CO)_{10}/C_2F_4$ and $Re_2(CO)_{10}/C_2F_4$ were found to be active photo-initiators of the polymerisation of common vinyl monomers at ambient temperatures.[9-11] Note that the polymeric products have metal atoms in their terminal groups, e.g. poly(methyl methacrylate) prepared by photo-initiation with $Mn_2(CO)_{10}/C_2F_4$ has end groups of the type shown in (I):

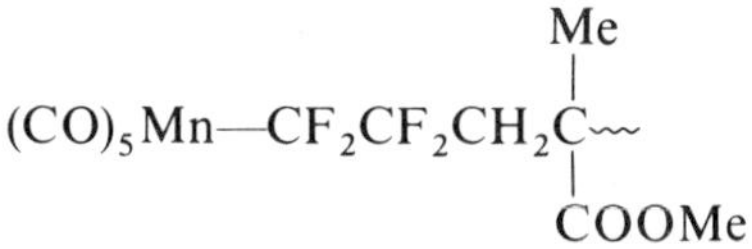

$$(CO)_5Mn\text{—}CF_2CF_2CH_2\overset{\overset{\displaystyle Me}{|}}{\underset{\underset{\displaystyle COOMe}{|}}{C}}\sim \qquad (I)$$

$$(CO)_5Re\text{—}\overset{\overset{\displaystyle R_1}{|}}{\underset{\underset{\displaystyle R_2}{|}}{C}}\text{—}\overset{\overset{\displaystyle R_3}{|}}{\underset{\underset{\displaystyle R_4}{|}}{C}}\text{—}CH_2\overset{\overset{\displaystyle Me}{|}}{\underset{\underset{\displaystyle COOMe}{|}}{C}}\sim \qquad (II)$$

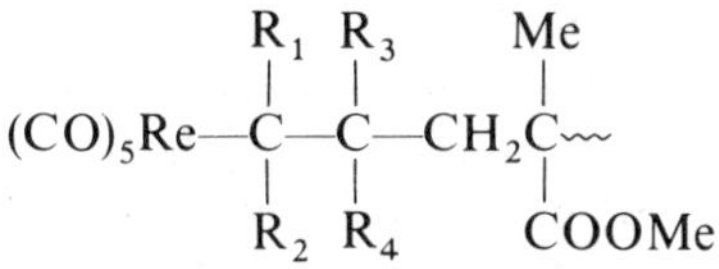

These, and the corresponding rhenium structures, are readily detected by their infra-red absorption bands near $2100\,cm^{-1}$.

Many fluoro-olefins behave as Type 2 co-initiators and other compounds active in this way have subsequently been found. Among these are olefins carrying electron-attracting substituents, acetylene and acetylene dicarboxylic acid and its esters. Some Type 2 co-initiators, with their activities are listed in refs 16 and 17.

For a co-initiator to possess high activity it must be able to form adequately strong metal–carbon bonds. This is achieved in the olefinic compounds by virtue of electronegative substituents such as carbonyl, nitrile, and, of course, fluorine. Steric factors are important.[17] For example, (II) shows the structure of terminal groups in poly(methyl methacrylate) prepared with a substituted ethylene as co-initiator. With all substituents carbomethoxy, the activity (with $Re_2(CO)_{10}$) is small, the quantum yield for initiation φ being 0·06, but if $R_1 = H$ (and $R_2 = R_3 = R_4 = COOMe$) a high activity is found, with $\varphi = 0{\cdot}74$. There is considerable steric strain in (II) with $R_1 = R_2 = COOMe$, which is relieved if one of these groups is hydrogen. The activities of co-initiators are explicable in terms of their electronic and steric influences.[17]

In general, $Re_2(CO)_{10}$ is more active than $Mn_2(CO)_{10}$. This is also likely to be a bond-energy effect, Re—C being much stronger than Mn—C. An additional contribution to $-\Delta H$ for radical formation exceeding 20 kcal mol^{-1} may arise from this source with $Re_2(CO)_{10}$.[17]

The dependence of the rate of radical generation on co-initiator concentration is similar to that described for Type 1 initiation. In active systems, the quantum yield for initiation under plateau conditions approximates to unity.

3. POLYMER MODIFICATION AND GRAFT AND BLOCK COPOLYMER SYNTHESIS

3.1. Type 1 Initiation

3.1.1. End-group Attachment

From the foregoing it will be evident that (with photo-initiation by $Mn_2(CO)_{10}$ or $Re_2(CO)_{10}$) the group in the co-initiator attached to halogen (R in eqn (4)) forms the initiating radical, and so becomes incorporated as a terminal group in the polymer chain. If termination occurs exclusively by combination, then, in the absence of chain transfer, both ends of each polymer chain carry R groups. The nature of R may be varied widely so that many different types of end group may be incorporated.

An example is provided by the work of Alimoglu *et al.*[18] who employed an initiating system composed of $Mn_2(CO)_{10}$ + N-bromoacetyldibenz[b,f]azepine (III) and so prepared poly(methyl methacrylate) and polystyrene with terminal photoactive units of N-acyldibenzazepine (IV). N-acyldibenzazepine cyclodimerises photochemically with the aid of appropriate triplet sensitisers, so that block polymers may be synthesized from a mixture of polymers such as (IV).

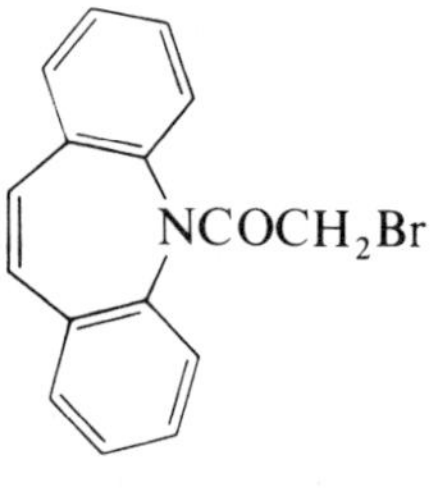

NCOCH$_2$Br (III) NCOCH$_2$〰 (IV)

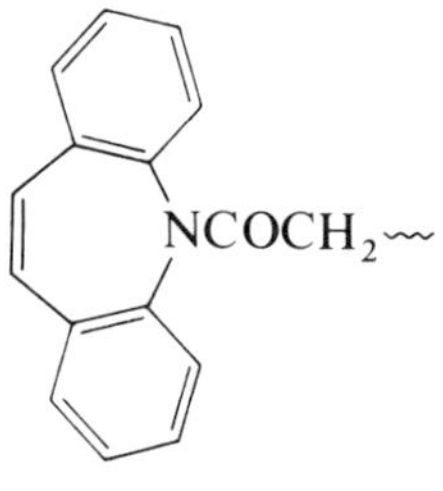

As a second example we may quote the work of Bamford *et al.*,[19] who used a similar route for attaching molecules of an antiplatelet agent to the ends of poly(N-vinylpyrrolidone) chains. This will be discussed below (see Biomedical Applications).

Terminal bonding of polymer chains to surfaces represents the ultimate extrapolation of the processes under discussion. Eastmond *et al.*[20] used Type 1 initiation for grafting to glass surfaces. Active halogen groups were bonded to glass surfaces by non-hydrolysable links using the reactions set out in (5).

$$\geq Si{-}OH \xrightarrow{Cl_2} \geq Si{-}Cl \xrightarrow{PhLi} \geq Si{-}C_6H_5 \xrightarrow[AlCl_3]{CCl_3COCl} \geq Si{-}C_6H_4{-}COCCl_3 \qquad (5)$$

Grafting of styrene and methyl methacrylate to the treated surface was carried out thermally with molybdenum carbonyl. Up to 3·8 %w/w of poly(methyl methacrylate) was attached to glass beads in this way, corresponding to a polymer layer of thickness 0·4 μm. Photochemical grafting with $Mn_2(CO)_{10}$ or $Re_2(CO)_{10}$ would seem to be an alternative in this work, preferably with the aid of the after-effect technique which we shall describe under Biomedical Applications.

3.1.2. Block Copolymer Synthesis

When the co-initiator is a polymer, Type 1 initiation results in the formation of block or graft copolymers, or networks. Polymeric co-initiators with active halogen groups at the ends of linear chains yield block copolymers. Suppose, for illustration, that the co-initiator is a polymer of monomer M_1 with terminal CBr_3 groups, such as may be prepared by polymerisation of M_1 in the presence of carbon tetrabromide.

$$\sim M_1 \sim CBr_3 \xrightarrow[h\nu]{Mn_2(CO)_{10}} \underset{(V)}{\sim M_1 \sim \dot{C}Br_2} \qquad (a)$$

$$\sim M_1 \sim \dot{C}Br_2 + nM_2 \longrightarrow \underset{(VI)}{\sim M_1 \sim CBr_2 \sim M_2 \sim \dot{M}_2} \qquad (b)$$

$$2(V) \nearrow \underset{(VII)}{\sim M_1 \sim CBr_2 \sim M_2 \sim CBr_2 \sim M_1 \sim} \qquad (c) \qquad (6)$$

$$2(V) \searrow \underset{(VIII)}{2 \sim M_1 \sim CBr_2 \sim M_2 \sim} \qquad (d)$$

Primary radicals (V) are formed as in (6a), and in the presence of a second monomer M_2 initiate block copolymerisation (6b). If the propagating

$\sim\dot{M}_2$ radicals terminate exclusively by combination (6c), the product is the three-block copolymer (VII). Disproportionation (6d) obviously leads to two-block structures (VIII).

With initial polymers carrying CBr_3 terminations at both chain ends reaction proceeds further. As before, the first product (assuming combination) is a three-block copolymer; however, this has CBr_3 end groups which are subject to further activation. Consequently the next product will be a seven-block copolymer, and so on. Generally, the product consists of alternating blocks of M_1 and M_2 linked by CBr_2 units with CBr_3 end units. Two general points are relevant here. First, CBr_2 is much less reactive than CBr_3 towards transition metal derivatives, so that complications (grafting) arising from activation of CBr_2 groups may not be significant until large molecules have been formed. *A fortiori*, CH_2Br groups are still less reactive, unless the bromine is activated by being in allylic or benzylic positions, or by proximity of electron-attracting groups. Secondly, since all primary radicals are macroradicals, the synthesis produces only minimal amounts of M_2 homopolymer (such as may arise by chain-transfer) and so meets requirement (i) set out in the Introduction.

This procedure was employed by Bamford and Han[21] to synthesise block copolymers in which each block is an alternating copolymer. These workers synthesised, purified and characterised four fully alternating block copolymers with structure (IX)

poly(A-*alt*-B)-*block*-poly(C-*alt*-D)-*block*-poly(A-*alt*-B) (IX)

A = St, B = MA, C = Ip, D = MMA
A = St, B = MA, C = Bd, D = MMA
A = Ip, B = MA, C = St, D = MMA
A = Bd, B = MA, C = St, D = MMA

(St = styrene, MMA = methyl methacrylate, MA = methyl acrylate, Bd = butadiene, Ip = isoprene).

The alternating character of the blocks was achieved by carrying out each polymerisation in the presence of a Lewis acid ($Al_2Et_3Cl_3$).[22] The preformed copolymer Brpoly(A-*alt*-B)-CBr_3 was first prepared by copolymerisation of monomers A,B containing $Al_2Et_3Cl_3$, with CBr_4 as transfer agent and $Mn_2(CO)_{10}$ as photo-initiator. CBr_4 also behaves as co-initiator, so that all polymer molecules formed have CBr_3 and Br terminations. The second alternating copolymerisation (of C and D, corresponding to (6b)) was carried out with this preformed polymer as co-initiator. Molecular weight data were consistent with the main product being three-block copolymers, showing that radical combination (6c) must occur. It is, of course, possible that minor amounts of two-block copolymers may have been formed and removed by the fractionation procedure.

In a synthesis of the kind we are considering the nature of the preformed polymer is unimportant so long as the material is soluble and carries suitable halide groups. There is certainly no need for the preformed polymer to be the product of a free-radical polymerisation. Thus the reactions may be employed in effecting transformations from anionic or cationic to free-radical polymerisation. An example of the former has been provided by Bamford *et al.*[23] These workers prepared polystyrene (M.Wt. ~2000) of narrow molecular weight distribution by anionic polymerisation, terminating the reaction via a Grignard compound to introduce a terminal benzylic bromine atom, according to the procedure of Burgess *et al.*[24] The product was used as co-initiator with $Mn_2(CO)_{10}$ in the photo-initiated polymerisation of methyl methacrylate. Since this monomer shows about 30 % of combination in termination at 25 °C, the major copolymer would be expected to contain two blocks with some three-block copolymer. It was concluded[23] that chromatographic and kinetic data provided strong evidence for the formation of block copolymers by the proposed mechanism.

The synthesis of block copolymers of polypeptides and vinyl polymers by Imanishi and Bamford[25] is another example.

3.1.3. *Graft Copolymer Synthesis*

An analogous series of reactions gives rise to graft copolymers when the polymeric co-initiator carries side chains with active halide groups. The reactions are summarised in (7); it is evident that crosslinks or simple grafts, or mixtures of the two may be formed, depending on the character of the termination reaction of $\sim\dot{M}_2$. The development of crosslinks leads ultimately to network formation and gelation; a very wide range of polymer networks may be constructed in this manner. Some novel materials, not readily synthesised by other routes, have been prepared. In early studies[26] poly(vinyl trichloracetate) was used as co-initiator with styrene, methyl methacrylate and chloroprene as crosslinking units. Polycarbonates, polystyrene and cellulose acetate, suitably functionalised, have also been employed as co-initiators.

Eastmond[2] and his colleagues have made detailed studies of these reactions and the properties of the products referred to as 'AB-crosslinked polymers'. Although the polymerisation of M_2 is 'near-normal', the

$$\text{CBr}_3 \ldots \text{M}_1 \ldots \text{Br}_3\text{C} \xrightarrow[h\nu]{\text{Mn}_2(\text{CO})_{10}} \dot{\text{C}}\text{Br}_2 \ldots \text{M}_1 \ldots \text{Br}_3\text{C} \xrightarrow{n\text{M}_2} \text{CBr}_2\text{~~M}_2\text{~~}\dot{\text{M}}_2 \ldots \text{M}_1 \ldots \text{Br}_3\text{C} \quad (\text{X})$$

$$2(\text{X}) \rightarrow \text{Br}_3\text{C} \ldots \text{M}_1 \ldots \text{CBr}_2\text{~~M}_2\text{~~Br}_2\text{C} \ldots \text{M}_1 \ldots \text{CBr}_3 \quad (\text{XI})$$

$$2(\text{X}) \rightarrow 2\ \text{CBr}_2\text{~~M}_2\text{~~} \ldots \text{M}_1 \ldots \text{Br}_3\text{C} \qquad (7)$$

termination coefficient for ~~$\dot{M}_2$ may depart from the normal value for two reasons. The attached chain of M_1 units impedes translational diffusion and so reduces k_t. On the other hand, with very high rates of initiation, non-random termination occurs, with an unusually high value of k_t for pairs of radicals originating from the same co-initiator chain. Both effects, which influence the lengths of the M_2 blocks, were suggested earlier,[27] and have now been verified experimentally. Eastmond[2] reports calculations on the populations of various types of species during the reaction. At moderately low degrees of grafting the products comprise M_1 homopolymer (the co-initiator) and relatively simple multicomponent species. When combination is exclusively by combination, it is probably adequate to assume that the products are blends of 'A_2BA_2 block copolymers' (i.e. (XI)) and the homopolymer. A knowledge of these and other related structural details is clearly a requisite before undertaking extensive property studies.

It appears that AB-crosslinked polymers show the same basic morphologies (microphase separation) as simple block copolymers. Further, in some systems macrophase separation may occur and produce complex morphologies.[28] A very interesting (and currently controversial) conclusion which has emerged from this work is that 'homopolymers are essentially incompatible with chemically identical blocks in multicomponent polymers . . . i.e. A-blocks in AB and A_2BA_2 block copolymers are incompatible with A-homopolymer of similar molecular weight'. A possible extension, with far-reaching consequences, is that 'polymer-chains which have one end located at an impenetrable interface are not miscible with chemically identical chains of comparable or greater molecular weight which are not so restricted'.

Variants of the crosslinking reactions under consideration have been investigated with other objectives; one of these is to formulate new (negative) photoresists and presensitised lithographic plates. This work has been described by Wagner and Purbrick.[29] Initial experiments involved the use of poly(vinyl trichloroacetate) and styrene, which together with manganese carbonyl were coated on to grained anodised aluminium foil. On exposure to light, the illuminated portions were crosslinked and insolubilised, so that an image developable in a mixture of ethanol and cyclohexanone was formed. In later work manganese carbonyl was replaced by benzene chromium tricarbonyl, which as already noted, shows Type 1 photo-initiation. Poly(vinyl trichloroacetate) does not have suitable mechanical properties for practical use in a photoresist or lithographic printing material, and the use of a liquid monomer is undesirable. A polymer was therefore designed to overcome these shortcomings; its chains, with the average composition shown in (XII), carried both CBr_3 groups and polymerisable methacrylate units.

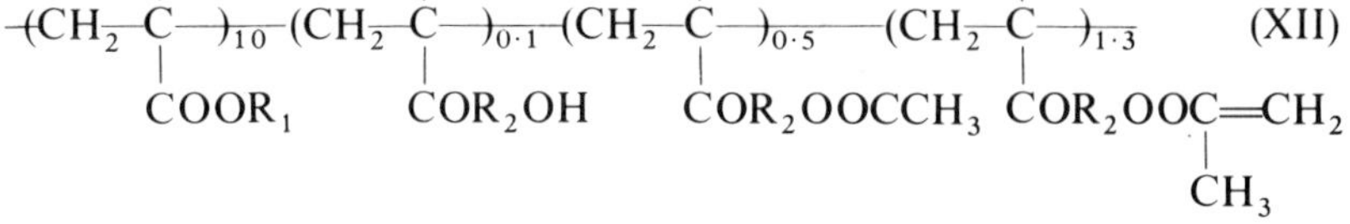

$$\left(CH_2-\underset{\underset{COOR_1}{|}}{\overset{\overset{CH_3}{|}}{C}}-\right)_{10}\left(CH_2-\underset{\underset{COR_2OH}{|}}{\overset{\overset{CH_3}{|}}{C}}-\right)_{0\cdot 1}\left(CH_2-\underset{\underset{COR_2OOCCH_3}{|}}{\overset{\overset{CH_3}{|}}{C}}-\right)_{0\cdot 5}\left(CH_2-\underset{\underset{COR_2OO\underset{\underset{CH_3}{|}}{C}=CH_2}{|}}{\overset{\overset{CH_3}{|}}{C}}-\right)_{1\cdot 3} \qquad \text{(XII)}$$

$$(R_1 = CH_3,\ C_2H_5 \text{ etc.},\ R_2 = -OCH_2CH_2- \text{ or } -NHCH_2\overset{|}{C}HCH_3).$$

This system gave very good results, which, however, were further improved by addition of a sensitiser (e.g. a thiapyrylium, pyrylium and selena-pyrylium salt) to bring the absorption into the visible region. The mechanism of sensitisation is not clear.

3.1.4. Biomedical Applications

We now discuss applications of Type 1 initiation which arose in connection with the development of a prosthetic artery of small bore (<4 mm internal diameter) fabricated from commercial poly(ether-urethane).[30] Poly(ether-urethanes) have good elastomeric and other mechanical properties and a relatively high compatibility with blood, consequently they have been much used for the manufacture of prostheses such as artificial hearts and arteries, and extracorporeal circulatory systems. Clearly the highest possible degree of haemocompatibility is essential for certain applications, including small-bore arterial prostheses. It is generally appreciated that prostheses of internal diameter 4 mm or less present specially difficult problems arising from thrombus formation. Two methods of improving polymer haemocompatibility are well known: the grafting of hydrophilic chains and the chemical attachment of antiplatelet agents. Each has been the subject of a voluminous literature.[19]

Poly(ether-urethanes) have no convenient functional groups (excluding end groups) for these purposes. Previous work on grafting, mainly by γ-irradiation or ceric ion techniques, appeared to us to suffer from several disadvantages[19,31] and we decided to develop a new method based on earlier work[32] with polypeptides and polyamides.

We found that these polymers could be readily (partially) converted to N-chloro or N-bromo derivatives by a short immersion in dilute aqueous sodium hypohalite solution at ambient temperatures. As already stated, the halogenated polymers are highly active Type 1 co-initiators so that grafting is simply carried out by the procedure described under sub-section 3.1.3 above. Further, work with low-molecular weight amides and imides,[30] showed that radicals of the type —CON̈— do not fragment under the operative conditions, so that as in the other systems considered, all the initiating centres are macroradicals.

This technique has been extended to poly(ether-urethanes) by using the reactions shown in (8) ($X = Cl$ or Br).

$$\mathrm{R'OOCNHR''} \longrightarrow \mathrm{R'OOCNXR''} \xrightarrow[h\nu]{\mathrm{Mn_2(CO)_{10}\ (etc.)}} \mathrm{R'OOC\dot{N}R''} \xrightarrow{n\mathrm{M}} \mathrm{R'OOC}\underset{\underset{\dot{\mathrm{M}}_n}{|}}{\mathrm{N}}\mathrm{R''} \qquad (8)$$

Conditions for the halogenation step have been elaborated to give minimal degradation of the polymers; halogenation may be carried out with solid polymer by immersion in aqueous hypohalite (preferably hypobromite, which gives much less degradation) or in solution in N,N-dimethylformamide (DMF) by bromine. Many monomers of different types have been grafted to several poly(ether-urethanes) in this way, including styrene, methyl methacrylate, methyl acrylate, 2-hydroxyethyl methacrylate(HEMA), 2-hydroxypropyl methacrylate, acrylic and methacrylic acids, acrylamide, N-vinylpyrrolidone (NVP). Either photo- or thermal-grafting is effective with the solid halogenated poly(ether-urethane) or the solution in DMF. Grafting to the solid has the advantage of leaving the interior and the mechanical properties effectively unchanged. Further, the inner surface of a tube such as a prosthetic artery may be halogenated by allowing the hypohalite solution to flow through the tube for a short time; grafting is subsequently confined to the halogenated surface.[31]

Photografting to a solid of irregular shape such as a vascular prosthesis has obvious disadvantages, since it is impossible to illuminate the whole area uniformly. To overcome this problem, recourse may be made to the photo-after-effects with $Mn_2(CO)_{10}$ and $Re_2(CO)_{10}$ described earlier. The procedure used in after-effect grafting consists of irradiating the metal carbonyl in monomer solution (with an additive such as acetylacetone for $Mn_2(CO)_{10}$) followed by pouring the resulting liquid (containing the long-lived active species) on to the halogenated polymer (co-initiator), all in vacuum. Grafting then proceeds thermally in the dark. This method has been very successful with arterial prostheses and produces little distortion.

Clearly any molecule with a polymerisable double bond may be bound to poly(ether-urethane) by these techniques. We have applied them to the antiplatelet agent 5-(6-carboxyhexyl)-1(3-cyclohexyl-3-hydroxypropyl) hydantoin (Burrows Wellcome 245C) (XIII). The carboxyl group of (XIII)

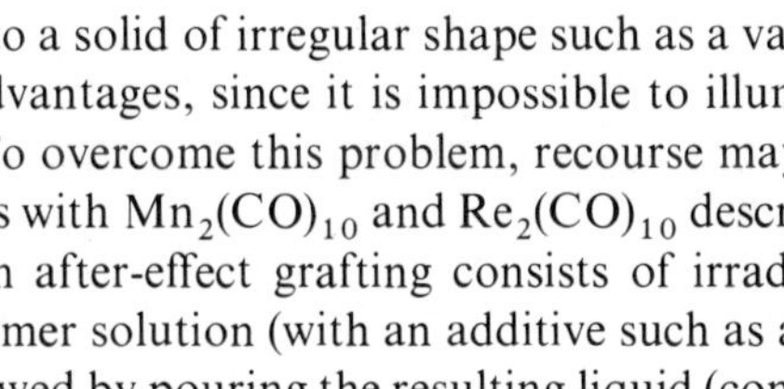

O C COOH HN N C O OH

(XIII)

was esterified with several hydroxy-methacrylate esters, and give spacer arms with different lengths and hydrophilicities. Care was taken to avoid reaction of the 15-OH in (XIII). After extensive purification, the esters were grafted to poly(ether-urethanes) as copolymers with HEMA, MMA and NVP.

The products of these reactions are poly(ether-urethane) molecules with (XIII) residues in side chains. Simple copolymers of the methacrylate esters of (XIII) with vinyl monomers also have (XIII) located in side chains. It was of interest to compare the antiplatelet activities of these species with polymers having only terminal groups of (XIII). Such polymers were synthesised by esterifying (XIII) with a halo-alcohol, e.g. CCl_3CH_2OH or $BrCH_2CH_2OH$, and using the product as co-initiator with $Mn_2(CO)_{10}$ in a simple free-radical polymerisation, as described in 'End-group attachment'.

Details of these operations and syntheses have been given by Bamford *et al.*[19]

3.2. Type 2 Initiation

3.2.1. Block Copolymer Synthesis

The synthesis of vinyl polymers such as (I) and (II) having terminal metal carbonyl residues has been described. Such polymers are normally prepared at or below room temperatures. At higher temperatures (e.g. 100 °C) they are unstable and decompose with generation of macroradicals:

$$\underset{\text{(I)}}{(CO)_5Mn{-}CF_2CF_2{\sim}M_1{\sim}} \longrightarrow Mn(CO)_5 + \underset{\text{(XIV)}}{\dot{C}F_2CF_2{\sim}M_1{\sim}} \tag{9}$$

If the decomposition is carried out in the presence of a second monomer M_2 a block copolymer is formed:[33]

$$(XIV) + nM_2 \longrightarrow \underset{\text{(XV)}}{{\sim}M_1{\sim}CF_2CF_2{\sim}M_2{\sim}} \tag{10}$$

The final copolymer molecule (XV) contains blocks of the two monomers joined by a unit of the original co-initiator (C_2F_4). As explained for Type 1 synthesis, copolymer molecules containing more than two blocks are built up if the monomer used in the first synthesis is such that termination occurs by combination; in this case each chain in the product carries two terminal $Mn(CO)_5$ residues. Note that the $Mn(CO)_5$ fragment arising in (9) does not initiate unless a co-initiator is present, but dimerises to the decacarbonyl. The desired copolymer (XV) is therefore uncontaminated by homopolymer of M_2, except insofar as this may be formed by chain transfer.

In view of these properties of macromolecules such as (I) we have called them *macro-initiators*.[33]

The technique is a versatile and general one for preparing block copolymers. However, some restrictions may arise with C_2F_4 as co-initiator on account of the high activity of perfluoroalkyl groups in transfer, for example with secondary and tertiary hydrogens in alkanes.

Among pairs of monomers used in synthesising block copolymers by this method are MMA/AN, C_2F_4/MMA, HEMA/DMA (DMA = decyl methacrylate).

Some macro-initiators of the type described are photo-active. Examples are shown in (XVI), based on platinum[35] and (XVII),[34] prepared with ADME as initiator.

$$\widehat{NN}Pt(CF_2CF_2CF_2)\text{—}CF_2CF_2\text{—MMA} \qquad \text{(XVI)} \quad \widehat{NN} = 2,2'\text{-bipyridyl}$$

$$(CO)_5MnC(COOMe){=}C(COOMe)\text{—MMA} \qquad \text{(XVII)}$$

Both absorb at $\lambda = 365\,nm$ and are photolysed with generation of macroradicals.

The concept has been generalised to include multifunctional macro-initiators suitable for the synthesis of graft copolymers. Bamford and Mullik[36] described a procedure based on the high reactivity of perfluoroalkyl radicals towards benzene. Propagating radicals formed by photolysis of $Mn_2(CO)_{10}$ in a solution of C_2F_4 in benzene enter into substitution and addition reactions with the latter, probably as a result of initial addition, to yield (XVIII) and (XIX).

$$(CO)_5Mn(CF_2CF_2)_n\text{—}C_6H_5 \qquad \text{(XVIII)}$$

$$(CO)_5Mn(CF_2CF_2)_n\text{—}C_6H_6\text{—}(CF_2CF_2)_mMn(CO)_5 \qquad \text{(XIX)}$$

Essentially similar reactions occur when benzene is replaced by a polymer containing phenyl groups. A solvent of low reactivity to perfluoro radicals, such as acetic acid, must be used. Results obtained with a copolymer of styrene and methyl methacrylate[36] showed that each Mn was associated with about 6 C_2F_4 units, so that the average size of radicals joining the benzene rings was approximately 6, and the polymer backbone carried about 20 such short side chains, each terminally functionalised by

$Mn(CO)_5$ units. As would be anticipated, such multifunctional macro-initiators initiate graft copolymerisation readily at 100 °C.[35]

Several extensions of these procedures have been described.[35]

REFERENCES

1. Sogah, Y., Hertler, W. R. and Webster, O. W., paper presented at symposium on Speciality Polymers, University of Birmingham, U.K., 1984; *ACS Polymer Preprints*, **25**, 3 (1984).
2. Eastmond, G. C., *Pure appl. Chem.*, **53**, 657 (1981).
3. Bamford, C. H. In: *Reactivity Mechanism and Structure in Polymer Chemistry* (Ed. A. D. Jenkins and A. Ledwith), Wiley, New York, 1974, Chap. 3.
4. Bamford, C. H. and Sakamoto, I., *J. chem. Soc., Faraday I*, **70**, 330 (1974).
5. Bamford, C. H. and Sakamoto, I., *J. chem. Soc., Faraday I*, **70**, 344 (1974).
6. Bamford, C. H., Crowe, P. A. and Wayne, R. P., *Proc. R. Soc.*, **A284**, 455 (1965).
7. Bamford, C. H., Crowe, P. A., Hobbs, J. and Wayne, R. P., *Proc. R. Soc.*, **A292**, 153 (1966).
8. Bamford, C. H., *Pure appl. Chem.*, **34**, 173 (1973).
9. Bamford, C. H. and Mullik, S. U., *Polymer*, **14**, 38 (1973).
10. Bamford, C. H. and Mullik, S. U., *J. chem. Soc., Faraday I*, **69**, 1127 (1973).
11. Bamford, C. H. and Mullik, S. U., *J. chem. Soc., Faraday I*, **71**, 625 (1975).
12. Church, P., Hermann, H., Grevels, F-W. and Schaffner, K., *J. chem. Soc., Chem. Comm.*, 785 (1984).
13. Bamford, C. H., Al-Lamee, K. G. and Konstantinov, C. J., *J. chem. Soc., Faraday Trans.*, I, **73**, 1406 (1977).
14. Bamford, C. H. and Al-Lamee, K. G., *J. chem. Soc., Faraday Trans.*, I, **80**, 2175 (1984).
15. Bamford, C. H. and Al-Lamee, K. G., *Faraday Trans.*, I, **80**, 2187 (1984).
16. Bamford, C. H., *Polymer*, **17**, 321 (1976).
17. Bamford, C. H. and Mullik, S. U., *J. chem. Soc., Faraday Trans.*, I, **73**, 1260 (1977).
18. Alimoglu, A. K., Bamford, C. H., Ledwith, A. and Mullik, S. U., *Macromolecules*, **10**, 1081 (1977).
19. Bamford, C. H., Middleton, I. P., Satake, Y. and Al-Lamee, K. G., paper presented at *International Symposium on Advances in Polymer Synthesis*, A.C.S. Annual Meeting, Philadelphia, 1984: *Polymer Preprints*, **25**, 27 (1984).
20. Eastmond, G. C., Nguyen-Huu, C. and Piret, W. H., *Polymer*, **21**, 598 (1980).
21. Bamford, C. H. and Han, Xiao-zu, *Polymer*, **22**, 1299 (1981).
22. Hirooka, M., Yabuuchi, H., Kawasumi, S. and Nakaguchi, K., *J. Polym. Sci. Chem. Ed.*, **11**, 1281 (1973); Bamford, C. H. In: *Alternating Copolymers* (Ed. J. M. G. Cowie), Plenum Press, New York, to be published.
23. Bamford, C. H., Eastmond, G. C., Woo, J. and Richards, D. H., *Polymer*, **23**, 643 (1982).
24. Burgess, F. J., Cunliffe, A. V., MacCallum, J. R. and Richards, D. H., *Polymer*, **18**, 719 (1977).

25. Imanishi, Y. and Bamford, C. H., in course of publication.
26. Ashworth, J., Bamford, C. H. and Smith, E. G., *Pure appl. Chem.*, **30**, 25 (1972).
27. Bamford, C. H., Dyson, R. W. and Eastmond, G. C., *J. Polym. Sci. C*, **16**, 2425 (1967).
28. Eastmond, G. C. and Phillips, D. G., *Polymer*, **20**, 1501 (1979).
29. Wagner, H. M. and Purbrick, M. D., *J. Photographic Sci.*, **29**, 230 (1981).
30. Annis, D., Bornat, A., Edwards, R., Higham, A., Loveday, B. and Wilson, J., *Trans. Am. Soc. Artif. Int. Organs*, **24**, 209 (1978).
31. Bamford, C. H. and Middleton, I. P., *Europ. Polym. J.*, **19**, 1027 (1983).
32. Bamford, C. H., Duncan, F. J., Reynolds, R. J. and Seddon, J. D., *J. Polym. Sci. C*, **23**, 419 (1968).
33. Bamford, C. H. and Mullik, S. U., *Polymer*, **17**, 94 (1976).
34. Alimoglu, A. K., Bamford, C. H., Ledwith, A. and Mullik, S. U., *Vysokomol. Soedin.*, **21**, 2403 (1979) (in Russian).
35. Bamford, C. H., Mullik, S. U. and Puddephatt, R. J., *J. chem. Soc., Faraday Trans.*, I, **71**, 2213 (1975).
36. Bamford, C. H. and Mullik, S. U., *Polymer*, **19**, 948 (1978).

9

New Aspects of the Photo-oxidation and Photostabilization of Polyolefins*

D. M. WILES and D. J. CARLSSON

Chemistry Division, National Research Council of Canada, Ottawa, Canada

1. INTRODUCTION

The susceptibility to oxidative degradation of polyolefins (polyethylene (PE) and polypropylene (PP)) has been the subject of vigorous research investigations for several decades. Much of this activity has arisen because of the reasonable assumption that a greater understanding of the chemical and physical reasons for the vulnerability of polyolefins to oxidation would lead to improvements in composition, fabrication, formulation and the like, in short to increased durability and service life. Indeed, polyolefins now represent the most widely used family of plastics, and additional applications will undoubtedly arise as further technology is developed in order to obtain even greater resistance to oxidation.

In this paper we discuss the photo-oxidation of PE and PP, especially the latter because it is more susceptible to UV degradation, because it has been more thoroughly investigated, and because incremental improvements in its photostabilization have been somewhat more dramatic. Consideration of the mechanism(s) of stabilization of PP by hindered amine light stabilizers (HALS) is included here as well because the widespread use of this type of additive has led to a major improvement in the outdoor performance of PP, and because elucidation of the relevant stabilization mechanism(s) has added to our knowledge of the fundamental photochemistry of this important plastic. Finally, an indication of the remaining unsolved problems is given.

* Issued as NRCC 23995.

2. PHOTO-OXIDATION (HISTORY)

The polyolefins, in common with a number of other vinyl, addition polymers, ought to be transparent to terrestrial sunlight, i.e. to wavelengths > 290 nm. Since significant photodegradation, as manifested by embrittlement, does occur in use, the early stages of photo-initiation must involve the absorption of photons by chromophoric impurities. Considerable research effort was expended during the 1960s and early 1970s in trying to identify the nature and source of these impurities on the reasonable assumption that known, conventional photochemical reactions would follow on from the electronic excitation of chromophoric groups fastened to the molecules of 'non-absorbing' polyolefins. The further assumption was made that it would then be relatively easy to select highly effective additives which should impart UV stability to polyolefins.

Light-absorbing impurities implicated in the photo-oxidation of PP[1] include hydroperoxide, and peroxide groups, catalyst residue (e.g. TiO_2), intrachain ketone groups, methyl ketone end groups, and oxygen/hydrocarbon charge transfer complexes. More than a decade ago in these laboratories, we proposed that this is the order of decreasing importance for photo-initiating chromophores in commercial PP articles. We continue to hold this view. There is a growing consensus that PP hydroperoxides dominate the photochemistry of this polymer during outdoor use, but some scientists are still convinced that carbonyl impurities are at least as important, or indeed, are more important than hydroperoxides.

A number of similarities exist in the photo-oxidation of PE, although both low and high density versions of this thermoplastic do not oxidize as rapidly as PP in sunlight. Owing to differences in structure (negligible tertiary C—H bonds) and chain conformation (non-helical), however, rates of propagation of photo-oxidation are lower and of termination are higher for PE than for PP. Indeed, initial attempts at detecting hydroperoxide groups on photo-oxidized PE were not successful, and many researchers concluded that ketone-group photolyses (Norrish I and II) dominated these systems.

Reviews of the research published up to 1970 are available[2-4] and, in comparison with a review article[5] from these laboratories in 1976, it is evident that extensive developments occurred during that period. In particular, kinetic and product data allowed for a reasonably comprehensive although qualitative and consistent picture of PP photo-oxidation 10 years ago.

3. PHOTOSTABILIZATION (HISTORY)

Preoccupation during the 1960s and early 1970s with identifying the initial chromophores in polyolefins and with relating these to classical (small-molecule) photochemistry led to an emphasis on two kinds of UV stabilizers: UV absorbers and excited state quenchers. Confusion arose because it was not widely appreciated that virtually all effective stabilizers operate by more than one mechanism. For example, additives based on hydroxybenzophenone provide protection in part because of UV absorption and in part because of radical scavenging. Some transition metal chelates absorb UV radiation, quench excited carbonyl triplets and singlet molecular oxygen, but a significant fraction of their effectiveness arises because they decompose hydroperoxides (Fig. 1). Owing to the sensitivity of the physical properties of polyolefins to surface degradation, thin articles especially cannot be stabilized by additives which operate solely as UV absorbers. Moreover, there is no definitive evidence that additives which operate solely as excited state quenchers provide commercially adequate UV stability.[7]

By the mid-1970s it was possible to discuss UV stabilizers for PP in the light of a reasonably comprehensive knowledge of the significant primary photochemical processes.[8] There was an improved appreciation of the mechanistic similarities between the liquid-phase oxidation of hydrocarbons and the photo-oxidation of solid polymers, correlations which had, in fact, been identified many years before. Moreover, some of the significant kinetic differences between oxidation in mobile liquid systems and in extremely viscous, heterogeneous solid polymers were becoming evident.

Up to this time, it was considered acceptable to classify UV stabilizing additives as either (a) UV screeners (absorbers or pigments), (b) excited state deactivators, (c) peroxide decomposers, or (d) radical scavengers. There was some understanding that a number of efficient protectors for polyolefins operate in more than one of these modes, and the concepts of synergism and antagonism between additives were being discussed.[7] Evidence that hindered piperidines provide excellent UV stability to PP had started to accumulate[9] and it had been reported[10] that piperidines are readily oxidized to the corresponding nitroxide in the presence, for example, of a peroxide undergoing photolysis. It was known that aliphatic nitroxides scavenge alkyl radicals[11,12] and that nitroxide radicals are regenerated during the thermal oxidation of PP.[13]

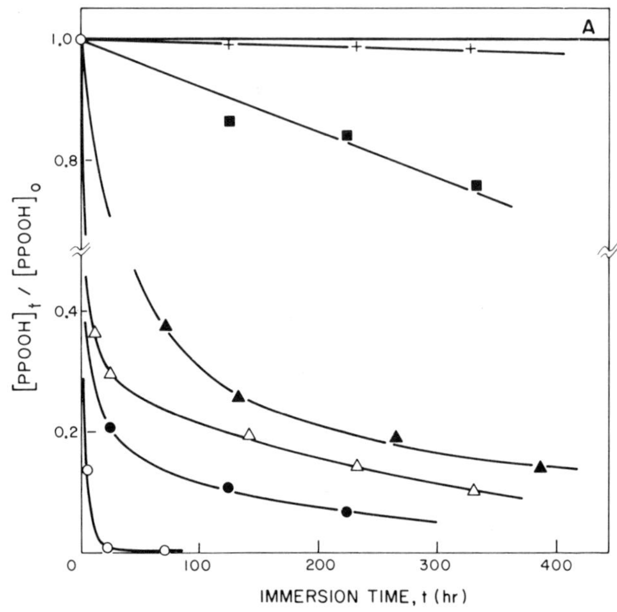

FIG. 1. Polypropylene hydroperoxide thermal decompositions by stabilizers and related compounds. Initial PPOOH concentration, 0·05*M*, 22 μm film. Films immersed in stabilizer solutions in iso-octane or ethanol at 25 °C; stabilizer concentration 2 wt % in solution. Dissolved stabilizers: (○) Ni(II) bis(di-n-butyldithiocarbamate); (△) Ni(II) bis(di-iso-propyl dithiophosphate); (+) Ni(II)[2,2′-thiobis-(4-tert.-octylphenolate)]-n-butylamine; (●) Zn(II) bis(di-n-butyldithiocarbamate); (■) 2,2′-methylene-bis(4-methyl-6-tert.-butylphenol); and (▲) 2,2′-thiobis(4-tert.-octylphenol). Ineffective decomposers (A): Ni(II) di(O-butyl 3,5-di-tert.-butyl-4-hydroxybenzyl phosphonate); octadecyl 3-(3′,5′-di-tert.-butyl-4′-hydroxyphenyl) propionate; dioctadecyl 3,5-di-tert-butyl-4-hydroxybenzyl phosphonate; 2,4-di-tert.-butyl-phenyl (4′-hydroxy-3′,5′-di-tert.-butylbenzene); 2-hydroxy-4-dodecyloxybenzophenone; 4-(1′,1′,3′,3′-tetramethylbutyl)phenyl salicylate; and 2-(2′-hydroxy-3′,5′-di-tert-pentyphenyl)benzotriazole. Reproduced from *J. Polym. Sci., Polym. Chem. Ed.*, ref. 6, by permission of John Wiley and Sons, Inc., New York.

4. PHOTO-OXIDATION (THE PAST DECADE)

Recent developments in our understanding of the photo-oxidation mechanisms and kinetics for polyolefins have been reviewed and summarized by Garton *et al.*[14] The effects of the inevitable non-uniformity in solid polymer oxidation, as these influence photo-initiation, propagation and radical termination, were discussed with particular emphasis on greater than predicted radical lifetimes. It was pointed out by these authors that, although there is still no detailed understanding of the oxidation of solid polymers, there is considerable merit in re-focusing attention on the simplified, universal scheme for hydrocarbon oxidation, i.e.

$$\text{Chromophore} \xrightarrow{h\nu} 2R^{\cdot} \tag{1}$$

$$R^{\cdot} + O_2 \longrightarrow RO_2^{\cdot} \tag{2}$$

$$RO_2^{\cdot} + RH \xrightarrow{k_p} ROOH + R^{\cdot} \tag{3}$$

$$ROOH \xrightarrow[\Delta]{h\nu \text{ or}} RO^{\cdot} + {}^{\cdot}OH \tag{4}$$

$$2RO_2^{\cdot} \xrightarrow{2k_t} \text{Non-radical products} \tag{5}$$

This scheme allows one to identify which rate constant values are needed, to account for the kinetic differences between liquid and solid systems, and to explain the differences among polyolefins. It also, however, indicates how much is still not known about the photo-oxidation of hydrocarbon polymers and emphasizes how difficult it will be to successfully address the major problem of macro- and micro-unevenness in oxidation processes.

An additional conclusion that arises from recent research in our laboratories and elsewhere is that effective UV stabilization of polyolefins requires that additives be used which cope with the ubiquitous hydroperoxide group and/or which deactivate alkyl and peroxy radicals, preferably both. Considerable research in several laboratories during the past 15 years has shown that hindered amine light stabilizers (HALS) based on 2,2,6,6-tetramethylpiperidines owe much of their effectiveness in photostabilizing polyolefins (and some other polymers) to just such characteristics (see below).

In a book published recently[15] some modern concepts are described in detail and several factors in addition to the chemistry of polyolefin photo-oxidation are discussed. For example, there is an excellent introduction[16] to the complexities of the degradation and stabilization of polyolefins which concludes with the admonition that 'the urgent need for better methods of predicting the actual service life of fabricated articles' remains.

A persistent controversy over the years has been the relative importance of hydroperoxides and carbonyl groups as chromophores during the photo-oxidation of polyolefins. Most scientists agree that hydroperoxides are involved in the photodegradation of many polymers, including all polyolefins, and that these groups dominate the photochemistry of PP. The controversy arises because PP ketones, etc., absorb more strongly in the near UV than do PP hydroperoxides although the quantum efficiency for radical generation from the former is a couple of orders of magnitude below that of the latter, in solid PP. Even in the case of low density polyethylene, hydroperoxide groups are considered to be the most important photo-initiating impurities[17] even though the level of carbonyl oxidation products is approximately 50 times higher.

Clearly both carbonyls and hydroperoxides play a role individually in polyolefin photo-oxidations and evidently the latter are more important in the case of PP. There may be, however, an additive effect since both groups are normally present simultaneously. A few years ago it was suggested[18,19] that energy absorbed by carbonyl groups (with the higher molar extinction coefficient) could be efficiently transferred to the more labile hydroperoxides (having lower molar extinction coefficients). Evidence was produced that this proposed sensitizer/energy transfer phenomenon does occur in certain model systems in the liquid phase. More recently it has been suggested[20] that the quenching of triplet aromatic chromophores by hydroperoxides observed in liquid systems may be relevant to polymer photo-oxidative degradation although evidence for this phenomenon in solid polyolefins is indirect.

Ng and Guillet proposed that the excited carbonyl-hydroperoxide interaction occurred via an exiplex to yield excited hydroperoxide which then dissociated:[18]

$$>C{=}O \xrightarrow{h\nu} >C{=}O^* \xrightarrow{ROOH} C{=}O + [ROOH]^* \longrightarrow RO^{\cdot} + {}^{\cdot}OH$$

Using nanosecond laser flash photolysis, we have shown that for excited ketones, including ketone copolymers, a more common process is the abstraction of the hydroperoxidic hydrogen:[20]

$$>C{=}O \xrightarrow{h\nu} >C{=}O^* \xrightarrow{ROOH} {}^{\cdot}C{-}OH + RO_2^{\cdot}$$

An energy transfer process may occur between excited carbonyl groups or polynuclear aromatics and dialkyl peroxides and azo compounds but at a rate appreciably slower than the hydrogen transfer reaction with hydroperoxides (Table 1). The hydrogen transfer reaction is, however, slowed in highly polar environments.[20]

Finally, Geuskens and Kabamba[21] have suggested a new chain scission mechanism to explain the photo-oxidation of polyolefins. From the results of their study using an ethylene/propylene copolymer, these authors

TABLE 1
Comparison of the Rates of Triplet Quenching by Several Substrates Capable of Generating tert-Butoxyl Radicals[a]

Sensitizer	*Triplet energy* (*kcal mole*$^{-1}$)	$k_q/M^{-1}s^{-1}$		
		Bu^tOOH	Bu^tOOBu^t	$Bu^tON_2OBu^t$
$RCOC_6H_5$	73·7	$7{\cdot}7 \times 10^{8}$ [b]	$9{\cdot}6 \times 10^{6}$ [c]	
$RCO(4\text{-}CH_3OC_6H_4)$	72·5	$1{\cdot}9 \times 10^{9}$ [d]	$7{\cdot}9 \times 10^{6}$ [c]	$1{\cdot}4 \times 10^{8}$ [d]
Benzophenone	68·6	$1{\cdot}8 \times 10^{8}$	$3{\cdot}4 \times 10^{6}$	$1{\cdot}7 \times 10^{7}$
Phenanthrene	61·9	$2{\cdot}3 \times 10^{7}$	$1{\cdot}8 \times 10^{6}$	$4{\cdot}3 \times 10^{6}$
Anthracene	42	$<2 \times 10^{6}$	$9{\cdot}7 \times 10^{4}$	$<5 \times 10^{5}$

[a] In benzene, at 300 K unless otherwise specified. Bu^t is the tert.-butyl group.
[b] In chloroform, $R = CH_3$—.
[c] Propiophenone, $R = C_3H_7$—.
[d] $R = CH_3$—.
(Reproduced from *J. Am. Chem. Soc.*, ref. 20, with permission of the American Chemical Society, Washington, DC.

conclude that there is reaction of the photolysis products from hydroperoxides with neighboring ketone groups. The specific reactions invoked are:

$$\underset{\text{(isolated)}}{ROOH} \xrightarrow{h\nu} {}^{\cdot}OH + {}^{\cdot}OR$$

$${}^{\cdot}OH + \sim\!\!C(=O)\!\!\sim\ \longrightarrow\ \sim\!C(=O)OH \qquad {}^{\cdot}OR + \sim\!\!C(=O)\!\!\sim\ \longrightarrow\ \sim\!C(=O)OR$$

and from hydrogen-bonded hydroperoxides

$$ROO^{\cdot} + \sim\!\!C(=O)\!\!\sim\ \longrightarrow\ \sim\!C(=O)\text{—}OOR$$

It is indicated that preliminary results suggest the occurrence of such reactions in photo-oxidizing PE and PP as well. The relatively high local concentrations required for these reactions to be significant are believed to occur in oxidized domains in solid polyolefins. However, the formation of acids, peracids and esters, detected by IR for example, can also be explained by the normal oxidation scheme (reactions 2–5) if extensive oxidation is occurring at primary and secondary C—H sites, as well as the accepted tert.-C—H sites.

5. PHOTOSTABILIZATION (THE PAST DECADE)

The effectiveness of low levels of HALS at providing UV stability to polyolefins and other polymers has resulted in much research to determine why. Work in this laboratory[22,23] indicates the following reactions to be important:

$$X\text{—}\langle\text{piperidine}\rangle\text{N—R} \xrightarrow[PO_2^{\cdot}]{PO^{\cdot}} {>}N\text{—}H \xrightarrow[PO_2^{\cdot}]{OOH} {>}NO^{\cdot} \tag{6}$$

$$>NO^{\cdot} + P^{\cdot} \longrightarrow\ >NOP \tag{7}$$

$$>NOP + PO_2^{\cdot} \longrightarrow\ >NO^{\cdot} + POOP \tag{8}$$

where P represents a polymer radical.

In the case where the initial HALS is a tertiary amine ($R = CH_3$) and the polymer is PP, we have shown that the stabilizer reacts with the free-radical products from the decomposition (photo or thermal) of polymer hydroperoxide groups to form the corresponding secondary HALS.[22] The rates of formation of the products in reactions (6–8) from the commercially important bifunctional HALS bis(2,2,6,6-tetramethyl-4-piperidyl) decanedioate in photo-oxidizing PP have been measured directly by FTIR and ESR spectroscopy.[22] These data confirm the validity of the above reactions, deduced initially from model compound experiments, and indicate the slow irreversible consumption of the HALS and its by-products.

Although there remains some uncertainty about the details of reaction (8), it is generally agreed that a major reason for the effectiveness of HALS is that the nitroxide derived from them leads to the catalytic scavenging of alkyl and alkyl peroxyl radicals. Owing to secondary cage effects in solid polymers like PP,[24,25] most of the polymer oxidation arises from the few peroxyl radicals that escape the cage, propagate with high kinetic chain lengths, and are easily scavenged.[26] We assume that this explains why HALS provide effective UV stabilization even though the rates for reactions (7) and (8) may be below those for (2) and (3), respectively.

It is well known that polymer additives which convert potential initiators to inactive products can impart stability to the polymer, e.g. the use of phosphites or thio-compounds in polyolefins to decompose hydroperoxides and increase thermal stability. Furthermore, the association between $>$NH and hydroperoxides in liquid systems is observed also in

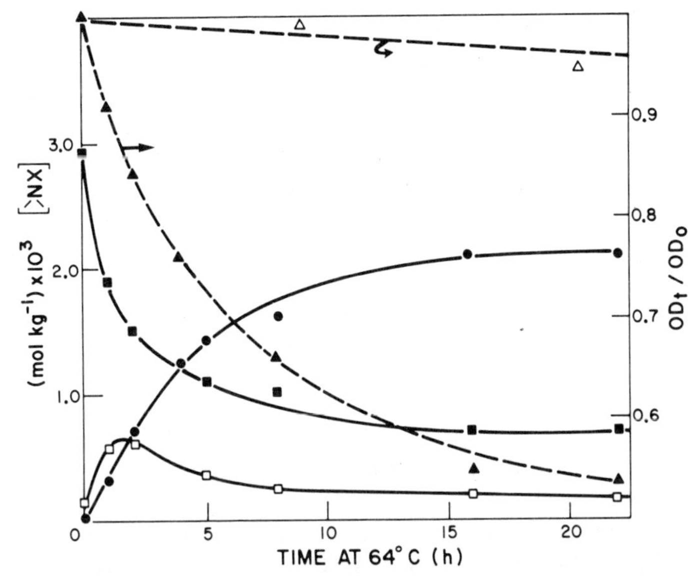

FIG. 2. PPOOH thermal decomposition by 1,2,2,6,6-pentamethyl octadecanoate (St—NH) under N_2 at 64 °C. Pre-oxidized PPH film (25 μm, $[PPOOH]_0 = 0{\cdot}03$ mol kg^{-1}). △. PPOOH concentration in the ABSENCE of St—NH by FTIR and iodometry. ▲, PPOOH concentration with St—NH present; ■, St—NH; □, St—NOH; ●, St—NO·. Reproduced from *J. Polym. Sci., Polym. Chem. Ed.*, ref. 29, by permission of John Wiley and Sons, Inc., New York.

solid PP films with a commercial HALS[27,28] and this allows for a dark reaction between >NH and PPOOH[29] at ambient temperatures (Fig. 2). The association between HALS or their nitroxides[27] and hydroperoxides should add to the effectiveness in all polyolefins since, in the fullness of time, the additive molecules will be located in the oxidizing regions of the polymer where they are required. The dark reaction to remove chromophores (—OOH) and generate active radical scavengers (>NO·) may apply primarily to PP with 'runs' of adjacent hydroperoxides since this reaction is observed in the liquid phase with model, mono-hydroperoxides only at temperatures > ~100 °C.

The unravelling over the years of the mechanisms by which HALS provide UV stability to polymers is important for several reasons. For example, this work has correctly refocused attention on the universality of the basic mechanism of the oxidation of hydrocarbons, i.e. reactions (1–5). This in turn has re-emphasized the necessity of deactivating radicals and of reducing the number of hydroperoxide groups in a variety of light-sensitive polymers in order to prolong their service life. The work has also reminded us all that, while there are many mechanistic similarities between liquid phase hydrocarbon oxidations and the photo-oxidation of solid polymers, there are critical kinetic differences as well, e.g. nonuniformity of oxidation, secondary cage effects, long kinetic chain lengths in extremely viscous solids. An additional important corollary of the HALS research is renewed interest in the concept of stabilizing species which can catalytically consume many reactive radicals because of cyclical regeneration, i.e. chain-breaking antioxidants. The principles involved here have been known for some time[30] and have been described recently in an authoritative chapter,[31] which covers both the photodegradation and photostabilization of polyolefins.

6. REMAINING CHALLENGES

It is worth repeating that we still cannot predict accurately the service life of articles fabricated from polyolefins. Accelerated testing protocols involve either more severe testing conditions (e.g. higher or UV-richer photon flux compared to terrestrial sunlight, above ambient temperature) or ultrasensitive tests for chemical or mechanical damage or combinations of both of these approaches. Persistent problems with using more severe testing conditions than those experienced during use are to be expected since degradation and stabilization mechanisms which occur in an accelerated test may not obtain during use, and temperature effects may not be amenable to extrapolation according to an Arrhenius relationship.

The chemistry of polymer photodegradation is frequently studied on additive-free samples (or those containing one or two additives only) because of the requirement to keep the analytical problems as simple as possible. Polyolefins in use usually contain several additives, however, which can be expected to complicate the chemistry and this may challenge the validity of the results obtained with simpler systems. For example, synergism or antagonism have been observed[31] with combinations of some stabilizers, and there are effects of pigments on the photostability of polyolefins.[32] In addition the melt processing of formulated polyolefins affects both the polymers and the additives irreversibly so as to influence the photostability of the articles that are produced. Some progress has been made in elucidating these effects[33] but much more information is required.

Despite the advances in photo-oxidative stability achieved by the use of HALS compounds, it seems very likely that the optimum performance of these additives is yet to be reached. Problems of efficiency being impaired by additive volatility have been partially overcome by the use of oligomeric additives containing either $>$N—R groups in the backbone or as pendant groups. Although optimum stabilization has been suggested to occur with piperidylacrylate polymers of ~2700 molecular weight, this optimum may result from compatibility and dispersion effects, rather than volatility and scavenging limitations.[34] Preliminary studies of piperidyl copolymers designed to have well separated piperidyl groups indicate that performance can be improved by molecular weights of ~8000.[35] Other structures, such as piperidyl groups spaced from the backbone by long, flexible chains may also be worth exploring.

Finally, the problem of the nonuniformity of polyolefin photo-oxidation persists. The chemical changes occur in tiny regions in the vicinity of the original photo-initiating chromophores and, owing to the severely restricted mobilities of polymer-bound reactive species, the degraded zones represent a very small volume fraction of the total polymer even after exposure to the point of brittle failure of a semi-crystalline polyolefin. On the other hand, chemical and physical measurements of degraded plastics necessarily produce average values of effects and fail to address the micro-heterogeneity that is known to exist. Vink[36] has noted at the conclusion of his recent review on the structural and morphological aspects of polyolefin photo-oxidation that additional research in this area is required in order to predict the long-term properties of plastic products.

REFERENCES

1. Carlsson, D. J., Garton, A. and Wiles, D. M., *Macromolecules*, **9**, 695 (1976).
2. Cicchetti, O., *Adv. Polym. Sci.*, **7**, 70 (1970).
3. Karpukhin, O. N. and Slobodetskaya, E. M., *Russ. Chem. Rev.*, **42**, 173 (1973).
4. Trozzolo, A. M. In: *Polymer Stabilization* (Ed. W. L. Hawkins) Wiley-Interscience, New York, 1972, Chapter 4.
5. Carlsson, D. J. and Wiles, D. M., *J. Macromol. Sci.-Rev. Macromol. Chem.*, **C14**, 65 (1976).
6. Carlsson, D. J. and Wiles, D. M., *J. Polym. Sci., Polym. Chem. Ed.*, **12**, 2217 (1974).
7. Wiles, D. M., *Pure appl. Chem.*, **50**, 291 (1978).
8. Carlsson, D. J. and Wiles, D. M., *J. Macromol. Sci.-Rev. Macromol. Chem.*, **C14**, 155 (1976).
9. Temchin, Y. I., Burmistrov, Y. F., Skripko, L. A., Burmistrova, R. S., Kokhanov, Y. V., Gushchina, M. A. and Rozantsev, E. G., *Vysokomol. Soedin., Ser. A*, **15**, 1038 (1973).
10. Janzen, E. G., *Top. Stereochem.*, **6**, 177 (1971).

11. Howard, J. A., *Adv. Free-Radical Chem.*, **4**, 49 (1972).
12. Murayama, K., Morimura, S. and Yoshioka, T., *Bull. Chem. Soc. Japan*, **42**, 1640 (1969).
13. Denisov, E. T., *International Symposium on Degradation and Stabilization of Polymers*, Brussels (Sept. 11–13, 1974), prepr. p. 137.
14. Garton, A., Carlsson, D. J. and Wiles, D. M. In: *Developments in Polymer Photochemistry*, Vol. 1 (Ed. Norman S. Allen) Elsevier Applied Science Publishers, London, 1980, Chapter 4.
15. Allen, N. S. (Ed.), *Degradation and Stabilization of Polyolefins*, Elsevier Applied Science Publishers, London, 1983.
16. Billingham, N. C. and Calvert, P. D., ref. 15, Chapter 1.
17. Scott, G., *Am. Chem. Soc. Symp. Ser.*, **25**, 340 (1976).
18. Ng, H. C. and Guillet, J. E., *Macromolecules*, **11**, 937 (1978).
19. Geuskins, G. and David, C., *Pure appl. Chem.*, **51**, 233 (1979).
20. Stewart, L. C., Carlsson, D. J., Wiles, D. M. and Scaiano, J. C., *J. Am. chem. Soc.*, **105**, 3605 (1983).
21. Geuskens, G. and Kabamba, M. S., *Polym. Deg. Stab.*, **4**, 69 (1982).
22. Wiles, D. M., Tovberg Jensen, J. P. and Carlsson, D. J., *Pure appl. Chem.*, **55**, 1651 (1983).
23. Carlsson, D. J. and Wiles, D. M., *Am. chem. Soc., Polym. Prepr.*, **25**, 24 (1984).
24. Garton, A., Carlsson, D. J. and Wiles, D. M., *Macromolecules*, **12**, 1071 (1979).
25. Garton, A., Carlsson, D. J. and Wiles, D. M., *Makromol. Chem.*, **181**, 1841 (1980).
26. Carlsson, D. J., Chan, K. H., Garton, A. and Wiles, D. M., *Pure appl. Chem.*, **25**, 389 (1980).
27. Grattan, D. W., Reddoch, A. H., Carlsson, D. J. and Wiles, D. M., *J. Polym. Sci., Polym. Lett. Ed.*, **16**, 143 (1978).
28. Durmis, J., Carlsson, D. J., Chan, K. H. and Wiles, D. M., *J. Polym. Sci., Polym. Lett. Ed.*, **19**, 549 (1981).
29. Carlsson, D. J., Chan, K. H., Durmis, J. and Wiles, D. M., *J. Polym. Sci., Polym. Chem. Ed.*, **20**, 575 (1982).
30. Scott, G., *Atmospheric Oxidation and Antioxidants*, Elsevier, London and New York, 1965.
31. Al-Malaika, S. and Scott, G., ref. 15, Chapter 7.
32. Allen, N. S., ref. 15, Chapter 8.
33. Al-Malaika, S. and Scott, G., ref. 15, Chapter 6.
34. Gugumus, F., *Res. Discl.*, **209**, 357 (1981).
35. Chmela, S. and Hrdlovic, P., 6th IUPAC Conference on Modified Polymers, Bratislava, Czechoslovakia, July 1984, paper P.43.
36. Vink, P., ref. 15, Chapter 5.

10

The Photochemistry of a Poly(ethylene-co-carbon monoxide)

R. GOODEN, M. Y. HELLMAN, D. A. SIMOFF and F. H. WINSLOW
AT&T Bell Laboratories, Murray Hill, New Jersey, USA

1. INTRODUCTION

Early in the development of polyolefins it was recognized that absorption of ultraviolet radiation of wavelengths exceeding 290 nm was a major cause of their deterioration during outdoor exposure. Since pure alkanes do not absorb sunlight, it was concluded[1,2] that traces of adventitious chromophores and/or contaminants function as sensitizers that promote degradation. The most important of these chromophores were thought to be the hydroperoxides and ketonic carbonyl groups formed during processing or contact with ozone in the atmosphere.

This paper describes the use of an ethylene–carbon monoxide copolymer (E/CO) for evaluating the effectiveness of main chain ketonic groups as initiators of polyethylene photo-oxidation. It seemed likely that an E/CO with a low CO content would be a particularly appropriate model[3,4] for study since Hartley and Guillet[5] had already explored its photochemical reactions in detail, and since others had shown that it had a branched structure[6] and a morphology[7] similar to that of a low-density polyethylene. But unlike polyethylene, photo-oxidized E/CO was completely soluble in the early stages of degradation—a distinct advantage for the determination of chain scission rates and especially for obtaining high resolution NMR spectra.[8]

Both the thermal oxidative and photo-oxidative behavior of E/CO and a low density polyethylene are compared under identical experimental conditions. Differences in oxidation rates of the polymers were small, and the peak ($3550\,cm^{-1}$) attributed to 'free' hydroperoxides was missing from the infrared spectrum of thermally oxidized E/CO.

2. EXPERIMENTAL

A noncommercial E/CO (1·4% CO by weight) was provided by the E.I. du Pont de Nemours and Co., and an unstabilized low density polyethylene (DYNK) was obtained from the Union Carbide Corp.

Compression-molded films (0·15–0·25 mm thick) of these polymers were exposed to >290 nm radiation at 30–32 °C in cells containing a 5 Å molecular sieve unless indicated otherwise. Net gas evolution during photolysis in argon and net gas consumption during irradiation in oxygen were measured volumetrically in conjunction with a reference cell.

Samples mounted on Nichrome screens were autoxidized in glass reaction tubes heated in a constant temperature bath maintained at 101 °C.

Changes in functional group concentrations were followed by transmission infrared spectroscopy, and by using the method of Mitchell and Perkins[9] for determining total hydroperoxides, and the SF_4 procedure of Heacock[10] for measuring carboxylic acids. All compositions of gaseous mixtures were analyzed with a gas chromatograph calibrated with standard gas mixtures. Molecular weight variations were observed by gel permeation chromatography and the resulting elution curves were compared with those obtained from standard linear polyethylene fractions. Details of these experimental methods have been reported previously.[8] The procedure for differential scanning calorimetric analysis is outlined in Table 2.

3. RESULTS

3.1. Gas Uptake or Evolution During Irradiation

Both weight loss and gas evolution during E/CO photolysis in argon involve two-step processes consisting of a rapid initial rate that shifts to a slower more prolonged rate as shown in Fig. 1. A related pattern is observed during photo-oxidation in a cell containing a molecular sieve absorbant. The initial rate of apparent oxygen uptake by E/CO is approximately four times the uniform rate exhibited by DYNK, as shown in Fig. 2. The second stage rate of E/CO, reached after about a 30-h exposure, is only twice as fast. Film weight was unchanged during the first 30 h of E/CO photo-oxidation, but decreased gradually as degradation continued. A slight weight gain occurred during irradiation of DYNK for 160 h under the same circumstances.

The mole ratios of volatile products from photolysis shown in Table 1 are

TABLE 1
Mole Percent of Volatile Products from Photolytic Reactions of E/CO and DYNK

Products	*E/CO*		*DYNK*
	Photolysis[a]	*Photo-oxidation*[b]	*Photo-oxidation*[c]
CO	68	17	48
CO_2	—	68	52
CH_4	2	<1	<4
C_2H_6	8	<1	<2
CH_3COCH_3	22	15	<2

[a] 20–40 h exposure under argon. Average of three determinations.
[b] 32 h exposure, 1·3 ml/g volume decrease.
[c] 145 h exposure, 3·4 ml/g volume decrease.

close to those reported by Li and Guillet[11] for an E/CO of similar composition. As might be expected, the main products obtained in an argon atmosphere were carbon monoxide, acetone and derivatives of acetone photolysis. Replacement of argon by an oxygen atmosphere reduced the CO/acetone ratio by threefold. Note also that the CO/CO_2 ratio was lower in E/CO than in DYNK photo-oxidation.

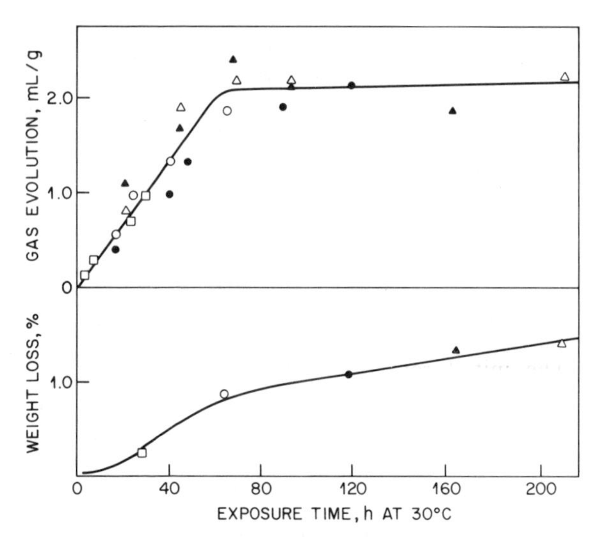

FIG. 1. Gas evolution and weight loss during photolysis of several E/CO films (□, ○, ●, ▲, △) in the absence of a molecular sieve absorbant. Note that weight loss was measured after maximum exposure of each sample.

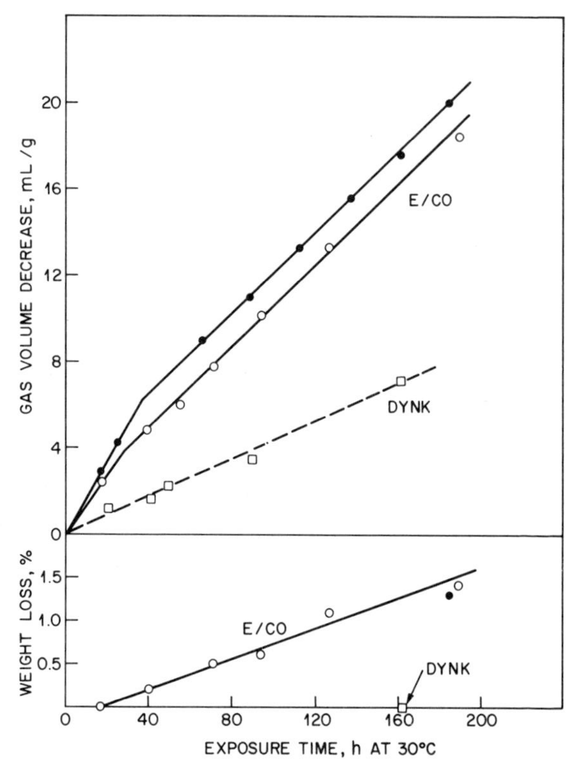

FIG. 2. Gas volume decrease and weight loss during photo-oxidation of two E/CO films (● and ○) in the presence of a molecular sieve absorbant. The photo-oxidation rate of DYNK (□) under identical experimental conditions is also shown.

3.2. IR and GPC Measurements

Changes in the infrared intensities of carbonyl (1720 cm^{-1}) and vinyl (908 cm^{-1}) absorbances by samples of E/CO exposed under argon and oxygen are compared in Fig. 3. The rate of decrease of carbonyl absorption appears to be somewhat slower for samples exposed under oxygen due to the formation of oxidation products, such as carboxylic acids and esters whose carbonyl peaks overlap ketonic carbonyls. A broad band centered at $\sim$3440 cm^{-1} increased with exposure under oxygen due to OH groups being formed.

Changes in number average molecular weights ($\bar{M}_n$) and average scissions per molecule are shown in Fig. 4. Oxidation enhances the scission rate significantly. At comparable losses of CO, a limiting value of four

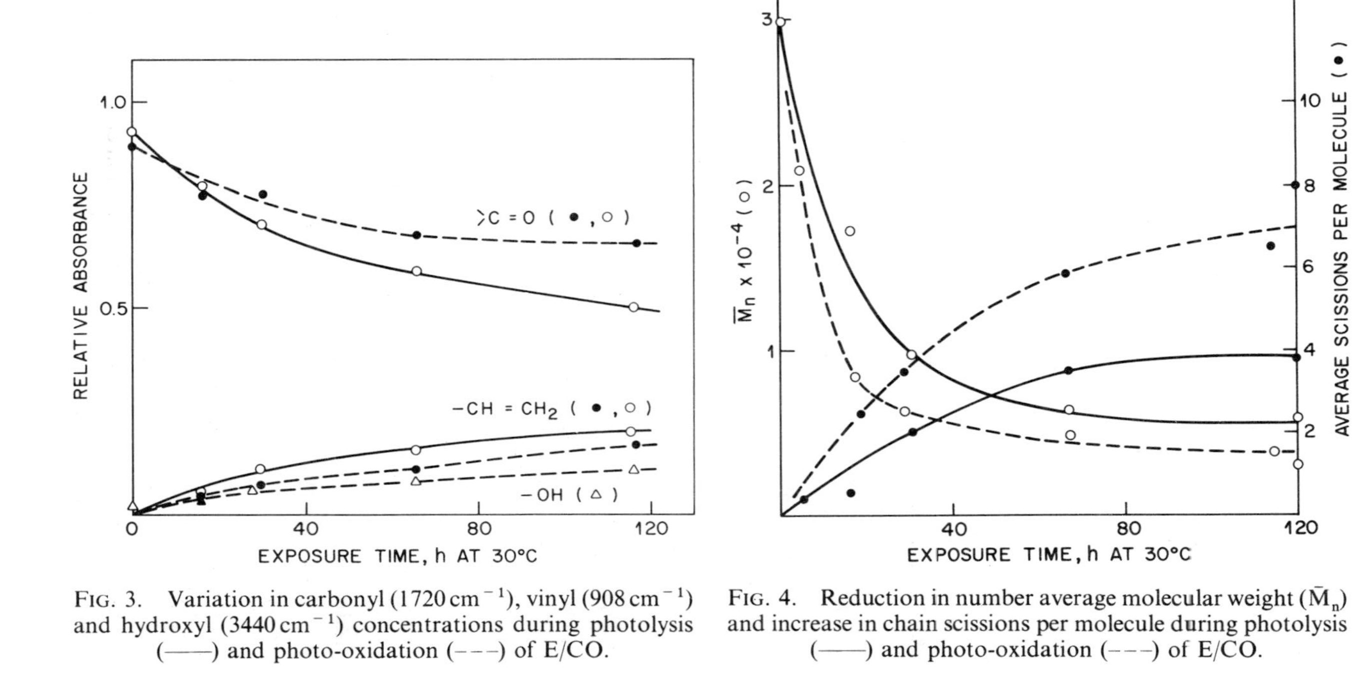

FIG. 3. Variation in carbonyl ($1720\,cm^{-1}$), vinyl ($908\,cm^{-1}$) and hydroxyl ($3440\,cm^{-1}$) concentrations during photolysis (——) and photo-oxidation (– – –) of E/CO.

FIG. 4. Reduction in number average molecular weight ($\bar{M}_n$) and increase in chain scissions per molecule during photolysis (——) and photo-oxidation (– – –) of E/CO.

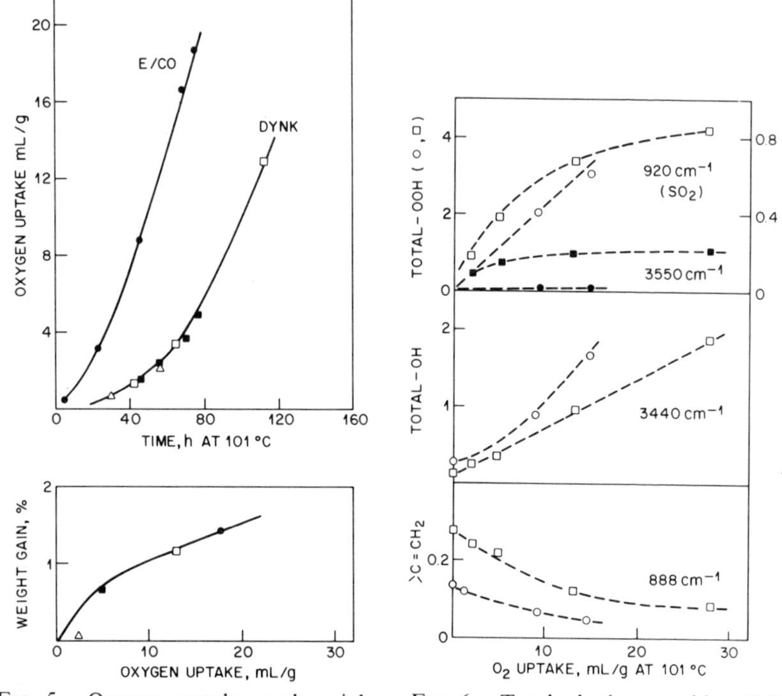

FIG. 5. Oxygen uptake and weight gain by E/CO (●) and DYNK (△, ■, □) during autoxidation at 101 °C. Weight gains were measured at the maximum oxidation time for each sample.

FIG. 6. Total hydroperoxide (SO_2 method), 'free' hydroperoxide, total hydroxyl, and vinylidene concentrations determined as a function of oxygen consumption during autoxidation of E/CO (○ and ●) and DYNK (□ and ■).

chain breaks is approached in argon, while the comparable value in oxygen is about seven scissions per molecule. A high molecular weight shoulder appeared in chromatograms of samples exposed under argon for more than 66 h, indicating the presence of some crosslinking not found in photo-oxidized films.[8]

3.3. Morphology and Molecular Structure

DSC measurements listed in Table 2 show that both polymers were nearly 50 % crystalline. During photolysis and photo-oxidation of E/CO the crystallinity increased, but T_m remained essentially unchanged.

NMR analysis[8] showed that E/CO, like DYNK, contained about two

TABLE 2
Differential Scanning Calorimetry of E/CO and DYNK

Polymer	*Treatment*	*Crystallinity* %	T_m, °C
DYNK	NONE	45 ± 1	107
E/CO	NONE	49 ± 1	109
E/CO	Photolyzed	57 ± 0	109
E/CO	Photo-oxidized	63 ± 4	110

Conditions: Mettler TA 2000 Z. Three samples per polymer heated at 10 °C/min from 5° to 135 °C. Crystallinity estimated using $\Delta H_f = 1\,800$ cal/mole as 100% crystallinity.

branches per hundred carbon atoms, and that CO groups were solely backbone ketonic carbonyls.

3.4. Autoxidation

The thermal oxidations of E/CO and DYNK are compared in Fig. 5. Although E/CO was reported to contain no inhibitor, it was stable in oxygen at 101 °C for 200 h. Consequently, the sample in Fig. 5 was preoxidized in air with >290 nm radiation. The oxidation rates of the polymers were then similar as shown in Fig. 5. Both polymers gained weight during oxidation.

Changes in hydroperoxide, hydroxyl and vinylidene concentrations are shown in Fig. 6 for E/CO and DYNK. Although comparable quantities of total hydroperoxides (920 cm^{-1}) were formed during oxidation, no 'free' or non-hydrogen bonded hydroperoxide (3550 cm^{-1}) was observed in E/CO.

4. DISCUSSION

Photolysis of E/CO has been attributed to a combination of Norrish scission reactions of ketones. Initially, the Type II process accounts for

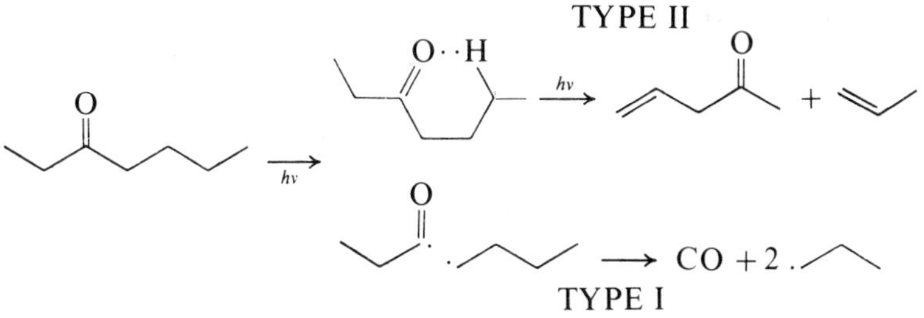

TYPE II

O··H

$h\nu$

O

O

+

$h\nu$

O

$\longrightarrow$ CO + 2 .

TYPE I

over 80% of the chain breaks.[5,8] Reactions in E/CO at 30 °C must be confined mostly to disordered regions because Type II fragmentation involves a six-member transition state that requires a high degree of chain mobility. According to theoretical[12] and experimental[7] evidence, the carbonyls should be distributed almost randomly throughout the polymer solid. This implies that only about half of the carbonyls (those in the amorphous phase) can readily decompose. Although the other half is constrained by the crystal lattice, some Type I scission can apparently still occur.† The severed chain ends that fail to recombine are then free to leave the lattice by diffusing an average distance of only half the lamellar thickness, about 25–30 Å.[13]

The effects of morphology are manifest in the two-stage mode of E/CO decomposition in Figs. 1–4. Incipient photo-oxidation is evidently initiated by radicals formed by Type I scission. But the reaction rate subsides as carbonyls in amorphous regions are decomposed. Continuing slow decomposition of residual carbonyls lodged in the lamellae could, in part, explain the second-stage E/CO rate which is still higher than that of DYNK. An oxidation mechanism involving a rather short kinetic chain length could further account for the fact that E/CO, with a thousandfold greater carbonyl content than DYNK, photo-oxidizes only twice as rapidly as the branched polyethylene.

Autoxidation of E/CO and DYNK was limited to amorphous regions accessible to oxygen.[14] The reaction was clearly autocatalytic in contrast to the nearly constant rates found for photo-oxidation. The main differences between the oxidized polymers were a more rapid hydroxyl build-up and the absence of 'free' hydroperoxide formation in E/CO.

Arnaud *et al.*[15] have proposed that free hydroperoxides are formed at vinylidene sites in polyethylene. Yet, the initial concentration of vinylidene groups in E/CO was only half that in DYNK and appeared to decrease at nearly the same rate. Hydroperoxides in thermally oxidized E/CO may be bound to carbons in the alpha positions adjacent to carbonyls, forming the following hydrogen bonded cyclic structure.

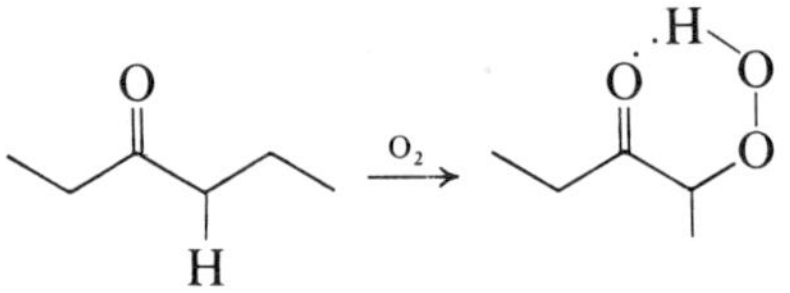

A more complete account of these studies will soon be published.

† More than 85% of the original CO was lost during photooxidation for 300 h, indicating CO loss from the crystal lattice as well as the amorphous regions.

5. CONCLUSIONS

Main chain ketonic carbonyls proved to be ineffective initiators of polyethylene photo-oxidation. Both radical and nonradical scissions took place in amorphous regions of the copolymer. Only a small fraction ($\sim 10\%$) of these groups in E/CO can decompose at 30 °C to form radicals by a mechanism with a low quantum yield. The degradation rate depended on morphology. A much slower radical scission process was presumably possible in the crystalline state.

Although branched polyethylene produced gel during photo-oxidation, E/CO remained soluble and consequently was conveniently analyzed by GPC and NMR techniques.

Thermal oxidation of E/CO presumably produced α-keto hydroperoxides which, in turn, formed hydrogen-bonded six-membered ring structures.

ACKNOWLEDGEMENTS

The authors thank Dr H. H. Hoehn of E.I. du Pont de Nemours and Co. for supplying the E/CO, and Drs H. D. Keith and A. J. Lovinger for helpful discussions.

REFERENCES

1. Pross, A. W. and Black, R. M., *J. Soc. Chem. Ind. (London)*, **69**, 113 (1950).
2. Burgess, A. R., *Natl. Bur. Standards (U.S.)*, Circ. 525, 149 (1953).
3. (a) Winslow, F. H., Matreyek, W. and Trozzolo, A. M., *Am. chem. Soc. Polym. Preprints*, **110**, 1271 (1969).
   (b) *ibid*, *S.P.E.J.*, **28**, 19 (1972).
4. Sitek, F. and Guillet, J. E., *J. Polym. Sci. Symp.*, **57**, 543 (1976).
5. Hartley, G. H. and Guillet, J. E., *Macromolecules*, **1**, 165 (1968).
6. Wu, T. K., Ovenall, D. W. and Hoehn, H. H. In: *Applications of Polymer Spectroscopy* (Ed. E. G. Brame, Jr), Academic Press, New York, 1978, pp. 19–40.
7. Wunderlich, B. and Poland, D., *J. Polym. Sci. A*, **1**, 357 (1963).
8. Gooden, R., Hellman, M. Y., Hutton, R. S. and Winslow, F. H., *Macromolecules* (in press).
9. Mitchell, J. and Perkins, L. R. In: *Weatherability of Plastics Materials* (Ed. M. R. Kamal), Interscience, New York, 1968, p. 167.
10. Heacock, J. F., *J. appl. Polym. Sci.*, **7**, 2319 (1963).
11. Li, S. K. and Guillet, J. E., *J. Polym. Sci. Polym. Chem. Ed.*, **18**, 2221 (1980).
12. Fuller, C. S., *Chem. Rev.*, **26**, 143 (1940).

13. Alfonso, G. C., Fiorina, L., Martuscelli, E., Pedemonte, E. and Russo, S., *Polymer*, **14**, 373 (1973).
14. Winslow, F. H. In: *Durability of Macromolecular Materials* (Ed. R. K. Eby), ACS Symposium Series, No. 95, American Chemical Society, Washington, D.C., 1979, 11.
15. Arnaud, R., Moisan, J.-Y. and Lemaire, J., *Macromolecules*, **17**, 332 (1984).

11

Some Photoreactions of Polystyrenes: Progress and Unsolved Problems

N. A. WEIR
Chemistry Department, Lakehead University, Thunder Bay, Canada

1. INTRODUCTION

While the manifestations of the photo and photo-oxidative degradations of polystyrene have long been recognized, and much research has been directed towards an understanding of the reactions involved, we find four decades later that not only are there large gaps in our knowledge of the fundamental processes, but also the mechanisms of several of the reactions are still the subjects of considerable controversy. Experimental evidence is available to support most of the plausible mechanisms. To some extent these observations are explicable. Apart from a desire to understand the deleterious effects of oxygen and light on PS, researchers have been attracted to the polymer, in preference to others, on account of its relative ease of preparation, and, more importantly, its ease of characterization. However, lack of standardization of preparation and reaction conditions has inevitably introduced a number of new variables, and consequently results obtained in one laboratory are not necessarily directly comparable with those from another; and varying reaction conditions can lead to different reaction pathways. The relatively low reactivity of PS has itself contributed. In the first place, detection of extremely low concentrations of intermediates (or supposed intermediates), has not been particularly successful, and hence the amount of analytical data available from conventional methods (e.g. NMR), has been limited. Application of ESCA appears to be promising, and it is hoped that the analytical situation can be ameliorated.[1] In order to circumvent these analytical problems, most researchers have studied what is essentially the 254 nm initiated photo-oxidation of PS. Under these conditions, appreciable (and detectable), changes take place; however, the reaction system bears little resemblance to the natural weathering environment to which PS is normally exposed. Short-wave photochemistry of PS is, however, a valid study in its own right.

The purpose of this article is to review some of the progress which has been made in our understanding of the photochemistry of styrene polymers and to point out some of the still unresolved problems.

The most important variable is the wavelength of the incident radiation, since it controls the type of photoreaction occurring; and it is convenient to classify photoprocesses into two general groups, i.e. those occurring on 254 nm irradiation (short-wave), and those occurring on long-wave ($\lambda \geq 300$ nm) irradiation. It should be emphasized that the reactions discussed occur in the absence of oxygen.

2. SHORT-WAVE PHOTOCHEMISTRY

254 nm radiation is strongly absorbed by films of styrene polymers, and from Fig. 1(A), it can be seen that the rapid attenuation of intensity has the effect of confining the effective reaction zone to the surface layers. Absorption becomes even less uniform, following comparatively small extents of photodegradation (Fig. 1(B)). However, the process conforms to Lambert's Law, in which the intensity at depth L, I_L, within a film is related to the incident 254 nm intensity (I_0), by the equation, $I_L = I_0 \exp(-\beta L)$ in

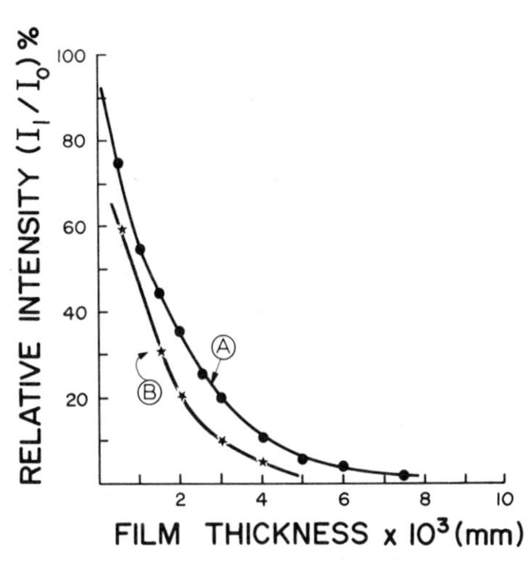

FIG. 1. Absorption characteristics of polystyrene films. A—before degradation. B—20 min, 254 nm irradiation.

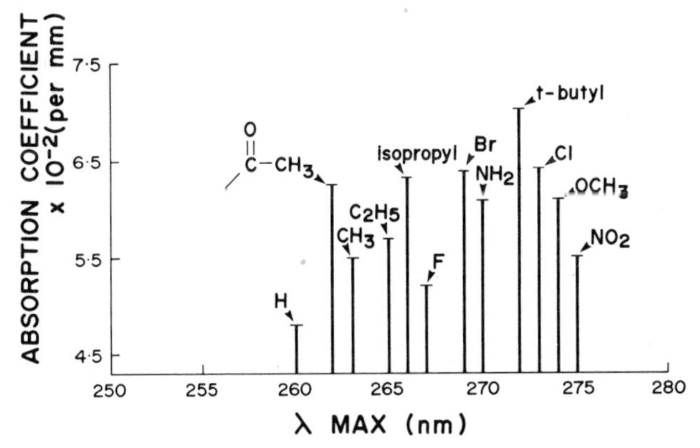

FIG. 2. Absorption characteristics of ring-substituted polystyrenes.

which β is an absorption coefficient (values are qualitatively similar to the corresponding extinction coefficient for solutions). In Fig. 2 the absorption characteristics of a number of ring-substituted polymers are compared. It can be seen that substitution leads to more intense 254 nm absorption, and hence smaller reaction zones, and to a general bathochromic shift in λ_{max} values. These observations are in line with previous findings,[2] and it can be concluded that perturbation of the energy levels in the phenyl chromophores is brought about by the varying electronic characteristics of the substituents.

Nonuniform absorption (and reaction), has lead to complications and to ambiguities in the quantitative interpretation of results of a number of studies, and a failure to recognize the effect of photodegradation (as yet another variable), has contributed to the noncompatibility of data from different sources. A major inhibition to the detection and unequivocal identification of reaction products has been their nonuniform distributions within photodegraded polymers, and unless very specific modifications are made in the spectroscopic techniques employed, e.g. the use of ATR, little useful information can be obtained. Indeed, degraded films often exhibit IR absorption characteristics which are almost indistinguishable from those of the pure polymers.

3. INITIAL PROCESSES

Absorption of 254 nm radiation by styrene polymers results in the formation of the excited singlet (S_1^*) state of the phenyl groups ($\pi \rightarrow \pi^*$

transition). The degree of vibrational excitation is relatively small. The energy equivalent of 254 nm quanta exceeding the S_1 energy level by less than 30 kJ mol^{-1}, and claims that the photochemistry of these polymers is derived entirely from highly excited species are not valid; at least for monophotonic processes. De-excitation of the excited phenyl singlet and the $S_1 \rightarrow S_0$ radiationless transition (internal conversion), occur very readily, with quantum yields in excess of 0·5 mol Einstein^{-1},[3] and further deactivation occurs by monomer and excimer fluorescence, decay lifetimes of which are 1 and 19 ns respectively.[4] Partition of the absorbed energy is very much in favour of photophysical processes, and quantum yields of photoprocesses (typically $10^{-5} \leq \phi \leq 10^{-3}$ mol Einstein^{-1}), reflect this.[5]

Initial photochemical processes are associated with fissions (α and β), of bonds adjacent to the phenyl groups. Possible fissions are as follows:

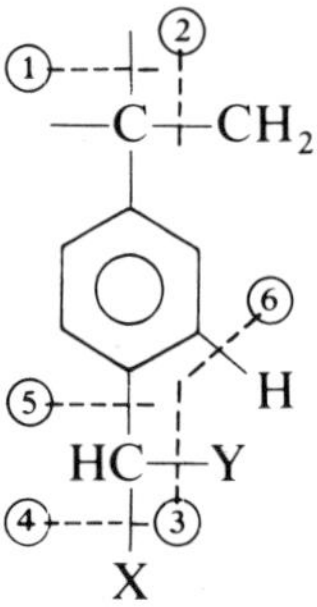

(1) ① —C—CH$_2$ ② ⑥ H ⑤ HC—Y ④ ③ X

(fissions (1), (2), (3) and (4) are β and (5) and (6) are α).

On energetic grounds alone, it can be predicted that all of these fissions occur, the energy equivalent of 254 nm radiation (470 kJ mol^{-1}) being greater than the dissociation energies of all bonds involved. However, this presupposes efficient and adequate transfer of energy from the excited singlet to the appropriate vibrational mode (or modes). No other activation is possible. The probability of energy transfer decreases rapidly as the bond dissociation energy value approaches that of the excitation level of donor. Thus it is by no means certain that the energy requirement for ring C—H fission (type (6)), (approx. 420 kJ mol^{-1}) can be satisfied in this way; and the failure to detect such fission so far would bear this out. (It is not suggested that fission (6) is impossible.) Other α-fissions (type (5)), involving C—C bonds in both aromatic hydrocarbons and in substituted polystyrenes have been established.[6,7] In the former case, predissociation has been invoked, but in both cases the dissociation energies are significantly lower than those of ring C—H bonds (by at least $100\,kJ\,mol^{-1}$). Quantum yields are low.

The most frequently observed fission in polymers and in aromatic compounds,[6,8,9] is the β-fission (i.e. types (1), (2), (3), (4)). The presence of a radical center on the α-C atom (type (1)), is well substantiated by direct ESR measurements.[10] Indeed, this is the only type of radical that has been detected by ESR. Others may be formed, but in insufficient concentrations. Chain scission (type (2)), has also been observed in both PS and in substituted PS,[11–15] molecular weight data being confirmed for PS by ESR measurements.[16] Although quantum yields are low (10^{-4}–10^{-5} mol Einstein^{-1}), they are sensitive to ring substitution, alkyl, NH_2 and OH substituted polymers being more susceptible to chain scission.

Analogous β-fission (types (3) and (4)), which have been frequently detected in the photolyses of hydrocarbons,[6–8] have also been shown to occur in substituted PS,[14–16] the quantum yields being slightly higher than those for chain scission.

The predominance of β-fission in both small and macromolecules can be rationalized in terms of the energetics of radical formation; the stability of the incipient benzylic-type radical being enhanced (delocalization), and the corresponding bond dissociation energy (BDE), lowered. The effect of ring substitution on, for example, chain scission, can be attributed to the stabilizing and destabilizing effects (inductive and conjugative), of the substituents on radical stabilities, and Table 1 shows a qualitative correlation between these factors.

While the photochemistry of small aromatics can be discussed largely in terms of energetics, additional factors, peculiar to the macromolecular environment (which have been neglected), must also be considered. Firstly, radicals formed on polymer photolyses are produced initially in cages, which are 'strong' on account of the very large bulk viscosity. Since the mobility of the polymers is severely limited at sub-T_G temperatures, diffusive escape (which is a prerequisite for many reactions), particularly that of macroradicals, is inhibited, and a β-fission in a small molecule cannot realistically be compared with chain scission in a polymer. It follows that small radicals and H atoms have greater probabilities of escape from cages, and hence the probability of forming a radical center on the α-C atom is greater. Larger radicals will undergo cage collapse reactions, so that observed quantum yields will be depressed. Although the separation of the phenyl radical (α-scission), suffers from an energy deficit (relative to β-fission), it is also subject to unfavourable cage effects, on account of the bulk of the radical.

TABLE 1
254 nm Photolyses of Films of Ring Substituted Poly(styrenes) High Vacuum, 25 °C

Ring substituent (para)	*Products in order of abundance*	*Relative* ϕH_2	*Relative rate of CS*	*Relative rate of CL*	*Reference*
H	H_2	1·0	1·0	1·0	13
CH_3	$H_2 > CH_4 > C_2H_6$	1·3	1·2	1·7	13
C_2H_5	$H_2 > CH_4 > C_2H_6$	2·0	1·5	1·9	15
iso-C_3H_7	$H_2 > CH_4 > C_2H_6 \gg$	2·3	1·7	2	14
tert.-C_4H_9	$H_2 > CH_4 > C_2H_6 \gg$ iso-C_4H_{10}	1·4	1·8	1·3	30
NH_2	H_2	3·0	1·3	10·0	12
NO_2	H_2	0·8	0·9	0·6	12
F	$H_2 >$ HF (trace)	1·0	0·8	0·8	31
Cl	$H_2 > HCl$	4·0	1·2	3·0	31
Br	$H_2 > HBr$	10·0	1·4	6·5	31
OCH_3	$H_2 > CH_4 \gg C_2H_6 > CH_3OH$	2·4	1·5	2·0	7
OH	$H_2 \gg H_2O$	4·8	2·0	6·0	28

CL = crosslinking, CS = chain scission and PS = 1·0 in all cases.

An additional energy factor, which has greater influence in the solid state, must also be considered. Formation of radical centres on the polymer is accompanied by a change of both hybridization (from sp^2 to sp^3 for β-fission), and geometry. In order to achieve this, at least two bonds adjacent to the incipient radical center must be displaced. This process may occur readily in small flexible molecules in the gas phase. However, mobility in a rigid polymer at sub-T_G temperatures is restricted to small amplitude oscillations of the phenyl groups.[17] A degree of segmental mobility is required to produce the appropriate geometry for a radical center on the α-C atom, for example; and it would appear to be feasible only after a net input of energy. This implies a higher than expected BDE.

Factors like these introduce considerable uncertainties in estimated BDE values for polymers, and it seems that simply matching BDE with singlet and triplet energy levels does not necessarily permit the identification of the excited precursor of, for example, β-fission. A S_1 or excited S_1 state would be the most likely precursor of α-fission, but β-fission could conceivably also occur from the excited T_1 state in PS. Little helpful information can be obtained from similar studies of aromatic hydrocarbons, since the reaction mechanisms are still the matter of considerable controversy, excited singlets, triplets and biphotonic processes being proposed.[18,19] β-fission in PS appears to be a monophotonic process.[20] These considerations ignore the possibility of activation of phenyl groups by intra and intermolecular energy transfer. The outcome of the reactions would presumably be the same as discussed above. Intermolecular energy transfer from excited phenyl groups to hydroperoxides, leading to their subsequent decomposition, has been proposed as an initiation step in photo-oxidation of PS,[20] and intramolecular energy transfer involving the migration of singlet energy down chains has also been suggested,[21] but challenged.[22]

Clearly much more work is required to resolve these, and some of the problems referred to above. It is unfortunate that so few advances have been made during the last decade in understanding the photochemistry of aromatic hydrocarbons.

4. PHOTOISOMERIZATION

The photoisomerization of benzene (on 254 nm irradiation), to fulvene and benzvalene is well established.[23] Substituted benzenes also undergo intramolecular isomerization, but there appears to be considerable uncertainty with regard to detailed mechanisms. In general, it would appear that benzvalenes are formed, and that these are the intermediates involved in the formation of positional isomers. 1, 2 and to a lesser extent, 1, 3 shifts involving migration of ring substituents in disubstituted benzenes have been observed,[24] and the excited singlet state (S_1^*), has been suggested as the precursor.[25] Quantum yields of these isomerizations are not insignificant.

It has been suggested that the phenyl groups in PS undergo photoisomerization to fulvenes and benzvalenes.[26] In the light of the above observations, the latter would appear more probable; however, it is difficult to distinguish between the species. As yet, no positive evidence for photoisomerization of PS has been obtained, although there are analytical difficulties, e.g. low concentrations and the presence of other species which exhibit similar spectral responses. Perhaps the greatest objection to fulvenes and benzvalenes as photoproducts lies in their instability to UV radiation, substituted benzvalenes having short life times,[24] and substituted fulvenes decomposing to give reactive unsaturated polymers under the reaction conditions used in the photolyses of PS.[27]

On the positive side, however, we have recently obtained evidence

(NMR), for 1,2 methyl shifts occurring during the photolyses of poly- o- and m-methylstyrenes. Whether these are examples of photoisomerization involving benzvalenes, or whether radical addition is involved, is yet to be established.[28]

5. SECONDARY REACTIONS

These consist of the interactions of the primary radicals with themselves and with the polymers. Hydrogen is the principal gaseous product of the photolyses of PS and substituted PS. While the main source of H_2 is associated with β-C—H bond fission in the main chain (i.e. the α-C atom), it has been shown (ESR), that β-type fissions within ring substituents are an additional source of H atoms.[29]

H_2 can be formed by combination of atoms (in presence of a third body), but recent work would suggest that this is a minor source, and that most of the H_2 evolved is the result of abstraction reactions between H atoms and the polymers.[28]

6. ABSTRACTION REACTIONS

On purely energetic grounds, abstraction would be expected to occur preferentially at α-C atoms, i.e.

$$H^{\cdot} + \text{Polymer} \longrightarrow H_2 + \sim\underset{\substack{|\\ \text{Ph}}}{\dot{C}}\sim CH_2\sim \tag{2}$$

and, indeed, direct evidence for this has been obtained.[28] However, abstraction from β-C atoms cannot be excluded, particularly when the very favorable relative orientations of the nascent H atom and β-C atoms are considered. Following dissociation of the C—H bonds, H atoms escape from cages with considerable translational energy, and the probability of collision with H atoms attached to β-C centres is also statistically higher. Thus a more favorable entropy of activation term compensates to some extent for the higher energy of activation of β-abstraction. The intramolecular β-abstraction can be shown as follows:

$$\sim CH_2-\underset{\substack{|\\ \text{Ph}}}{\overset{\dot{H}\ +}{\dot{C}}}-\overset{\substack{H\\ |}}{CH}-\sim \longrightarrow \sim CH_2-\underset{\substack{|\\ \text{Ph}}}{C}=CH\sim \tag{3}$$

Recent work,[28] using isotopically labelled PS has established that β-abstraction does occur, and while there are correlations between this and the formation of unsaturated groups in the polymers, it cannot be concluded that all abstractions from β-positions take place by the mechanism shown above. In addition, there are other explanations for unsaturation.[11]

Table 1 shows the effect of substitution on H_2 quantum yields. On the basis of earlier work,[29] it appears that the augmented H_2 yields are directly attributable to fission of β-C—H (and O—H and N—H), bonds within the substituents, dissociation of these bonds being as energetically favorable as that of the α-C—H bonds. Additional molecular H_2 can then be formed by abstraction, including abstraction from the substituent groups.[30] Substituents probably also influence primary H_2 atom yields indirectly, by stabilizing the incipient radical centres on α-C atoms (inductive and conjugative effects). It is impossible to quantify these effects. Abstraction (by H), of halogen atoms from halopolymers also occurs, but it is possible that abstraction by halogen atoms from C—H bonds also contributes to the observed yields of the hydrogen halides as shown in Table 1.[31]

7. ADDITION REACTIONS

Addition of H atoms to phenyl groups and to other unsaturated species competes favorably with abstraction, and cyclohexadienyl radicals are formed in detectable quantities.[32] Recent studies of deuterated PS suggest that approximately 50 % of the primary H atoms may be involved in such additions, and this finding has considerable implications for some quantum yield determinations, which generally assume that this figure is negligible. Addition of H atoms to unsaturated species formed on photolysis may account for the apparent deceleration observed in their rates of formation.[5]

8. ELIMINATION AND COLORATION REACTIONS

An alternative source of H_2 has been proposed in which molecular elimination occurs from a terminally unsaturated species, which is activated by energy transfer from excited phenyl chromophores, energy presumably migrating along the chains.[33]

$$\left[\begin{array}{c}CH_2{=}\underset{\displaystyle Ph}{\underset{|}{C}}H{-}CH_2{-}\underset{\displaystyle Ph}{\underset{|}{C}}H\sim\end{array}\right]^* \longrightarrow H_2 + CH_2{=}\underset{\displaystyle Ph}{\underset{|}{C}}H{-}CH{=}\underset{\displaystyle Ph}{\underset{|}{C}}\sim \quad (4)$$

Energy transfer has been shown to occur between PS and styrene,[34] but it is questionable whether such a transfer would provide enough energy to bring about the elimination of H_2 from what is essentially a neighboring unit in the chain, bearing in mind the competition from photophysical processes and the specific activation pattern required.

Coloration is always observed on photodegradation of styrene polymers (both in vacuum and in presence of O_2), and the high vacuum coloration has been attributed to the presence of a number of unsaturated species. Rånby has identified fulvene derivatives, formed by phenyl group isomerization, with coloration, and these species show intense UV and visible absorptions in the same regions as degraded PS.[26] While such fulvenes have not yet been positively identified, there are other objections, principally on the basis of stability and irreconcilability with experimental observations. The lack of stability of fulvenes and benzvalenes (in the case of substituted PS), has been discussed, and we have demonstrated that when PS films containing diphenylfulvene are irradiated in vacuum, the intensity of the visible absorption decreases, and at the same time that in the UV region increases.[28] Spectral changes and decreases are accelerated on irradiation in presence of O_2. These results are quite contrary to the experiences with pure PS, in which it has been found that while coloration is subject to an apparent inhibition after about one hour reaction (probably explicable in terms of the data in Fig. 1 (B)), the intensity does not decrease and the spectral characteristics do not change noticeably on irradiation. Moreover, subsequent irradiation in O_2 of a film, which has been colored by vacuum irradiation leads to an increase in, or at least a persistence of color. Whether or not fulvenes are involved is extremely difficult to determine, on account of their spectral similarities to other plausible structures. The structure most frequently associated with coloration is a polyene of the type:

$$\sim CH_2-\underset{\substack{|\\ Ph}}{C}=CH-\underset{\substack{|\\ Ph}}{C}=CH\sim \qquad (5)$$

Original assignments were made on the basis of IR and UV spectral data,[35] but more convincing evidence has been obtained by Geuskens *et al.*,[33] who showed that emission spectra of irradiated PS films (consisting of broad absorption bands extending from 400 to 550 nm, and having discrete maxima superimposed), were consistent with the presence of polyenes containing up to 4 conjugated double bonds. These measurements have been refined by Fox and Price,[36] who confirmed that emission spectra of

α,ω-diphenylpolyenes are qualitatively similar to those of photodegraded PS, and also showed that maxima due to dienes, trienes, etc., coincided with the maxima in the PS spectra.

Recent detailed analyses of the colored material in PS films have revealed the presence of a structure whose ^{13}C NMR spectrum corresponds very closely with that of a triphenylhexatriene.[28] Experiments so far have failed to detect species whose spectra are consistent with polyenes having more than 4 double bonds in conjugation, and it is possible that the number is limited by the limited mobility of the polymer along with the steric effects introduced by the large phenyl groups. These in turn reduce the probability of attainment of an extended conjugated (coplanar), configuration, by inhibiting large scale bond displacements. It is possible, as in the case of PVC, that the concentration of double bonds is reduced by radical additions, H atoms likely being involved.[37]

Although there is strong evidence for the formation of phenyl polyenes, the mechanisms of their formation are still far from being understood. Two general approaches have been made: one involving molecular elimination[33] (Scheme 4), and the other favoring free radical abstractions[35] (Scheme 3). It is not only extremely difficult to distinguish between them, but also to establish unequivocally that either is involved (if at all).

9. CHAIN SCISSION AND CROSSLINKING

Chain scission involving β-fission has been suggested and discussed above. Relative rate data for substituted polymers are summarized in Table 1. Another chain breaking process, involving the decomposition of the tertiary radical has been proposed,[33] i.e.

$$\sim CH_2-\dot{C}(Ph)-CH_2-CH(Ph)\sim \longrightarrow \sim CH{=}C(Ph) + \dot{C}H_2-CH(Ph)\sim \qquad (6)$$

Scheme 6 also provides for the development of chain end unsaturation, which has been implicated in H_2 formation (Scheme 4). While the analogous decomposition is well known in alkoxy radical chemistry, it has been observed that quantum yields for this type of reaction with aromatic hydrocarbons are small. In the case of PS, in which reactions such as 6 are likely to be adversely influenced by the general lack of mobility and the presence of strong cages within the polymer, quantum yields will certainly be lower, and while the quantitative data available are subject to some uncertainty, it would appear that 6 alone could not account for the observed extent of chain scission. The critical (and undefinable) factor is the degree of excitation of the decomposing radical.

Crosslinking occurs readily in styrene polymers, and for some time it was thought that it occurred to the total exclusion of chain scission. This conclusion was based largely on experience, in which excessive crosslinking was found to obscure the effects of chain scission. Despite rapid insolubilization of these polymers, it is possible to obtain useful data for both chain scission and crosslinking, but measurements must be made in the initial phases. Quantum yields for crosslinking are about twice those for chain scission,[11] in PS, but this ratio is considerably greater for substituted PS (Table 1). It is emphasized that these data reflect reaction rates in the very early stages of the photoprocesses, and while they are of limited use in the absolute sense, they are self-consistent and comparisons are valid. Rates of crosslinking increase rapidly with increasing reaction time, chain scission being overwhelmed and inhibited by the presence of the highly crosslinked polymer matrix.

Crosslinking has frequently been envisaged as the union of two tertiary radicals (as in Scheme 6). However, there are other, equally plausible radical interactions involving the tertiary radicals, cyclohexadienyl radicals and those formed by H atom addition to main chain unsaturated groups. The sensitivity of quantum yields for crosslinking to ring substitution supports the view that radical centres formed within the substituents are also involved.

On the basis of ESR data, it can be concluded that a considerable proportion of the radicals become immobilized, and while they may eventually participate in secondary cage reactions,[38] they are unable to contribute to crosslinking on account of the greatly diminished mobility of the systems.

10. LONG-WAVE PHOTOCHEMISTRY

Long wave in this context refers to wavelengths $\lambda \geq 295$ nm. Long-wave photochemistry as applied to PS would appear to be a misnomer, since the pure polymer contains no groups which have significant absorptions in this spectral region. Absorption is confined to the short-wave region and is entirely due to the phenyl moieties.

PS does, however, undergo degradation on exposure to terrestrial sunlight (which has a UV component extending to about 295 nm), in air, conditions which are frequently experienced in commercial applications. Chain scission, O_2 absorption and the formation of a number of oxygenated structures in the polymers have been cited as evidence for photo- and photo-oxidative degradations, and while the details of the overall weathering processes are far from understood, it is generally recognized that little degradation occurs in the absence of light; which implies that the initiation steps are photolytic in nature. However, the photochemistry is that of adventitious impurities, which are incorporated in the chains during the polymerization and/or subsequent processing.[39]

Attempts have been made to isolate the constituent reactions, and a number of studies of the vacuum photolysis ($\lambda > 300$ nm) of PS have been made. The principal observations, along with the most important conclusions are summarized below.

Lawrence and Weir[40] showed that radically prepared PS undergoes random chain scission on photolysis, and this is accompanied by incorporation of scavenger molecules into the polymer, suggesting that radicals are involved in the initial fission. Studies of molecular weight changes on long-wave photo-oxidation show that at least two chain scission processes are involved, one a very rapid initial decomposition of photolabile bonds, and the other associated with decomposition of oxygenated products. Figure 3 shows typical data for an AIBN initiated PS, and the degradation may be described in terms of degradation, α where

$$\alpha = \gamma + \theta t \tag{7}$$

in which θ is a constant related to the rate of light absorption, t is the reaction time, and γ is another constant, the intercept on the degradation axis, and is identified with the rapid initial fission of photolabile species (γ depends on the preparative method used, and varies from 0 for an anionic PS to about 2×10^{-5} for a redox initiated PS).

Qualitatively similar behaviour has been demonstrated for the thermal degradation of radically prepared PS, and an equation of the same general form as (7) has been derived to account for both the rapid initial rupture of thermally labile bonds, and the less rapid dissociation of main chain bonds, which are considerably more thermally stable.[41] The magnitude of the intercept associated with the thermal decomposition of labile bonds is considerably greater than that of γ, and it is suggested that there are a number of anomalous links present in PS chains, some of which can be selectively broken by the absorption of light, and that these and a number of others are thermally unstable at subdegradation temperatures.

Significantly, anionically prepared PS behaves completely differently on

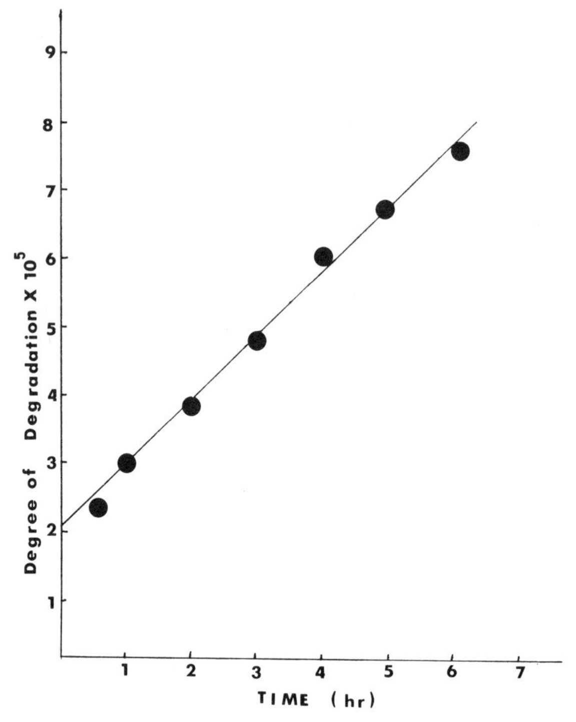

FIG. 3. Degree of degradation as a function of oxidation time for radically prepared polystyrene.

both thermal and photodegradation, and have enhanced stabilities to both modes of degradation, as reflected by the absence of intercepts (such as that in Fig. 3), on the degradation axes. These observations would imply that, whatever their nature, photo- and thermally labile bonds are not present in the chains of anionically prepared PS.

The differential stability of anionic PS is further exemplified by recent ESR investigations, in which nitroxide radicals were used as scavengers. No evidence for the production of radicals on the photolysis of anionic PS could be obtained.[42]

Cameron *et al.*[43] has shown that the thermal degradation of radically initiated PS is much more complex than that of both thermally and anionically prepared samples. Three concurrent chain scission processes are involved, but the nature of the weaker bond and the highly labile bonds was not established.

The nature of the photolabile structures has been, and still is, the subject of much discussion and controversy, and as yet no single species has been positively identified. (It should be emphasized that identification involves the detection of one anomalous link in perhaps 10^4 or 10^5 normal bonds.)

Lawrence and Weir[40] attributed the initial rapid molecular weight decrease to the photolysis of in-chain peroxides, which absorb in the $310 \leq \lambda \leq 350$ nm region; and although their concentrations and extinction coefficients are low, the near unity value of the quantum yield for O—O bond fission ensures that a finite concentration of alkoxy radicals could results. Peroxy units can be incorporated into the chains by the random copolymerization of O_2, present as an impurity and apparently impossible to eradicate from the usual free radical polymerization setup. Subsequent photolysis of O—O bonds would lead to chain scission, i.e.

$$O_2 + CH_2{=}\underset{\mid\atop Ph}{CH} + \sim CH_2{-}\underset{\mid\atop Ph}{\dot{C}H} \longrightarrow \sim CH_2{-}\underset{\mid\atop Ph}{CH}{-}O{-}O{-}CH_2{-}\underset{\mid\atop Ph}{\dot{C}H}$$

$$\sim\underset{\mid\atop Ph}{CH}{-}O{-}O{-}CH_2{-}\underset{\mid\atop Ph}{CH}\sim \xrightarrow{h\nu} \sim\underset{\mid\atop Ph}{CH}{-}\dot{O} + \dot{O}{-}CH_2{-}\underset{\mid\atop Ph}{CH}\sim \qquad (8)$$

Anionic polymerization is generally carried out under more hygienic conditions, and it is also likely that the residual O_2 is eliminated by reaction with the catalysts. Consequently, the probability of inclusion of peroxy links is reduced. That polymers containing such links actually undergo chain scission on irradiation has been established[44,45] by subjecting PS that had been synthesized in presence of known amounts of O_2 to vacuum photolysis ($\lambda \geq 300$ nm). According to George and Hodgeman,[44] however, the initially formed alkoxy radicals undergo a cage collapse reaction to form a ketonic and a hydroxyl species. These conclusions do not conflict with results of the scavenger experiments,[40] since it is quite possible that the scavenger diffuses into the cages, from which diffusive escape of two macroradicals is impeded by the high local viscosity of the system.

Rånby was unable to reproduce these results using a commercial PS sample.[26] However, George and Hodgeman[44] were able to reconcile the two apparently different results; and showed that samples of PS which had been polymerized at temperatures of up to 200 °C do not contain appreciable concentrations of peroxides. Such species have limited lifetimes at temperatures in excess of about 150 °C, and the apparent photostability of commercial PS can be attributed to the prior thermal decomposition of potential weak links.

Ketonic impurities have also been detected. Klöpffer[46] showed that radically prepared PS contains terminal keto groups (similar to acetophenone), and suggested that they are formed from hydroperoxides during the polymerization or subsequent processing. George and Hodgeman,[44] using similar phosphorescence techniques, showed that PS containing in-chain O—O bonds dissociated to give ketonic products, disproportionation of alkoxy radicals being involved (Scheme 8).

Such ketones absorb in the long-wave region, but a Norrish type I decomposition would not account for the observed chain scission. The non-terminal ketone suggested by Geuskens *et al.*[33] would undoubtedly undergo chain scission. However, the photo-oxidation sequence in which it is formed is unlikely to be encountered during exposure of PS to long-wave UV radiation.

The above discussion has centred around the two most frequently cited (and plausible) impurities. There are, in addition, a number of photosensitization processes involving ketones, molecular oxygen, and adventitious impurities; and while it is extremely unlikely that any (or all), of these makes a contribution to the controlled degradation,[40,44] it is conceivable that they play a part in the overall weathering of PS, and a brief discussion is justified.

The triplet ($n \rightarrow \pi^*$), states of aromatic ketones participate in photoreduction reactions, and abstraction of tertiary H atoms from PS is at least thermodynamically favourable.[47] Such a reaction would not in itself cause chain scission (unless β-fission occurs), but could be the initiation step of an autoxidation, which would in turn produce hydroperoxides, provided that the polymer had the appropriate exposure to O_2 and UV.

It has also been suggested that triplet states of such ketones could interact with O_2 to produce singlet O_2 ($^1\Delta g$),[48] which in turn can lead to hydroperoxidation of PS. It is difficult to reconcile such reactions with the vacuum degradation conditions,[40] in which the probability of forming singlet O_2 must be extremely small, and at the same time the possibilities for its quenching are numerous. The findings of Geuskens and David,[49] in which PS, irradiated in presence of benzophenone and O_2, underwent negligible degradation due to singlet O_2, would tend to bear this out.

It is known that PS and O_2 form a charge transfer complex which absorbs in the long-wave region. According to Rånby, the photo-excited form of the complex is capable of reacting with O_2 to form singlet O_2.[26] Recently, Allen and Fatinikun[50] have proposed that such complexes are the precursors of hydroperoxides in polyolefins. While it has been established that PS undergoes more rapid photo-oxidation under conditions which are conducive to the formation of charge transfer complexes,[51] the role played by these has yet to be determined.

Polycyclic aromatic hydrocarbons are known to accumulate in measurable quantities on the surfaces of polymers exposed to urban atmospheres, and a number of studies have shown that anthracene is capable of photosensitizing degradation of polymers.[52,53]

In summary, the long-wave photodegradation of PS is very sensitive to, and ultimately controlled by impurities.

REFERENCES

1. Peeling, J. and Clark, D. T., *Polym. Degrad. Stab.*, **3**, 97 (1981).
2. Lukac, I., Pilka, J., Kulickova, M. and Hrdlovic, P., *Europ. Polym. J.*, **15**, 1645 (1977).
3. Cundall, R. B., Pereira, L. and Robinson, D. A., *J. chem. Soc., Faraday II*, **69**, 701 (1973).
4. Ghiggino, K. P., Wright, R. D. and Phillips, D., *J. Polym. Sci., Polym. Phys. Ed*, **16**, 1499 (1978).
5. Weir, N. A. In *Developments in Polymer Degradation*, Vol. 4 (Ed. N. Grassie), Applied Science Publishers, London, 1982, p. 143.
6. Wilzbach, K. E. and Kaplan, L., *J. Am. Chem. Soc.*, **86**, 2307 (1964).
7. Weir, N. A. and Milkie, T. H., *Polym. Degrad. Stab.*, **1**, 105 (1979).
8. Porter, G. and Strachan, E., *Trans. Faraday Soc.*, **54**, 1595 (1958).
9. Hofer, H. and Heusinger, H., *Z., Phys. Chem., Neue Folge*, **69**, 147 (1970).
10. Lucki, J. and Rånby, B., *Polym. Degrad. Stab.*, **1**, 1 (1979).
11. David, C., Baeyens-Volant, D., Delanois, G., Lu Vinh, Q., Piret, W. and Geuskens, G., *Europ. Polym. J.*, **14**, 501 (1978).
12. Weir, N. A. and Milkie, T. H., *Makromol. Chem.*, **180**, 1729 (1979).
13. Weir, N. A. and Milkie, T. H., *Polym. Degrad. Stab.*, **1**, 181 (1979).
14. Weir, N. A. and Milkie, T. H., *Europ. Polym. J.*, **16**, 141 (1980).
15. Weir, N. A. and Milkie, T. H., *Polym. Photochem.*, **1**, 205 (1981).
16. Szocs, F. and Placek, J., *J. Polym. Sci., Polym. Phys. Ed.*, **13**, 1789 (1975).
17. Wetton, R. E. In: *Dielectric Properties of Polymers* (Ed. F. Karasz), Plenum Press, New York, 1972.
18. Brocklehurst, B., Gibbons, W. A., Lang, F. T., Porter, G. and Savadatti, M. I., *Trans. Faraday Soc.*, **62**, 1793 (1966).
19. Schwartz, F. P. and Albrecht, A. C., *J. phys. Chem.*, **77**, 2808 (1973).
20. Geuskens, G., Baeyens-Volant, D., Delaunois, G., Lu Vinh, Q., Piret, W. and David, C., *Europ. Polym. J.*, **14**, 299 (1978).
21. David, C., Baeyens-Volant, D. and Geuskens, G., *Europ. Polym. J.*, **9**, 533 (1973).
22. MacCallum, J. R., *Ann. Reports* (*A*), 99 (1978).
23. Bryce-Smith, D. and Gilbert, A., *Tetrahedron*, **32**, 1309 (1976).

24. Noyes, W. A. and Hunter, D. A., *J. phys. Chem.*, **75**, 2741 (1971).
25. Anderson, D., *ibid.*, **74**, 1686 (1970).
26. Rabek, J. F. and Rånby, B., *J. Polym. Sci., Polym. Chem. Ed.*, **12**, 273 (1974).
27. Rentsch, C., Slongo, M., Schonholzer, S. and Neunschwander, M., *Makromol. Chem.*, **181**, 19 (1980).
28. Weir, N. A. and Arct, J., unpublished data (1984).
29. Chernova, I. K., Golikov, V. P., Leshchenko, S. S. and Karpov, V. L., *Khimiya Vysokikh Energii*, **8**, 265 (1974).
30. Weir, N. A., Milkie, T. H. and Nicholas, D., *J. appl. Polym. Sci.*, **23**, 609 (1979).
31. Weir, N. A. and Milkie, T. H., *J. Polym. Sci., Polym. Chem. Ed.*, **17**, 3735 (1979).
32. Wilske, J. and Heusinger, H., *J. Polym. Sci. A1*, **7**, 995 (1969).
33. Geuskens, G., Baeyens-Volant, D., Delaunois, G., Lu Vinh, Q., Piret, W. and David, C., *Europ. Polym. J.*, **14**, 291 (1978).
34. Basile, L. J., *Trans. Faraday Soc.*, **60**, 1702 (1964).
35. Grassie, N. and Weir, N. A., *J. appl. Polym. Sci.*, **9**, 975 (1965).
36. Fox, R. B. and Price, T. R., *Stabilization and Degradation of Polymers*, Am. Chem. Soc. Symp. Ser., **196**, 96 (1978).
37. Abbås, K. B. and Sörvik, E. M., *J. appl. Polym. Sci.*, **19**, 2991 (1975).
38. Garton, A., Carlsson, D. J. and Wiles, D. M., *Makromol. Chem.*, **181**, 1841 (1980).
39. Rånby, B. and Rabek, J. F., *Photodegradation, Photooxidation and Photostabilisation of Polymers*, New York, Wiley, 1975.
40. Lawrence, J. B. and Weir, N. A., *J. Polym. Sci., Polym. Chem. Ed.*, **11**, 105 (1973).
41. Cameron, G. G. and Kerr, G. P., *Europ. Polym. J.*, **6**, 423 (1970).
42. Gerlock, J. L. and Bauer, D. R., *J. Polym. Sci., Polym. Lett. Ed.*, **22**, 447 (1984).
43. Cameron, G. G., Bryce, W. A. J. and McWalter, I. T., *Europ. Polym. J.*, **20**, 563 (1984).
44. George, G. A. and Hodgeman, D. K. C., *Europ. Polym. J.*, **13**, 63 (1977).
45. Weir, N. A. and Milkie, T. H., *Makromol. Chem.*, **179**, 1989 (1978).
46. Klöpffer, W., *Europ. Polym. J.*, **11**, 203 (1975).
47. Weir, N. A., Rujimethabas, M. and Clothier, P. Q., *Europ. Polym. J.*, **17**, 1394 (1979).
48. Trozzola, A. M. and Winslow, F. M., *Macromolecules*, **1**, 98 (1968).
49. Geuskens, G. and David, C., *Pure appl. Chem.*, **51**, 233 (1979).
50. Allen, N. S. and Fatinikun, K. O., *Polym. Degrad. Stab.*, **4**, 59 (1982).
51. Nowakowska, M. In: *Singlet Oxygen* (Ed. B. Rånby and J. F. Rabek), Wiley, Chichester, 1978.
52. Aspler, J., Carlsson, D. J. and Wiles, D. M., *Macromolecules*, **9**, 691 (1976).
53. Rämme, G. and Rånby, B., *Europ. Polym. J.*, **13**, 855 (1977).

12

Unwanted Photochemical Reactions in Polymer Coatings

BENGT LINDBERG and CHARLES M. HANSEN
Scandinavian Paint and Printing Ink Research Institute, Hørsholm, Denmark

1. UNWANTED REACTIONS IN PAINT FILMS: SOME EXAMPLES

Coatings in use both indoors and outdoors are exposed to radiation which can cause unwanted photochemical reactions. Most of these reactions have never been studied in detail. (Therefore the coatings industry customarily reverts to extensive exterior exposure series or hopes for serendipic events to lead to improving coatings performance.) The lack of fundamental knowledge is the reason the formulation of coatings is still very much an art.

The undesirable effect of unwanted photochemical reactions can be reduced or eliminated by changing any of a number of formulating parameters. Some polymeric binders are more stable to these effects than others, and some pigments and additives screen or reflect incoming radiation better than others. There are many possibilities.

In this paper we will briefly consider a number of unwanted reactions before proceeding to demonstrate how evaluations of coatings are most often done. These reactions can affect a substrate, a primer layer, given pigments, polymers, or additives and even the customary crosslinking of alkyds by oxidation reactions involving double bonds.

1.1. Protecting a Substrate

An undesirable photochemical reaction of some commercial significance is the yellowing of wooden furniture, for example. Lignin derivatives cause the colour. Clear coatings can delay this effect to varying degrees by screening out the damaging parts of incoming radiation. White or coloured pigmentation can also be employed to reflect or absorb the radiation but many customers prefer clear coatings for this purpose. Various types of additives to screen damaging radiation while not contributing colour are available commercially. Other additives for clear coatings interfere with photochemical reactions by scavenging radicals. There are signals that significant progress is being made in this area.

Clear coatings for exterior wood are a special case where superior performance is required. Only in recent years have exterior clear coatings been used to any extent for wood. They have traditionally flaked off after a year or two since the wood substrate degraded to such an extent that adhesion was lost. Even pigmented coatings can have this type failure and radiation well into the visible region can be responsible for degradation of a wooden substrate leading to coatings failure. The degradation of wood substrates by light prior to painting has also been pointed out as a significant problem for the exterior durability of all coatings.[1,2] The longer the exposure time of the bare wood to light prior to coating, the shorter the durability of the coating. A coating which may last longer than 8 years if applied to a fresh wooden substrate may only last 2 years, if the wood has been exposed to light for as much as 12 months.

The light fastness of printing inks might also be mentioned here as well as the yellowing of paper. Both are of course the result of resistance (or the lack of it) to unwanted photochemical reactions.

1.2. Delamination from Primers

In recent years more light stable coatings have been developed. In some instances (polyvinylidine fluoride based) these can allow significant UV transmission even though they appear to hide and suitably protect. These have often been used in connection with epoxy primers. This combination can lead to peeling off of the topcoats as sheets because of photochemical degradation of the epoxy primer. Epoxy coatings have not been traditionally accepted as having good chalking resistance during exterior exposure.

An interesting corollary to this type behaviour is the introduction of clear coatings as topcoats over base coats with colour in the automotive industry. This will not be discussed here however.

The topcoat must screen for radiation which can damage the primer to attain long-term system durability. If the primer is exceptionally stable to radiation degradation these problems may not be as significant.

1.3. Yellowing

A major problem with many potentially durable coatings is that they yellow on exposure to light, either indoors or outdoors. This single factor has led to many headaches for organic chemists trying to develop the perfect polymeric binder. Customer acceptance of such coatings which yellow is low. These yellowing reactions can be very elusive. One coating can yellow on exposure to light and regain whiteness when stored in darkness. This behaviour continues with continued cycling. The reactions are reversible. Another coating can react in the opposite manner. Such reactions have never been studied in detail, but demonstrate how difficult product development can be. A typical approach to solve this type problem in practice might be to mix both types of coatings and see what happens.

Yellowing has also been caused by such an elusive effect as retained solvent. Solvent is retained in many coatings for many years. In large companies formulations can be subject to alteration from plant location to plant location. Among other things different solvents and solvent combinations can be used in otherwise identical formulas. It is known that such variations unexpectedly led to yellowing where the retained solvent apparently took part in an unwanted photochemical reaction, and a large area had to be repainted for this reason. In this case as in many others no one studies the reactions, but a quick solution had to be found based on available knowledge.

1.4. Radiation Curing

Radiation curing is one of the newer coatings technologies designed to reduce environmental disturbances. This controlled radiation exposure leads of course to a wanted reaction and to a desired result.

It has been known for many years,[3] however, that radiation levels common in laboratories can significantly affect the drying of ordinary alkyd paints. This has caused many problems where comparative tests are carried out, for example, and drying conditions must be maintained constant to make the necessary comparisons. In one instance a variety of alkyd type paints were studied systematically to these effects.[3] These coatings dry by a catalysed oxidation process involving double bonds in the alkyd. These are found in various types of oils or fatty acids derived from them, from which the alkyds are made. Some of these types of paints dry faster in a room with fluorescent lamps which are on, than had they not been on. Such variations can be quite disturbing in a climatised room where drying is intended to occur uniformly according to standards, for example. The only way to solve the problem of irregular drying rates because of personnel irregularly turning the lights off and on was to require that the lights were on continuously. Reduction in hardness of up to 50% after 28 days drying was found in one series of tests to study this phenomenon where coatings dried in darkness.

2. EVALUATION OF WEATHER DURABILITY OF COATINGS

Having mentioned a number of disturbing effects caused by unwanted (and largely unstudied) photochemical reactions, we will now proceed to show how weather durability of coatings actually is evaluated.

A very important property of paint films used outdoors is their resistance to unwanted photochemical degradation. In this respect we can talk about a technical life time based on the protection of the underlaying material. Many research workers have written papers and discussed weathering of paint films from different backgrounds.[4–6] At our Institute we also have studied the photochemical degradation of paint films during exposure both by natural ageing outdoors and by accelerated ageing in different Atlas WeatherOmeters.[7–10] We have only studied the paint film as such during weathering and consciously excluded the influence of the substrate. Some results from these investigations with focus on photochemical degradation are presented in the following to demonstrate how testing is done.

2.1. Degradation of Paint Films by Natural and Accelerated Weathering

Six very different paints with respect to binder were exposed at three outdoor test stations and in seven different weathering cycles. The following materials and procedures were used:

2.1.1. Paints

- A Alkyd, air drying (linseed oil alkyd 68 %, Plexal P 68 H), PVC 25 %
- C Chlorinated rubber (Pergut S10), plasticiser chloroparaffin 40, PVC 23 %
- E Epoxy/polyamide, two-component (Epikote 1001/Versamid 115), PVC 19 %
- F Commercial fluorocarbon polymer, polyvinylidene fluoride PVF_2 (Kynar 500), coil-coat product
- L Acrylic latex (Rhoplex AC-34), PVC 40 %
- U Polyurethane, aliphatic two-component (Demmodur N/Desmophen 651), PVC 18 %

All the paints except F were prepared at the Institute and were all white, pigmented with titanium dioxide (Kronos RNCX) rutile grade. The paints were applied without primer to anodised aluminium panels by spiral applicator to a dry film thickness of approximately 50–60 μm (except paint F which was industrially applied by the supplier; epoxy primer + 25 μm top coat).

2.2. Weathering Cycles/Outdoor Exposure

The outdoor weathering tests were carried out at three different test stations:

W 1: Lejre, 40 km outside Copenhagen rural climate.
W 2: Bohus Malmön, 100 km north of Gothenburg, coastal climate.
W 3: Florida, USA (Sub-Tropical Testing Services), subtropical climate.

The test panels were exposed at 45° towards south.

The laboratory weathering was carried out in different Atlas WeatherOmeter models under various cycles. The exposure time (total exposure time) was approximately 1500 h except for two cycles with much UV-light (540 and 900 h). The following WeatherOmeter (WO) cycles were studied:

Xenon Lamps

WO 1: Xenon 6000 W, cycle 102/18 (min light/min water without light), borosilicate filters, 63 °C, 40–45 % RH, model 600 DMC-WRC, 1500 h.
WO 2A: The same as WO 1, but light also during the water period (in accordance with ASTM G 26-70), 1500 h.
WO 2B: The same as WO 1 but with double quartz filters instead of borosilicate filters, 540 h.
WO 3: Xenon 6000 W, cycle 60/60 (min light/min cold water on the back of the panels without light), type Dew-cycle, borosilicate filters, 63 °C, model DMC-WRC, 1500 h.

Carbon Arcs

WO4: Twin enclosed carbon arc, cycle 102/18 (min light/min water without light), 63 °C, model DL-TS (newer description DMC-W), 1500 h.
WO 5: Sunshine carbon arc with Corex D filters, cycle 1-2-2-6 (1 h water without light/2 h light/2 h water without light/6 h light), model XW-W, 1500 h.
WO 6: Sunshine carbon arc without filter, original Dew-cycle 60/60 (min light/min cold water on the back of the panels without light), model XW-W, 900 h.

As the original data is extensive (about 50 diagrams and 100 SEM photographs) the results can only be presented by a selected number of typical diagrams and tables.

2.3. Gloss

The loss of gloss during ageing was measured in accordance with ASTM-D523 with a 60° Gardner gloss meter. Panels exposed outdoors (with dirt pick-up) were measured before and after washing with a detergent solution (1 % Tween at 50 °C). Panels exposed in WeatherOmeters were not washed as no dirt exists in this case.

The same type of gloss must be measured when talking about correlation between different exposures. If outdoor panels are washed then panels in laboratory weathering should be washed. On the other hand it is desirable not to wash, thus avoiding severe variation in the gloss curve. The effect of washing on the gloss value is illustrated by results from the Florida exposure (Table 1).

Two of the paints studied, namely the acrylic latex and the fluoropolymer gave no information about gloss changes. The latex paint had too low a gloss value from the beginning (approx. 5) and the chosen exposure periods were not long enough to really affect the gloss of the fluoropolymer. Already here there are some limitations with gloss measurements.

The loss of gloss for two of the paints in outdoor and laboratory weathering is shown in Fig. 1 (alkyd) and Fig. 2 (epoxy paint).

TABLE 1
Gloss Variations due to Washing

Paint		A	C	E	F	L	U
Unexposed		85	89	100	29	5	96
6 months' exposure	I	21	49	5	26	2	85
	IIb	18	53	5	26	3	87
	IIa	43	63	36	28	5	93
12 months' exposure	I	9	32	0	25	0	70
	IIb	12	36	3	25	3	72
	IIa	23	53	25	27	4	89
24 months' exposure	I	5	15	0	20	0	25
	IIb	5	22	0	20	0	30
	IIa	11	38	25	23	0	35

I = gloss without washing. IIb = gloss before and IIa = gloss after repeated washing every month on the same panel. Exposure in Florida.

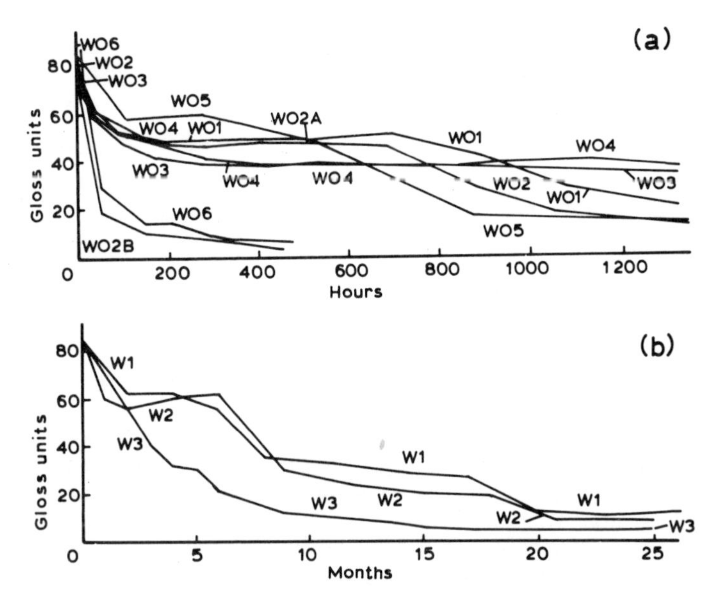

FIG. 1. Gloss of alkyd paint (A). (a) WeatherOmeters. (b) Outdoor exposures.

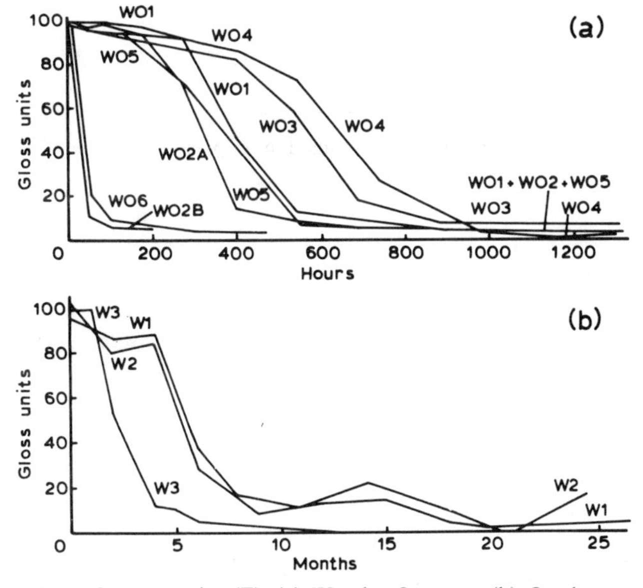

FIG. 2. Gloss of epoxy paint (E). (a) WeatherOmeters. (b) Outdoor exposures.

TABLE 2
Gloss Correlation between Outdoor Exposure W 1–3 and Laboratory Cycles WO 1–6

	Xenon lamp				*Carbon arc*			
	WO 1	WO 2A	WO 2B	WO 3	WO 4	WO 5	WO 6	Σ
W 1, Lejre	4	4	1	4	3	4	1	21
W 2, B-M	3	3	1	4	3	4	1	19
W 3, Florida	3	3	2	3	3	4	2	20
Points	10	10	4	11	9	12	4	

From similar diagrams showing the six paints for each exposure, the correlation between outdoor and WO-exposure was judged visually by four research workers at the Institute (Table 2) using the following scale:

5 = very good correlation, 4 = good correlation, 3 = slight correlation, 2 = poor correlation, 1 = very poor correlation.

Conclusions from the gloss studies are:

1. A high amount of UV light as in WO 2B and WO 6, gives poor correlation with outdoor exposures, especially for the polyurethane paint.
2. The best overall correlation related to the three outdoor climates was obtained with WO 5 (Sunshine Carbon Arc with Corex D filter), but the difference between the seven cycles is in fact relatively small when WO 2B and WO 6 are excluded.
3. No accelerated degradation of the paint films was achieved by spraying cold water on the back of the panels. Compare WO 3 with WO 1.
4. The chlorinated rubber paint gave the largest deviation in gloss for the different exposures. The reason for this may be explained by variation in film homogeneity at application of the wet paint. Outdoors there were also problems with severe dirt pick-up on this paint.
5. Compared with the two Scandinavian climates (W 1, W 2), the Florida deterioration is much more severe, especially for the epoxy and polyurethane paints. This is reasonable as the total global irradiation for one year is approx. 5700 MJ/m^2 in Florida compared to approx. 3600 MJ/m^2 in Scandinavia.

Loss of Weight

During the exposure to UV light and water, the organic binder in the paint film is chemically degraded and the more resistant inorganic pigments and fillers are gradually exposed. This phenomenon is called film erosion or chalking. The loss of weight for the painted Al panels was measured with an analytical balance. Before weighing, the panels were stored in a constant climate room for 24 h. Panels exposed outdoors were washed before weighing while WO-panels were not washed. In order to check that the aluminium substrate does not corrode, an unpainted aluminium panel was exposed outdoors (W 1, Lejre). The change in weight of this panel was very small, namely an increase of 13 mg during one year of exposure.

Quite a different picture of weather durability for the six paints was obtained by this method compared with gloss measurements. To illustrate the typical type of curve Fig. 3 is shown.

From the studies of weight loss following observations were noted:

1. The WO 2B and WO 6 cycles did not differ very much from the other WO cycles with respect to weight loss (contrary to gloss measurements).
2. Alkyd and acrylic latex gave the highest loss of weight, namely approx. 0·6 mg/cm^2 after 1500 h in WO-meter and about the same figure after 24 months outdoors (W 1). The film erosion of 0·6 mg/cm^2 corresponds to approx. 3 μm decrease in film thickness.
3. The weight loss method seems to be useful for evaluating low gloss paints and where the paint films contain water soluble components, e.g. latex paints. On the other hand the method is somewhat limited for very durable paints, e.g. fluoropolymers and 2-component polyurethane.
4. The epoxy paint gave a significantly low weight loss, although the gloss had disappeared rapidly. This indicates that the film degradation occurs just in the surface region to a depth of few μm.
5. Measuring loss of weight is not such a sensitive method as gloss measurements (for glossy paints).

2.5. SEM Studies

With a Scanning Electron Microscope (SEM) it is possible to study the surface structure at high magnification with large depth of focus. More than 100 SEM photographs were taken at various stages during the different exposures. Some comments on the SEM photographs are given in the following. On each photo, exposure time in months, 60° gloss value and chalking value (ASTM-D659) are given in this order.

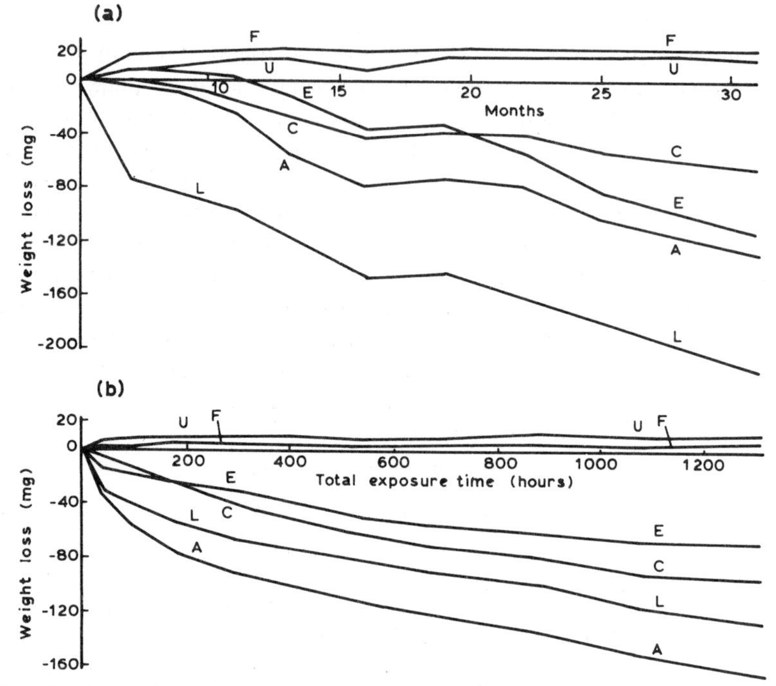

FIG. 3. Loss of weight for the paints on Al-panels. Size 7·5 × 23 cm. (a) Outdoor, Lejre (W1). (b) WeatherOmeter WO.

2.5.1. *Alkyd (A)*

After 3–4 months typical outdoor chalking could be noted. In the WO 1 surface defects could be observed after 280 h and after 1300 h the paint film surface was completely eroded (Fig. 4).

2.5.2. *Chlorinated Rubber (C)*

Outdoor deterioration could clearly be seen after 12 months in Florida (W 3) and approximately after 24 months at W 1 (Lejre). Relatively good resistance to degradation was noted in the WO-cycles.

2.5.3. *Epoxi (E)*

The film degradation started after 3 months outdoor exposure and already after 100 h in WO 1. After 700 h exposure in WO 1 the erosion was complete. The pigment particles could very easily be washed off (Fig. 5).

2.5.4. *Fluoropolymer (F)*

This paint changes very little during WO-exposure and outdoor exposure, though some degradation could be observed after 21 months in Florida. The degradation started from small defects (craters) already existing in the unexposed paint film. These defects seem to be related to imperfect film formation.

2.5.5. Acrylic Latex (L)

The thin layer of binder covering the pigment and filler particles in the unexposed paint film disappeared at an early stage of the exposure and the inorganic particles could clearly be seen.

2.5.6. Aliphatic Polyurethane (U)

Very little change in surface structure after 1300 h in WO 1, but a rather high degree of deterioration was found after 21 months in Florida. The degradation morphology is quite different from the alkyd to the epoxy. Areas of binder are left between rather deep craters or erosion spots (Fig. 6).

From the SEM photographs it can be concluded that the surface deterioration is very dissimilar for the six paints studied. In some cases the pigments (titanium dioxide and fillers) are exposed as individual particles, as for alkyd and epoxy paints, while for fluoropolymer and polyurethane, holes and 'islands' are formed in the binder matrix.

The same type of deterioration was found in the WO-exposure WO 1 (xenon lamp with borosilicate filters) as during outdoor exposure, while WO 2B (xenon lamp with quartz filters) gave a somewhat different deterioration compared to outdoor exposure.

2.6. Influence of UV Light of Varying Wavelength

2.6.1. Measuring Technique

As an extension of the studies with white pigmented paints, we have carried out investigations with changes of the type of pigment in the same alkyd binder (Alftalat AT 846). Eight different organic and inorganic coloured pigments were studied both in full-tone and half-tone 1:1 with titanium dioxide. The pigment volume concentration (PVC) was 17 %. The influence of certain wavelengths in UV light was studied by filtering the light from a xenon lamp into six UV areas by use of 'UV-kantenfilter' in glass (Glaswerk Schott & Gen., Mainz). These six filters were mounted at a distance of 4 cm in front of test panels of anodised aluminium (Fig. 7). The weathering was carried out in an Atlas xenon WeatherOmeter with double quartz filters in order to obtain a high degree of light energy also in the lowest UV range.

The transmission curves for the six glass filters were determined with a

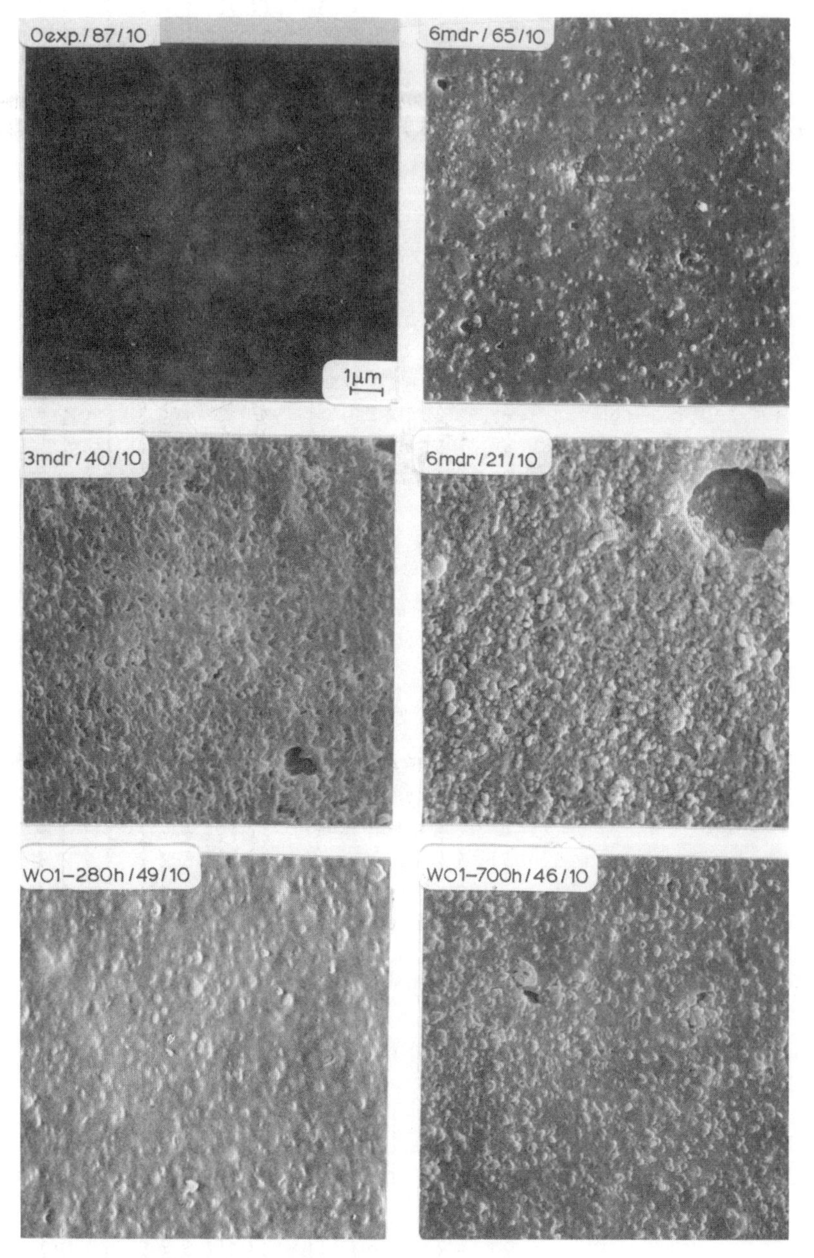

FIG. 4. SEM-photo of alkyd paint exposed outdoor W1 + W3 and WO1. 7000×.

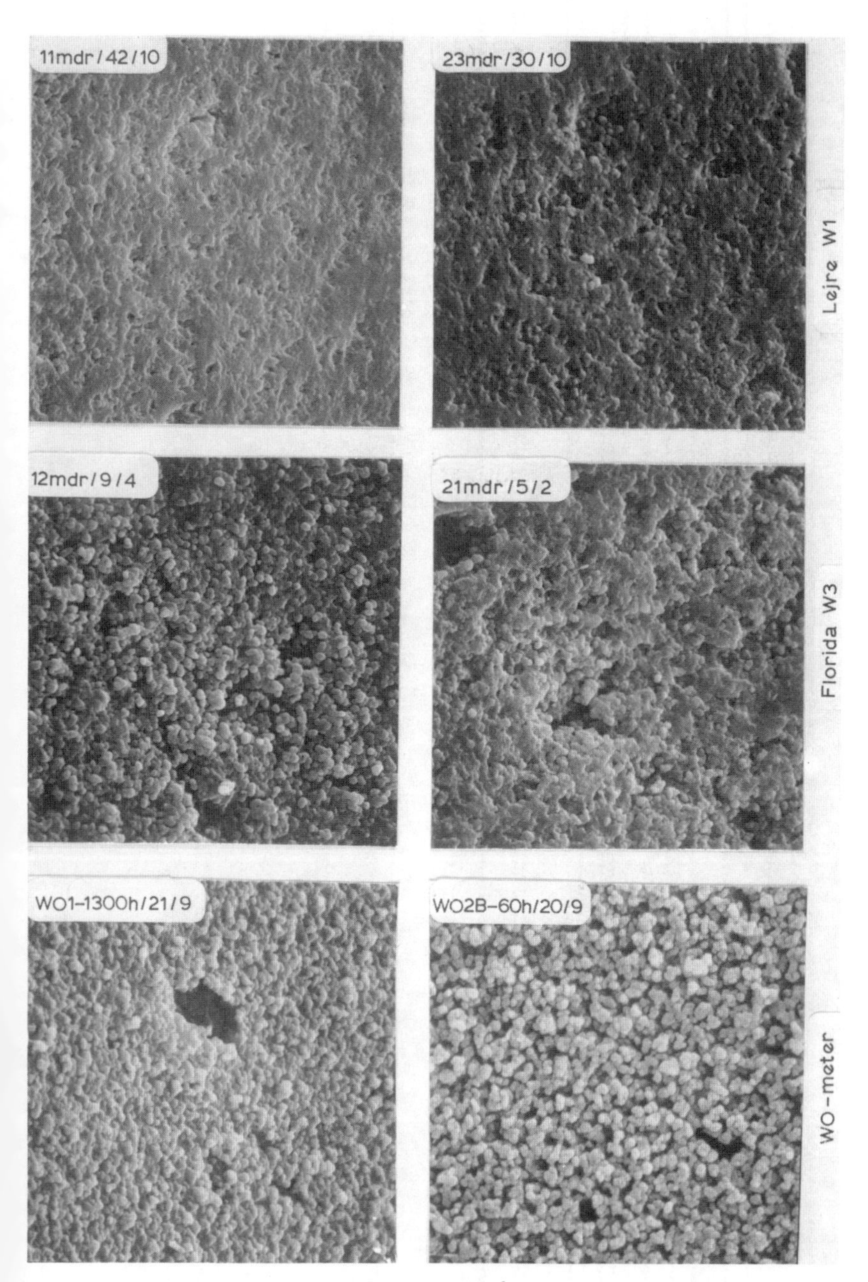
11mdr/42/10
23mdr/30/10
Lejre W1
12mdr/9/4
21mdr/5/2
Florida W3
WO1-1300h/21/9
WO2B-60h/20/9
WO-meter

FIG. 4.—*contd.*

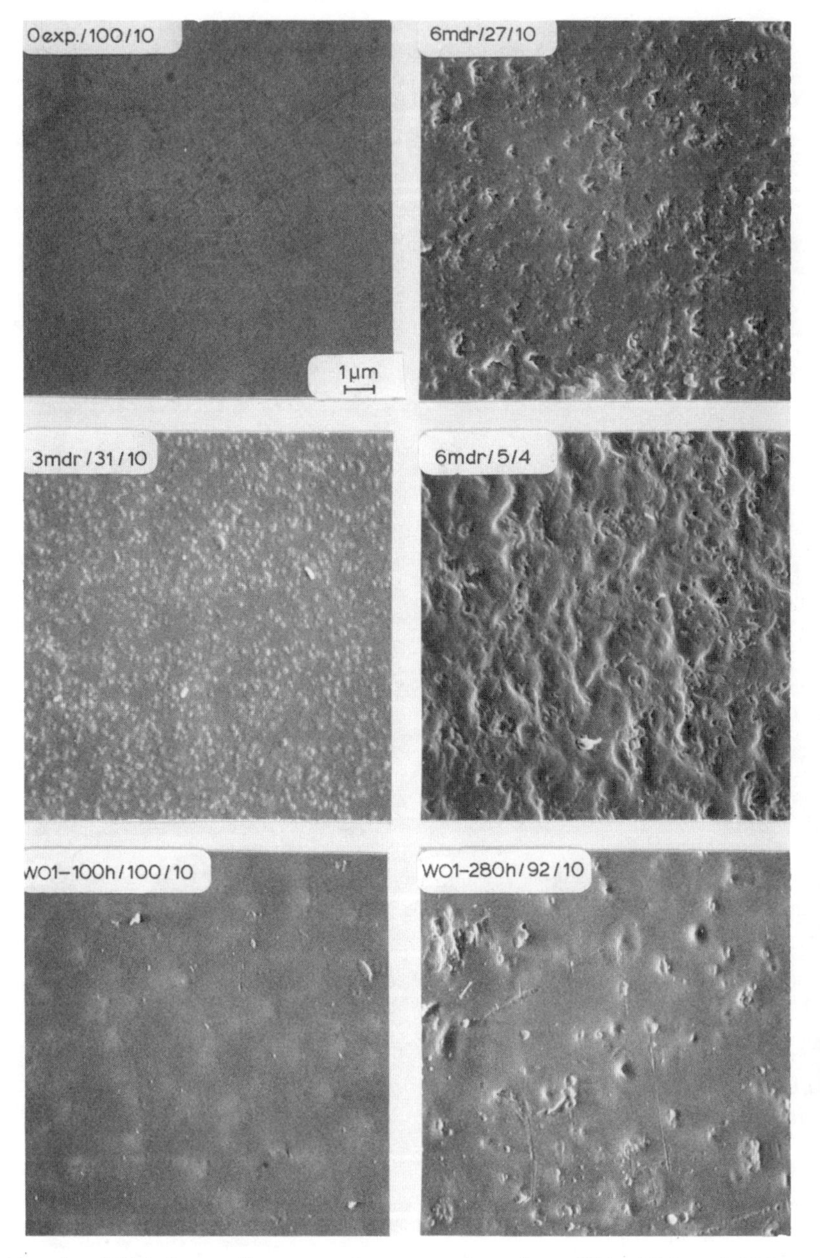

FIG. 5. SEM-photo of epoxy paint exposed outdoor W1 + W3 and WO1. 7000×.

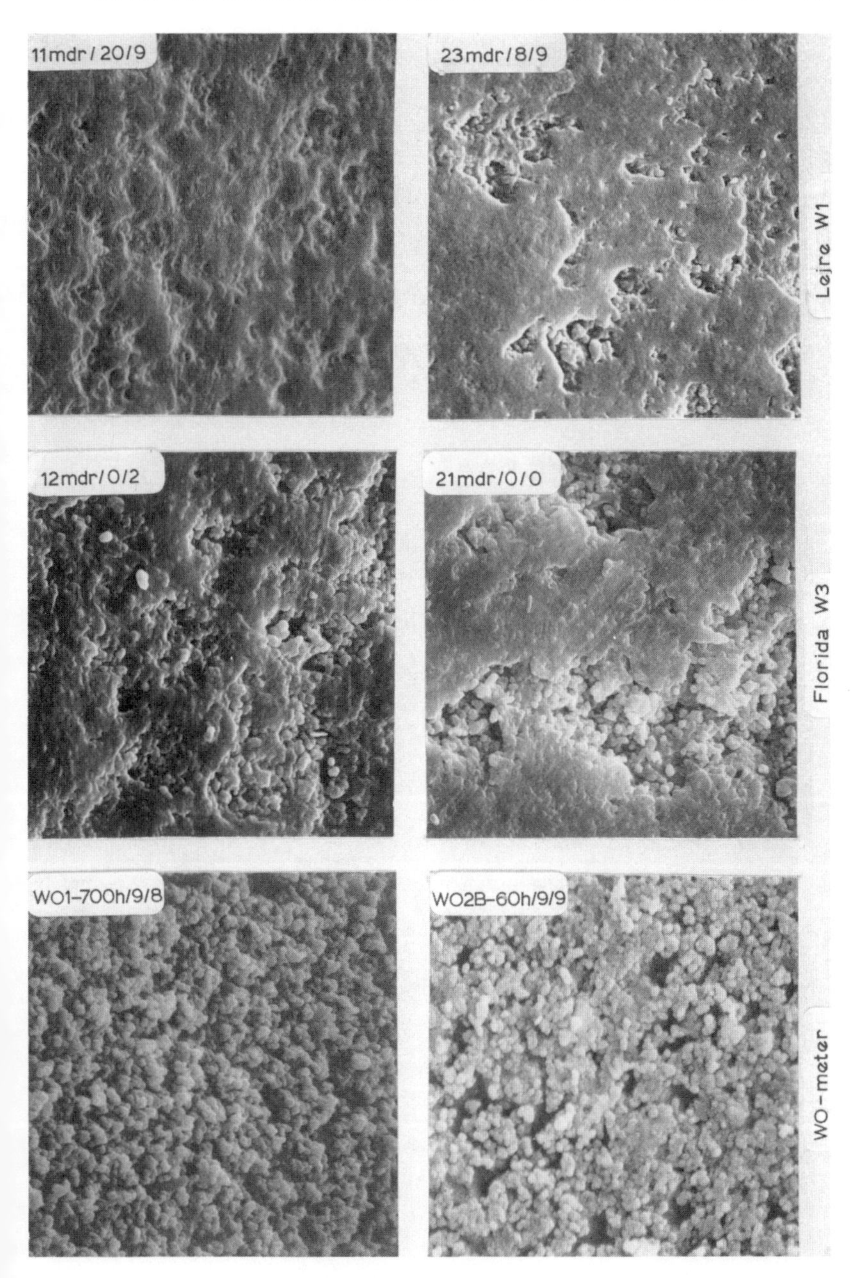
11mdr/20/9
23mdr/8/9
Lejre W1
12mdr/0/2
21mdr/0/0
Florida W3
WO1-700h/9/8
WO2B-60h/9/9
WO-meter

FIG. 5.—*contd.*

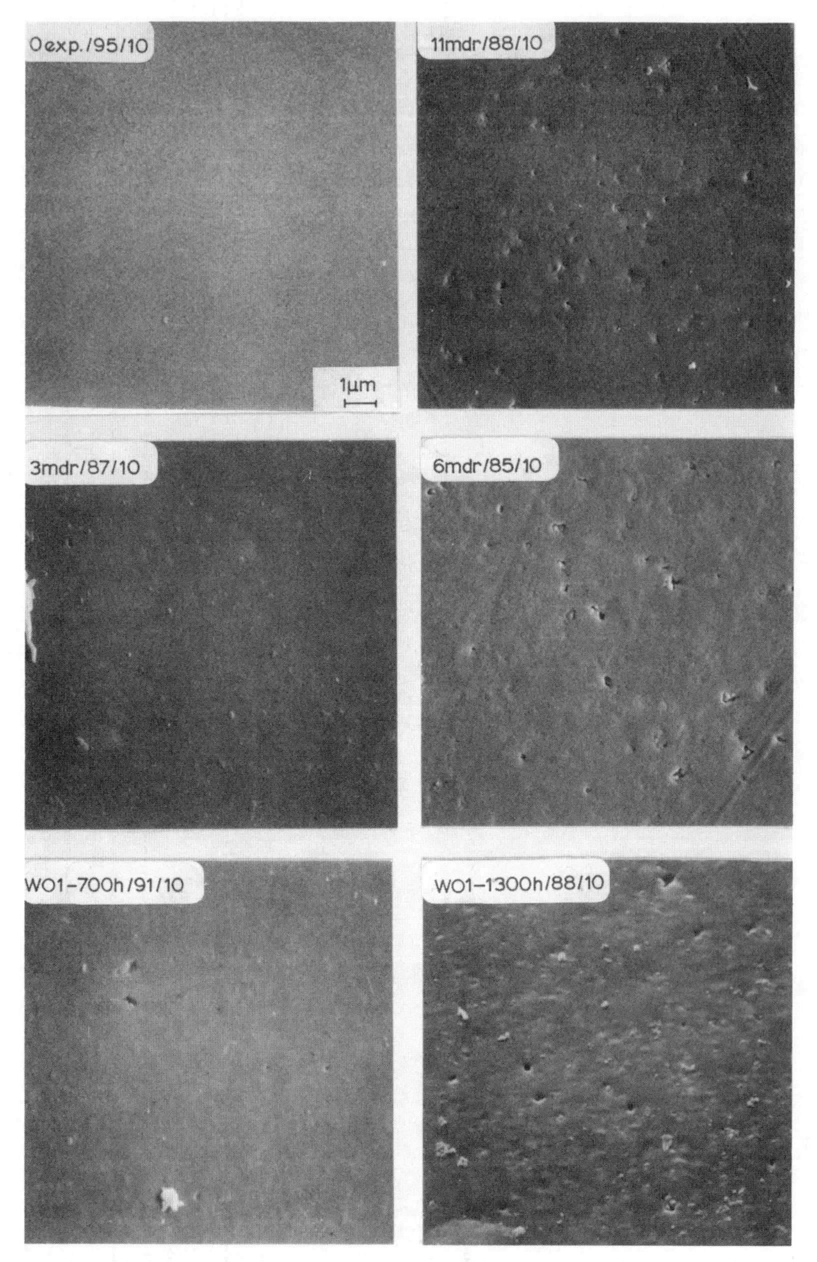

FIG. 6. SEM-photo of 2-component aliphatic polyurethane paint exposed outdoor W 1 + W 3 and in WO 1. 7000 ×.

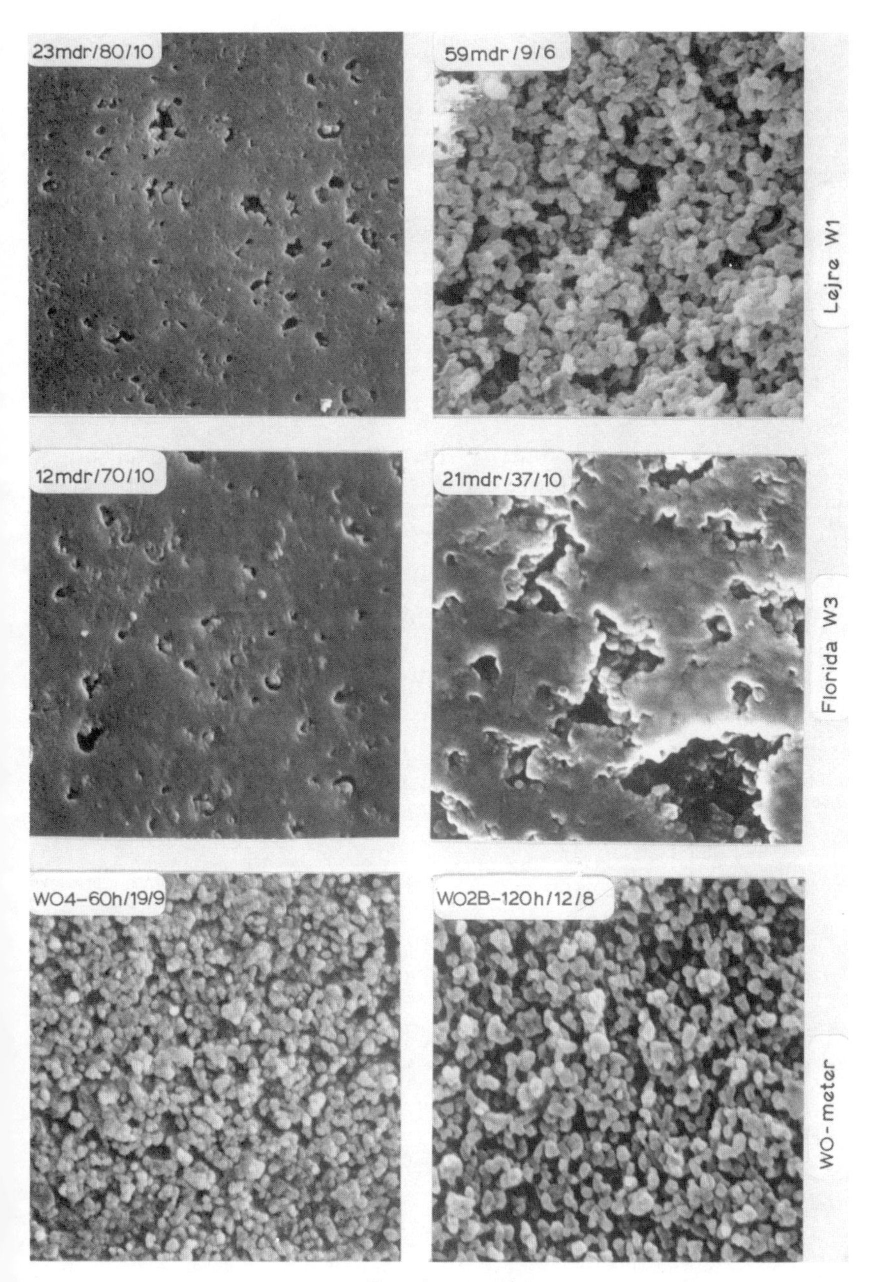
23mdr/80/10
59mdr/9/6
Lejre W1
12mdr/70/10
21mdr/37/10
Florida W3
WO4–60h/19/9
WO2B–120h/12/8
WO-meter

FIG. 6.—*contd.*

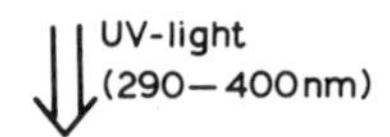

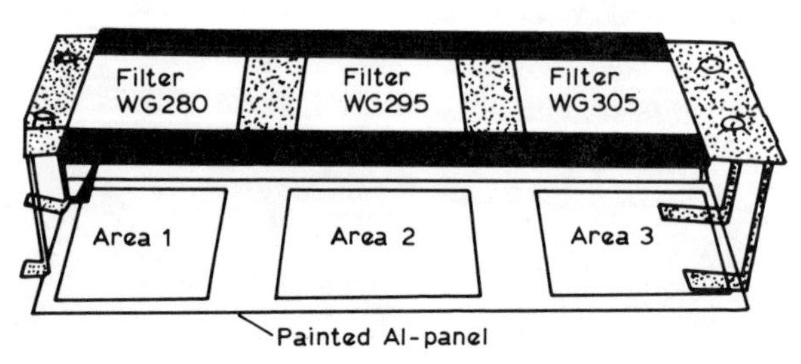

FIG. 7. Test panel with filter holder.

Zeiss PMQ spectrophotometer (Fig. 8). The curve for filter no. 4 corresponds rather well to the end of UV wavelengths in sunlight spectra.

2.7. Effect on Colour (ΔE-value)

Weathering evaluation properties such as gloss, colour change, chalking, and micromorphology (SEM) were studied for each of the six different UV-exposed areas on the painted Al panels. Only a few diagrams are presented here to illustrate some of the results. In Fig. 9 the colour change (ΔE) for a yellow iron oxide pigment (Eisenoxid gelb 3910, Bayer) in the alkyd is

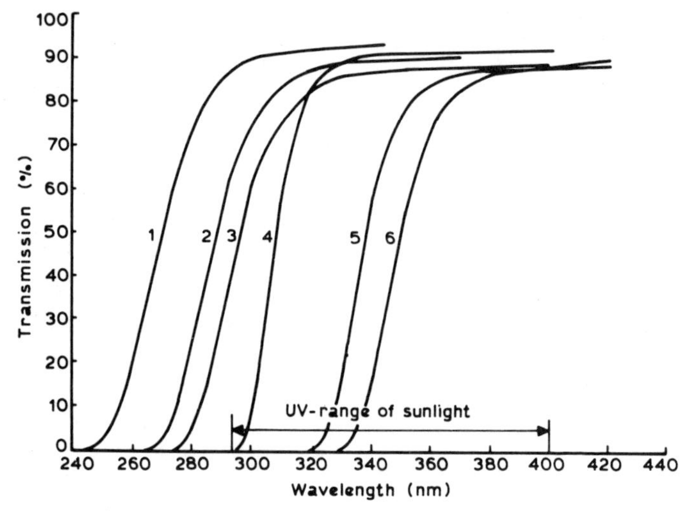

FIG. 8. Transmission curves for the six UV cut-off filters. Filter numbers: 1, WG 280; 2, WG 295; 3, WG 305; 4, WG 320; 5, WG 345; 6, WG 360.

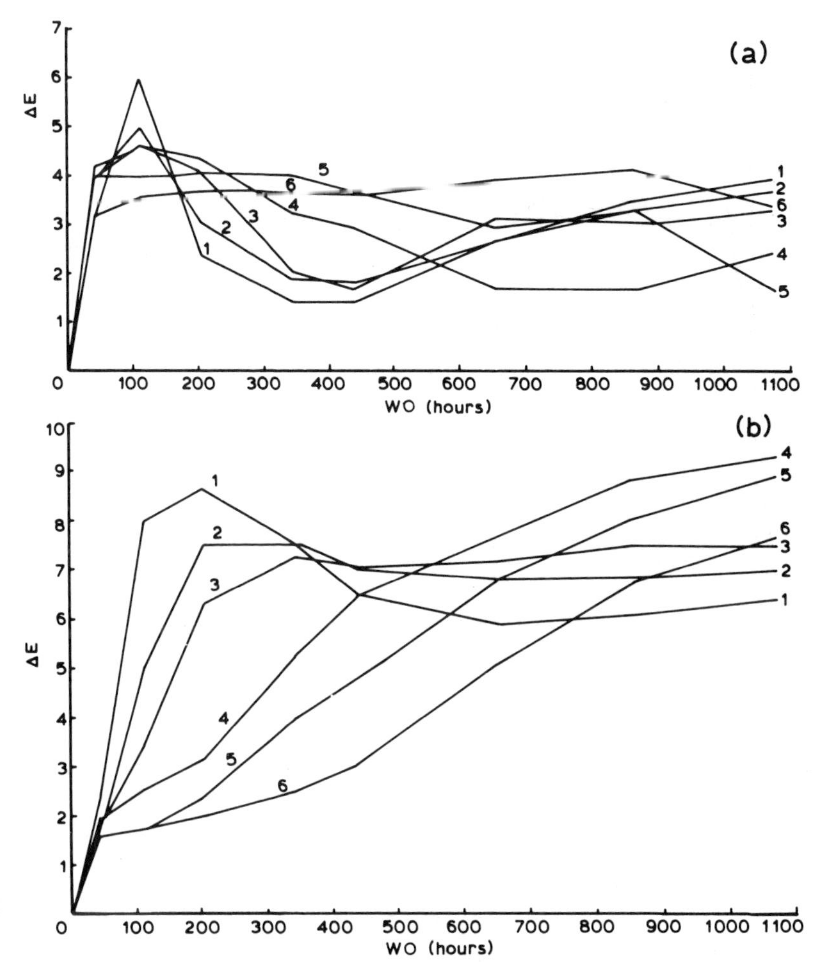

FIG. 9. Influence of UV range (1–6) on total colour change ΔE for yellow iron oxide in alkyd during WO-exposure. (a) Full-tone. (b) Half-tone, 1:1 with TiO_2.

shown in full-tone (without TiO_2) and half-tone (1:1 with TiO_2). The total colour difference (ΔE) was determined on a Hunterlab three filter spectrophotometer (25-2Δ) in accordance with ASTM D 2244-68.

From Fig. 9 it is clear that a large difference exists between pigmentation in full-tone (only yellow pigment) and in half-tone with respect to colour change during weathering. This is a rather well-known phenomenon and is explained by deterioration of the coloured pigment, while the more light stable titanium dioxide pigment is unaffected. This means that the paint

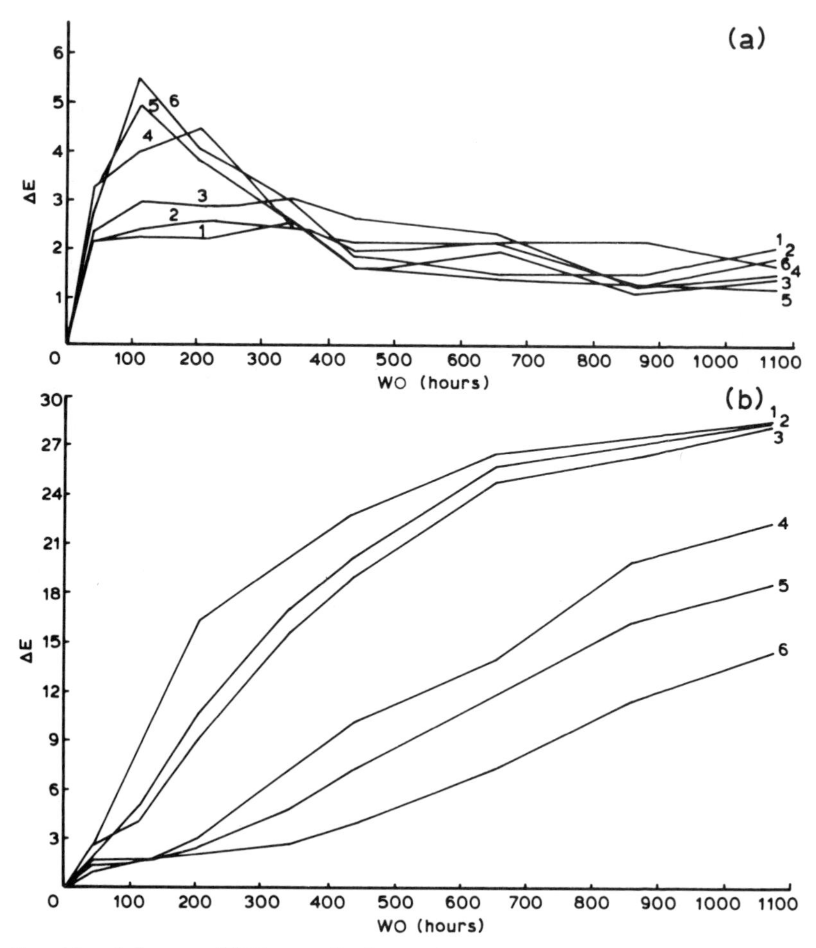

FIG. 10. Influence of UV range (1–6) on colour change ΔE for aryl yellow in alkyd during WO-exposure. (a) Full-tone. (b) Half-tone, 1:1 with TiO_2.

film gets lighter or whiter. It is also clear that the dominating colour change occurs in the beginning of the exposure period and that the paint in half-tone is especially sensitive for short, highly energetic UV irradiation.

The same testing for an organic yellow pigment (aryl ISOL 10 G, KVK) was carried out and the results are presented in Fig. 10. In this case, with an organic yellow pigment, it is even more drastic how large a difference there is between full- and half-tone. For the pigment in full-tone the colour change ΔE is almost negligible after 400 h of exposure (the human eye cannot observe colour differences of 1–2 ΔE-units). The pigment in half-tone is very sensitive to colour change and also here it is the lowest UV range, which affects the colour most.

3. CONCLUSION

There are many unwanted photochemical reactions in coatings. These are rarely studied even though they often are the most significant factor the formulator must contend with. Degradation, yellowing, colour and gloss change, etc., all are significant factors in coatings performance and all are the result of such photochemical reactions.

Some of the methods and problems of finding rapid and reliable weathering data have been described. There are no truly reliable, generally valid methods to accelerate weathering. With the many techniques available, experience, and great care, the coatings formulator can contend with unwanted photochemical reactions in a reasonably systematic manner. In spite of a lack of fundamental understanding coatings are steadily improving in quality as time goes on based on a combination of accelerated and natural exposure weathering data.

REFERENCES

1. Kleive, K. *et al.*, Wood surface degradation and consequences for the durability of coating systems. XVI FATIPEC Congress book 1982, Vol. 3, p. 121.
2. Miller, E. R. and Derbyshire, W., The photodegradation of wood during solar irradiation. *Holz als Roh- und Werkstoff*, **39**, 341–50 (1981).
3. Frost, M. and Sørensen, H. K., The importance of light for the cure of oxidatively drying coatings (in Danish). Report T 6-73, Scandinavian Paint and Printing Ink Research Institute, Copenhagen, 1973.
4. Dunderdale, J. and Colling, J. H., *Progress in Organic Coatings*, **1**, 47 (1981).
5. Blakey, R. R., XV FATIPEC Congress book 1980, part I, p. 244.
6. Kämpf, G., XVII FATIPEC Congress book 1984, Vol. 2, p. 81.
7. Lindberg, B., NIF-report T 11-82 M. Evaluation of weather durability of paint films.
8. Rosenqvist, L. and Lindberg, B., NIF-report T 15-82 M. Determination of light transmission through thin materials, Nord-test method.
9. Lindberg, B. and Rosenqvist, L., NIF-report T 17-83 M. Long term durability of painted surfaces. Effect of UV-absorbers in pigmented paint systems (in Swedish).
10. Lindberg, B. and Rosenqvist, L., NIF-report T 19-83 M. The UV-resistance of paint films (in Swedish).

13

Current Understanding of Photostabilising Mechanisms in Polyolefins

NORMAN S. ALLEN

Department of Chemistry, John Dalton Faculty of Technology, Manchester Polytechnic, UK

1. INTRODUCTION

The photostabilisation of polymers continues to be a rapidly advancing area of scientific and technological interest.[1,2] Unfortunately, the latter was often rarely concurrent with or occurred as a result of the former. Some of the most effective stabilisers were in fact developed with no firm understanding of their mode of action. However, our current deeper understanding of the complexity of both photo-oxidation and photo-stabilisation mechanisms has gradually resulted in the 'scientific' design of effective stabilisers for specific as well as more general applications. Nevertheless, the complexity of photo-oxidation processes and their dependence on thermal history still complicates the problem of stabilisation.

This paper discusses our current understanding of photostabilising mechanisms in polyolefins with reference to certain classes of stabiliser molecules, namely, the ortho-hydroxyaromatics and hindered piperidines. The interactions between light stabilisers and anti-oxidants will also be mentioned since this is a current area of industrial interest.

2. PHOTO-OXIDATION PROCESSES

The photo-oxidative degradation of polyolefins has been a long standing subject of some controversy.[3] Several photo-initiators are believed to be responsible and their relative importance still remains uncertain. Manufacturing and processing history plays a major role here. After the former, residues of unsaturation,[4,5] metal catalysts,[6] hydroperoxides,[7-11] and carbonylic groups[12,13] all appear to be important to varying extents. Taking the groups in no order of priority, unsaturation and or polymer (P—H) are known to give weakly absorbing complexes with ground-state molecular oxygen. On exposure to UV light they supposedly generate hydroperoxides by the following mechanism:

$$\mathrm{P{-}H} + \mathrm{O_2} \longrightarrow [\mathrm{P{-}H}{-}{-}{-}\mathrm{O_2}] \xrightarrow{h\nu} \mathrm{P{-}\overset{+}{H}}{-}{-}{-}\mathrm{O_2^-} \longrightarrow \mathrm{P^{\cdot}} + {}^{\cdot}\mathrm{O_2} \longrightarrow \mathrm{POOH} \quad (1)$$

Hydroperoxides undergo direct photolysis to give initially alkoxy and hydroxy radicals. The alkoxy radicals may then propagate further reactions by either hydrogen atom abstraction to give alcohols or cleavage to produce carbonyl groups:

$$\mathrm{R{-}C(CH_3)(OOH){-}CH_2{\sim}} \xrightarrow{h\nu} \mathrm{R{-}C(CH_3)(O^{\cdot}){-}CH_2{\sim}} + {}^{\cdot}\mathrm{OH} \quad (2)$$

$$\mathrm{R{-}C(CH_3)(O^{\cdot}){-}CH_2{\sim}} + \mathrm{R'H} \longrightarrow \mathrm{R{-}C(CH_3)(OH){-}CH_2{\sim}} + \mathrm{R'^{\cdot}} \quad (3)$$

$$\mathrm{R{-}C(CH_3)(O^{\cdot}){-}CH_2{\sim}} \longrightarrow \begin{cases} \mathrm{RC({=}O){-}CH_2{\sim}} + {}^{\cdot}\mathrm{CH_3} \\ \mathrm{R{-}C(CH_3){=}O} + {}^{\cdot}\mathrm{CH_2{\sim}} \end{cases} \quad (4)$$

Differences in chemical structure of the hydroperoxides in polyolefins results in different kinetic chain lengths. In polyethylene, for example, they are much shorter than in polypropylene.

Metallic catalyst residues in polyolefins, particularly the newer gas-phase processed types, also play a major role and good correlations between Ti and photo-oxidative stability have been found.[6] Here transition metal ions are known to catalyse the decomposition of hydroperoxides to alkoxy or peroxy radicals by the following processes:

$$PO_2H + M^{2+} \longrightarrow PO^{\cdot} + OH^- + M^{3+} \qquad (5)$$

$$PO_2H + M^{3+} \longrightarrow PO_2^{\cdot} + H^+ + M^{2+} \qquad (6)$$

In this way they can accelerate the photochemical oxidation of the polymer.

Carbonyl groups, which are the strongest light absorbing impurity, undergo the well-established Norrish Type I and II processes resulting in chain cleavage:

Norrish Type I

$$\sim CH_2{-}CH_2{-}\underset{\underset{O}{\|}}{C}{-}CH_2\sim \xrightarrow{h\nu} \sim CH_2CH_2^{\cdot} + {}^{\cdot}\underset{\underset{O}{\|}}{C}{-}CH_2\sim \qquad (7)$$

Norrish Type II

$$\sim CH_2{-}CH_2{-}\underset{\underset{O}{\|}}{C}{-}CH_2{-}CH_2{-}CH_2\sim \xrightarrow{h\nu}$$

$$\sim CH_2CH_2C\begin{matrix} /\!\!/ O{-}{-}{-}H \backslash \\ \backslash CH_2{-}CH_2 / \end{matrix}CH\sim \qquad (8)$$

$$\downarrow$$

$$CH_2{=}CH{-}CH_2\sim + \sim CH_2CH_2C\begin{matrix} /\!\!/ O \\ \backslash CH_3 \end{matrix}$$

The latter occurs via a six-membered cyclic intermediate involving intramolecular hydrogen-atom abstraction.

On processing, the further development of hydroperoxide and carbonyl groups results in both species playing a major role, the former being by far the most important initiator in terms of free radical production and 'chain scission'.[2,8,9] However, it is known that these processes are significantly reduced in the absence of carbonyl groups.[14,15] It is still not clear what the reason is for this phenomenon but some type of intermediate complex is proposed as one possibility wherein light absorption and energy transfer from the carbonyl group to the hydroperoxide may occur.[17] Other possibly less important species are trace metallic residues such as iron incorporated during processing and fabrication and polynuclear aromatics from the atmosphere.[16]

One other process which has attracted interest over the years is the involvement of singlet oxygen generated by photo-excited carbonyl groups transferring their triplet energy to ground-state molecular 'triplet' oxygen. In polyolefins the efficiency of this process is considered to be negligible,[3,8] whereas in rubbers such as polybutadiene[19] it may be more important.

2.1. Mechanisms of Photostabilisation

Over the years, four different classes of stabilising systems have been developed and these rely for their stabilising action on the presence of (a) an ultra-violet screener, (b) an ultra-violet absorber, (c) an excited state quencher and (d) a free radical scavenger and/or hydroperoxide decomposer. Of these, (d) is now clearly the most effective system but even within this category the mechanisms involved may be quite complex.

2.2. Ortho-hydroxyaromatic Compounds

Ortho-hydroxyaromatic compounds based on benzophenone and phenylbenzotriazole still find widespread application for the photoprotection of many polymer systems.[19,20] Early theories on the mode of operation of these stabilisers showed that they absorb the incident harmful sunlight, thereby preventing it from being absorbed by the photoactive impurities or structural units in the polymer.[19,20] These compounds all have high absorbance in the wavelength range which is most harmful to commercial polyolefins (300–350 nm) and are capable of harmlessly dissipating the absorbed energy by one or more non-radiative processes, such as internal conversion. The 2-hydroxybenzophenones dissipate their absorbed energy by a mechanism that involves the reversible formation of a six-membered hydrogen bonded ring.[1,20] The following two tautomeric forms in equilibrium provide a facile pathway for deactivation of the excited state induced by the absorption of light.

'Keto'-form $\underset{\text{OH}}{\overset{h\nu}{\rightleftharpoons}}$ 'Enol'-form (9)

The result of this mechanism of light absorption and energy dissipation thus leaves the stabiliser chemically unchanged and still able to undergo a large number of these activation–deactivation cycles, provided of course no other processes interfere. Until the advent of modern photochemical techniques direct evidence for such a mechanism was difficult to obtain because the role of the stabiliser is essentially that of a 'passive' nature. However, kinetic studies on the excited states of these stabilisers using nano- and picosecond laser flash photolysis studies have confirmed the above mechanism.[20,21] On the basis of this mechanism one might conclude that the polymer should last indefinitely, whereas in fact this is clearly not the case. Whilst such a mechanism may be important for protecting the stabiliser molecule, it does not contribute in any major way to the photoprotection of the polymer. Both theoretical and practical arguments have been put forward to discount this theory. In thin films, for example, the 2-hydroxybenzophenones offer much higher protection than would be expected on the basis of a 'pure' filter effect. Also, since photo-oxidation is predominantly confined to the surface then UV absorption alone is unlikely to be the major mechanism of stabilisation. The ortho-hydroxyphenylbenzotriazoles operate in a similar fashion.

These stabilisers are also known to quench the excited states of photoactive chromophores in a number of polymer systems and whilst this mechanism may have seemed an appropriate explanation no correlations between excited-state quenching and stabilising efficiency have ever been observed and, in many cases, the simple observation of a reduction in emission lifetime or intensity of a photoactive chromophore is not sufficient evidence to justify the implication of energy transfer in photostabilisation.

It is now well known that the photostabilising action of light stabilisers depends very much on prior processing history.[1,2] The ortho-hydroxy-aromatics are no exception and the results in Table 1 clearly demonstrate this effect for a number of ortho-hydroxybenzophenones (Cyasorb UV 531).[22] Here it is seen that the stabilisers photodecompose more rapidly in processed polypropylene compared with unprocessed polymer. It is also noted that the rates of photodecomposition correlate exactly with initial hydroperoxide concentrations in the polymer. These results are associated with the following mechanism whereby alkoxy and hydroxy radicals produced in the photolysis of hydroperoxides abstract the ortho-hydrogen atom on the hydroxyl group. The radical product (I) is no longer capable of intramolecular hydrogen bonding:

$$\begin{matrix}\mathrm{PO^{\cdot}}\\ + \\ \mathrm{{}^{\cdot}OH}\end{matrix}\ \mathrm{C_6H_5{-}C({=}O){-}C_6H_3(OH){-}OR} \longrightarrow \begin{matrix}\mathrm{POH}\\ + \\ \mathrm{H_2O}\end{matrix}\ \underset{\text{(I)}}{\mathrm{C_6H_5{-}C({=}O){-}C_6H_4{-}O^{\cdot}}} \qquad (10)$$

TABLE 1
Initial Hydroperoxide Concentration in Polypropylene Films and Rate of Change of Absorbance of 2-Hydroxybenzophenones in Polypropylene Films Irradiated with 365 nm Light

Additive	*[POOH](μg/g)*		*ΔABS/SEC*	
	0 min[a]	*10 min*[a]	*0 min*[a]	*10 min*[a]
Control	21	80	—	—
4-*n*-octoxy	21	70	$8{\cdot}3 \times 10^{-8}$	$2{\cdot}2 \times 10^{-7}$
4-*n*-octoxy + Weston 618	14	80	$1{\cdot}4 \times 10^{-8}$	$1{\cdot}7 \times 10^{-7}$
4-methylthio	27	64	$5{\cdot}2 \times 10^{-6}$	$9{\cdot}6 \times 10^{-6}$
4-hydroxy	50	35	$8{\cdot}6 \times 10^{-6}$	$4{\cdot}4 \times 10^{-6}$
4-methoxy	5	20	$6{\cdot}9 \times 10^{-6}$	$6{\cdot}3 \times 10^{-6}$
4-dodecyloxy	12	75	$5{\cdot}6 \times 10^{-8}$	$9{\cdot}7 \times 10^{-8}$

[a] Processing time.

The protective effect of Weston 618 (distearyl pentaerythritol diphosphite) (Table 1), a hydroperoxide decomposer, also supports this mechanism. A further interesting feature of the data in Table 1 is that the light source used was monochromatic and emitted mainly 365 nm light. Under these conditions light absorption by the stabilisers was very low, but the hydroperoxides would still be susceptible to photolysis. The synergistic behaviour of these light stabilisers with metal chelates such as nickel dialkyldithiocarbamates which are also hydroperoxide decomposers, confirms the above mechanism.[1,2,30]

One other feature of the data in Table 1 is that the stabilisers with long n-alkoxy groups in the 4-position in the ring are more photostable than those stabilisers with shorter groups. This is reflected by the actual light stability data shown in Table 2. Here embrittlement times are compared for all the polymer films using polychromatic light (Microscal λ's > 300 nm) and monochromatic 365 nm light. The interesting feature here however, is that if the embrittlement data are simply ratioed it is noted that the stabilisers with large 4-n-alkoxy groups (octoxy/dodecyloxy) give higher ratios, i.e. significantly greater stabilising effects under monochromatic light. Theoretically, this behaviour cannot be accounted for on the basis of any differences in inductive effects, by the 4-substituents, in enhancing the strength of the intramolecular hydrogen bond, since they should be similar. Differences in migration rates were also ruled out by controlled oven ageing experiments and also the fact that under monochromatic light the films were at room temperature and the effects were predominantly due to

TABLE 2
Time to Embrittlement (0·06 Carbonyl Units) for Polypropylene Films Stabilised with 2-Hydroxy Benzophenones During Exposure in Microscal Unit and to 365 nm Light Source at Processing Times of 0 and 10 min

Additive	*Microscal*		*365 nm light*		*Ratio*	
	0 min	*10 min*	*0 min*	*10 min*	*0 min*	*10 min*
Control	100	70	150	100	1·5	1·3
4-*n*-octoxy	1 872	740	4 750	2 400	2·5	3·2
4-*n*-octoxy + Weston 618	1 584	600	4 300	2 500	2·7	4·2
4-methylthio	280	170	260	200	0·9	1·2
4-hydroxy	360	450	300	450	0·8	1·0
4-methoxy	350	240	450	240	1·3	1·0
4-dodecyloxy	1 600	630	4 400	2 200	2·8	3·5

photosensitised oxidation. The stability of the 4-alkoxy groups may be an important factor here. During irradiation these groups may photolyse to give corresponding alkyl radicals thus:

$$\begin{array}{l} \text{—}\langle\bigcirc\rangle\text{—}OCH_3 \\ \text{—}\langle\bigcirc\rangle\text{—}OC_8H_{17} \end{array} \xrightarrow{h\nu} \left[\begin{array}{l} \text{—}\langle\bigcirc\rangle\text{—}O^{\cdot} + CH_3^{\cdot} \longrightarrow \\ \text{—}\langle\bigcirc\rangle\text{—}O^{\cdot} \longleftrightarrow {}^{\cdot}C_8H_{17} \end{array} \right]_{\text{CAGE}} \qquad (11)$$

The smaller methyl radicals may move more easily out of the polymer cage whereas the larger octyl radical may not. In the latter case radical recombination could therefore increase the light stabilities of the 4-methoxy and 4-n-octoxy derivatives of 2-hydroxybenzophenone in solution and these are compared in Fig. 1. Here it is seen that in the absence of added hydroperoxide there is some cage effect even in solution and oxygen is having a strong effect in quenching the photo-excited triplet state of the stabilisers. In the presence of added hydroperoxide however, both stabilisers have the same reduced stability confirming mechanism (11) above.

The results of some screening experiments are shown in Table 3 for both polypropylene and low density polyethylene containing the 4-methoxy and

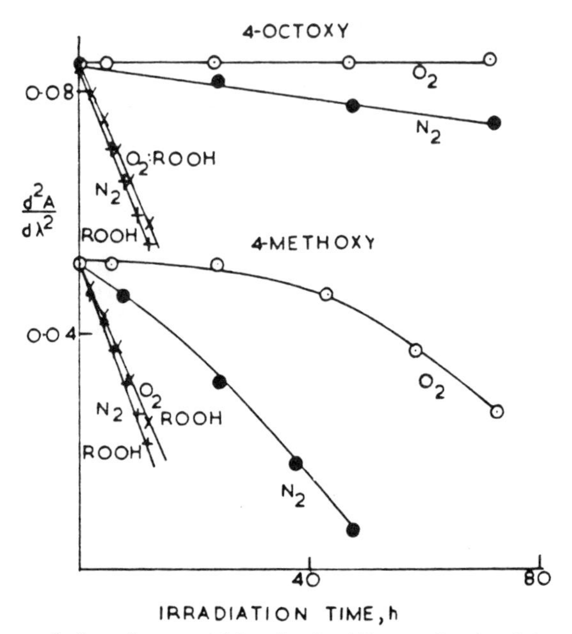

FIG. 1. Rates of photodecomposition in the Microscal unit of 4-n-octoxy and 4-n-methoxy substituted 2-hydroxybenzophenones (10^{-4} M) in ○: aerobic and ●: anaerobic propan-2-ol and ×: aerobic and +: anaerobic propan-2-ol containing 10^{-3} M cumene hydroperoxide.

n-octoxy stabilisers and these demonstrate a number of interesting features.[23,24] Here stabilised polymer films were exposed directly and their embrittlement times recorded as well as placing the stabilised films in front of control unstabilised films and measuring the embrittlement times of the latter only. In the case of polypropylene it is seen clearly that screening is not an efficient mechanism whereas in polyethylene it does appear to be important. Also, at low concentrations (0·01 % w/w) both stabilisers are weak prodegradants and this would be consistent with mechanism (10) above being dominated by hydroperoxides. In polyethylene the poor solubility of both stabilisers results in their migration to the surface where they may concentrate and operate as both screens and radical scavengers. In fact, in polyethylene higher concentrations of the stabilisers photodecomposed more rapidly than lower concentrations.

2.3. Hindered Piperidines

These are the newest class of light stabilisers and, in fact, are now widely established in many applications. Compared with other classes of light stabiliser they are highly efficient in polyolefins.[2,25] A range of different commercially acceptable systems for polyolefins are now available and are listed below:

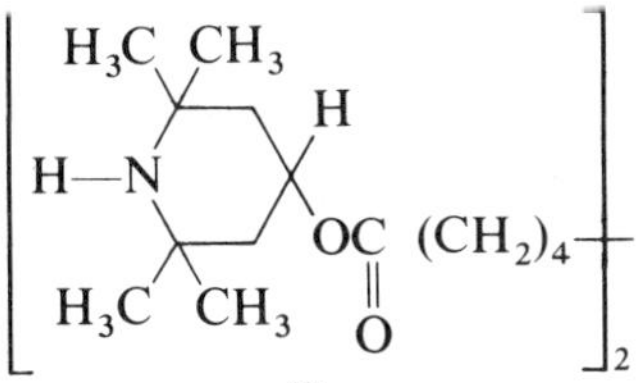

(I)

Tinuvin 770 (Ciba-Geigy Corp.)
Bis[2,2,6,6-tetramethyl-4-piperidinyl]sebacate

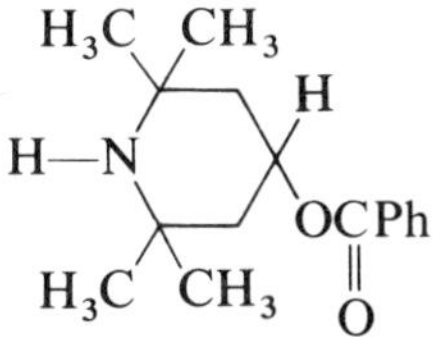

(II)

Sanol 774 (Sankyo)
4-benzoyloxy-2,2,6,6-tetramethylpiperidine

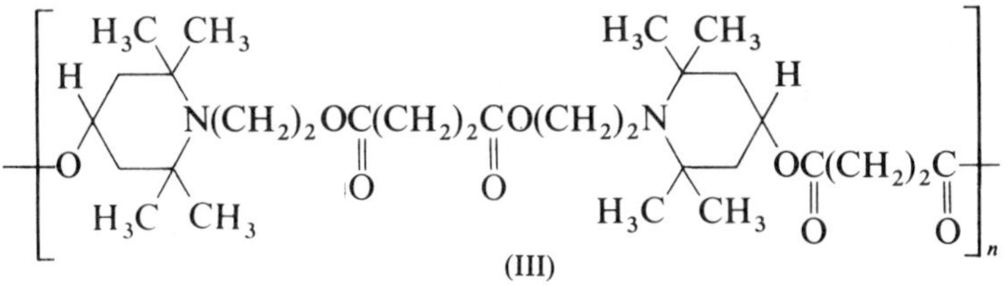

(III)

Tinuvin 622 (Ciba-Geigy Corp.) Polyester of succinic acid with N-β-hydroxyethyl-2,2,6,6-tetramethyl-4-hydroxypiperidine

(IV)

R = —$NH(CH_3)_2CCH_2C(CH_3)_3$

Chimassorb 944 (Ciba-Geigy Corp.) Poly[2-N,N'-di(2,2,6,6-tetramethyl-4-piperidinyl) hexanediamine]-4-[1-amino-1,1,3,3-tetramethylbutane]-symtriazine

R = —N O V [N-morpholino]

Cyasorb UV 3346 (American Cyanamid Comp.)

TABLE 3
UV Embrittlement Times for Polypropylene and Polyethylene Films Containing Different Concentrations of 2-Hydroxy-benzophenone Stabilisers ~300 μm thick (Microscal Exposure)

Additive %*w/w*	*Polypropylene*	*Polyethylene*
	No Screening	
Control	70	130
0·01 4-methoxy	60	105
0·10 4-methoxy	198	120
1·00 4-methoxy	294	165
0·01 4-n-octoxy	50	130
0·10 4-n-octoxy	1 140	210
1·00 4-n-octoxy	>5 000	890
	Screening	
Control	61	119
0·01 4-methoxy	—	172
0·10 4-methoxy	—	172
1·00 4-methoxy	—	170
0·01 4-n-octoxy	63	135
0·10 4-n-octoxy	58	190
1·00 4-n-octoxy	185	1 230

These compounds exhibit no absorption in the UV region >300 nm and are ineffective 'excited-state' quenchers,[25,26] although there is some evidence to suggest that in rubbers they quench singlet oxygen.[27,28] The stabilising efficiency of these compounds is associated with the following cyclic mechanism:[1,2,25,26]

$$>N{-}H \xrightarrow[\Delta H,\ [O]]{h\nu} >N{-}O^{\cdot} \xrightarrow{-C(CH_3){-}CH_2{\sim}/P^{\cdot}} >N{-}O{-}P \ \text{ or } \ >N{-}OH + {-}C(CH_3){=}CH_2{\sim}$$

$$>N{-}O{-}P \xrightarrow{PO_2^{\cdot}} >N{-}O^{\cdot} + PO_2P/PO_2H \longrightarrow >N{-}O^{\cdot} \qquad (12)$$

Here the amine is initially oxidised through to the nitroxyl radical by reaction with hydroperoxides[1,2] and/or the possible formation of an amine–oxygen complex:[29]

$$\begin{array}{ccc} >\text{N—H} + \text{POOH} & \xrightarrow{\Delta H/h\nu} & >\text{N}^{\cdot} + \text{PO}^{\cdot} + \text{H}_2\text{O} \\ & & \big\downarrow {\scriptstyle \text{P—H/O}_2} \\ >\text{N—O}^{\cdot} + {}^{\cdot}\text{OH} & \longleftarrow & >\text{N—OOH} + \text{P}^{\cdot} \end{array} \qquad (13)$$

There is much evidence in the literature in favour of this mechanism occurring in the polymer at elevated temperatures[30–32] despite room temperature model system studies[33] to the contrary. During irradiation the concentration of nitroxyl radicals grows rapidly but thereafter falls to a very low steady-state value of 10^{-4} M.[1,2,25,26] This low level of nitroxyl radicals is suggested to be insufficient to account for the high photoprotective efficiency of hindered piperidines. Much of the nitroxyl is converted into either the hydroxylamine (>N—OH) or substituted hydroxylamine (>N—OP) produced by reaction of the nitroxyl radicals with different types of macroalkyl radicals.[1,2,25,26] These reactions are now well established[35–37] and, in fact, may occur thermally or photochemically. The hydroxylamine or substituted hydroxylamines are now clearly established as the most effective stabilising intermediates and act as reservoirs for nitroxyl radicals by reacting with peroxy radicals. This latter reaction is somewhat controversial since for each nitroxyl radical generated a molecule of hydroperoxide is produced for the hydroxylamine. Recent work by Carlsson and Wiles using Fourier-transform infra-red indicates that the substituted hydroxylamine is the most important intermediate produced on irradiation.[38] No hydroxylamine (>N—OH) could be found in sufficient concentration. However, the latter appears to be more important thermally and this is indicated by the results shown in Table 4.[37] Here it is seen that prior oven ageing of polypropylene containing the hindered piperidine compound (I) and its bis-nitroxyl derivative results in a marked improvement in light stabilising activity. It is interesting to note that the bis-nitroxyl derivative is a much better light stabiliser compared with a mono-nitroxyl compound and is also much more efficient than the starting amine. This might at first suggest that the nitroxyl radical generated from the amine during thermal oxidation is responsible for the improvement in light stability. However, little bis-nitroxyl has been found in the polymer; it is mainly the mononitroxyl radical[39] and the stability of polymer containing the bis-nitroxyl radical

TABLE 4

UV Embrittlement Times (0·6 Carboxyl Index) and Unsaturated Index at 1 640 cm^{-1} for Stabilised Polypropylene Films Oven Aged at 140 °C in Air (0·1 % w/w)

Heating time (*h*)	*Tinuvin 770*		*Mono-nitroxyl*		*Bis-nitroxyl*	
	Embrittlement (*h*)	[>C=C<]	*Embrittlement* (*h*)	[>C=C<]	*Embrittlement* (*h*)	[>C=C<]
0	1 375	0·032	2 020	0·037	3 050	0·032
1	2 750	0·044	360	0·040	3 875	0·043
4	1 875	0·044	900	0·042	3 925	0·046
8	2 000	0·044	2 200	0·045	3 375	0·046

TABLE 5
Influence of Prior Thermal Oxidation at 130 °C on Photostability of Polymeric Hindered Piperidines in Polypropylene Film (300 μm)

Heating times (*h*)	*UV Embrittlement times* (*h*)	
	Chimassorb 944	*Cyasorb UV 3346*
0	920	1 300
10	1 110	2 270
25	1 200	1 920
50	1 160	1 860

increases anyway after ageing. The concurrent increase in vinyl absorption at 1650 cm^{-1} (Table 4) seems to suggest that the hydroxylamine product is important here. A similar effect to that in Table 4 is seen in Table 5 for the two polymeric hindered piperidines (IV) and (V).[40] These two compounds differ only by the nature of the 4-substituent on the triazine ring. Stabiliser (V) is seen to be a much more effective light stabiliser in polypropylene before oven ageing but, in fact, after just 10 h at 130 °C the efficiency of the system is nearly doubled. For stabiliser (IV) maximum stability is evident only after 25 h of heating but the effect is only marginal. Both effects are however consistent with maximum nitroxyl radical formation shown by the ESR data in Fig. 2.[40] However, this could also imply a high concentration

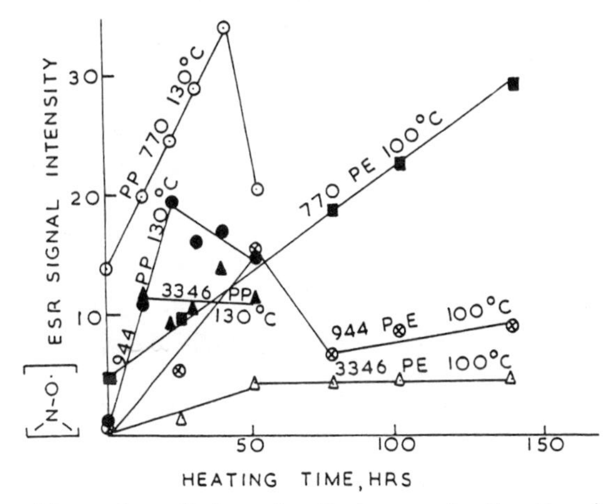

FIG. 2. Rate of formation of nitroxyl radical versus heating time in hours for polypropylene at 130 °C and polyethylene at 100 °C containing ○, ■ Tinuvin 770, ○, ⊗ Chimassorb 944 and △, ▲ Cyasorb UV 3346 respectively at 0·1 %w/w concentration.

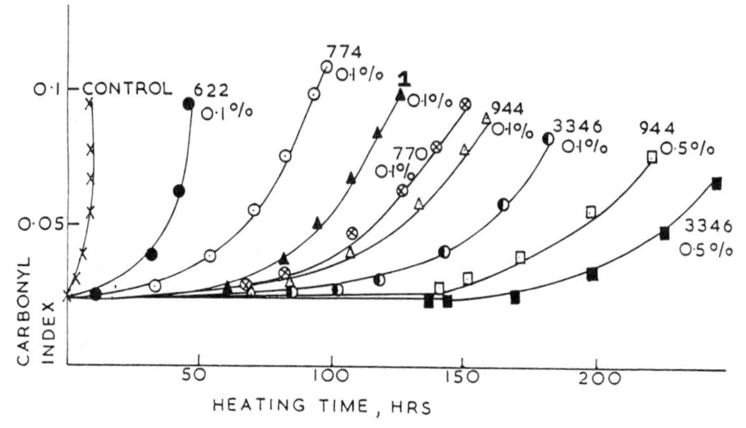

FIG. 3. Rates of oxidation of polypropylene films during oven ageing at 130 °C containing ×: no additive, ○: Tinuvin 774, ●: Tinuvin 622, ⊗: Tinuvin 770, △: Chimassorb 944, ▲: Bis 1,8 2,2,6,6-tetramethylpiperidinyl 4-carboxy octanedioate, ◐: Cyasorb UV 3346 at 0·1 % w/w concentration and □: Chimassorb 944, ■: Cyasorb UV 3346 at 0·5 % w/w concentration.

of hydroxylamine. The enhanced stabilising effect of (V) may well be due to the better radical oxygen scavenging ability of the morpholino group compared with the tetramethylbutylamine.

The above cyclic mechanism has been carefully scrutinised by many workers[1,2,25,26] in the last few years and, in fact, it has been concluded that it cannot alone fully account for the high photoprotective efficiency of the parent amine molecule. The nitroxyl radical itself is a radical scavenger but is not as effective as hindered phenols in competing with oxygen for the radicals. The hindered amine itself is not very effective although the results in Fig. 3 seem to suggest that this obviously depends on the structure of the light stabiliser.[40] These oven ageing results show that in polypropylene film at 130 °C the stabiliser (V) Cyasorb UV 3346 is the most efficient system.

To account for this deficiency in the cyclic mechanism it has been suggested and indeed confirmed by many workers that hindered piperidine stabilisers and their derived nitroxyls form weakly bonded localised complexes with hydroperoxides in the polymer, e.g.[1,2,25,26,35]

$$\rangle N—O^{\cdot} --- P—H + POOH \rightleftharpoons \rangle N—O^{\cdot} --- POOH + P—H \quad (14)$$

This mechanism raises the local concentration of nitroxyl radicals in regions where alkyl radicals are generated after the photocleavage of hydroperoxide groups. Under these conditions the nitroxyl radicals would then effectively compete with oxygen for the alkyl radicals and stabilise the polymer. Evidence of this mechanism is based on infra-red studies[1,2,25,26] and the observation of an increase in stabiliser adsorption from solution by oxidised polypropylene films containing higher concentrations of hydroperoxide groups.[35] One recently observed effects of this mechanism is to reduce the quantum yield of hydroperoxide photolysis.[41]

Other mechanisms include the reaction of acyloxy radicals with the hindered amine by:

$$2\ \rangle N\text{—}H + 2RC(=O)O^{\cdot} \longrightarrow \rangle N\text{—}C(=O)OR + \rangle \overset{+}{N}H_2RCOO^- \quad (15)$$

and acylperoxy radicals by the following scheme:[42,43]

$$+OOC(=O)C(=O)OO+ \xrightarrow{\Delta H} 2CO_2 + 2\dot{O}+ \quad (16)$$

$$+O^{\cdot} + \rangle\!\!-\!C(=O)H \xrightarrow{O_2} \rangle\!\!-\!C(=O)O_2^{\cdot} \xrightarrow[O_2]{\rangle N\text{—}H} \rangle N\text{—}O^{\cdot} \quad (17)$$

Reaction of the parent amine with transition metal ions is also a possibility as well as the ability of the nitroxyl radical to oxidise Fe^{2+} ions over to Fe^{3+} ions. The latter are apparently inert and will not then catalyse the breakdown of hydroperoxides.[26]

2.4. Interactions of Hindered Piperidine Light Stabilisers

Despite the high efficiency of these stabilisers they often interact unfavourably with many other additives used in commercial polymers for various purposes such as anti-oxidants, and fire-retardants.[25,26] Hindered phenolic anti-oxidants antagonise the photostabilising effect of hindered piperidine compounds although their effect depends very much on the stabiliser/anti-oxidant combination and processing history. The antagonism may be associated with the three following processes:

(1) Oxidation of the phenol to an active quinone by:[31,32]

$$\rangle N\text{—}O^{\cdot} + HO\text{—}C_6H_2(t\text{-}Bu)_2\text{—}R \rightleftharpoons \rangle N\text{—}OH + O\!=\!C_6H_2(t\text{-}Bu)_2\dot{}\,R \quad (18)$$

(2) Inhibition of hydroperoxide formation by the anti-oxidant thus preventing mechanism (13) above.[31,32]

(3) Reaction of the nitroxyl radicals with radical intermediates from the phenol such as (II) in reaction (18) above:[44]

$$>N—O^{\cdot} + O{=}C_6H_2(tBu)_2(R)^{\cdot} \longrightarrow O{=}C_6H_2(tBu)_2(R)(O—N<) \quad (19)$$

This would effectively remove nitroxyl radicals from the cyclic mechanism (12) above.

With thioesters the antagonism is highly detrimental and is associated with reaction of nitroxyl radicals with sulphenyl radicals to give inactive sulphonamides:[26]

$$>N—H,\ >S \xrightarrow{O_2} >N—O^{\cdot},\ >SO \longrightarrow >N—SO_2^- \quad (20)$$

With ortho-hydroxyaromatic compounds the effects are not synergistic but nevertheless greater than that of the original hindered piperidine alone.[45] The reason for this is uncertain but may, in fact, be due to the formation of nitroxyl and stabiliser radical adducts as shown in (19) above. With pigments the effects quoted above are often enhanced, i.e. worsened, and this is associated with the ability of the pigment to absorb the additives into its surface thereby increasing the effect of any potential interaction.[46,47]

REFERENCES

1. Allen, N. S. (Ed.), *Degradation and Stabilisation of Polyolefins*, Elsevier Applied Science Publishers Ltd, London, 1983.
2. Scott, G. (Ed.), *Developments in Polymer Stabilisation*, Vols 1–6, Elsevier Applied Science Publishers Ltd, London, 1979–83.
3. Vink, P., in ref. 1, Chapter 5, p. 213.
4. Allen, N. S., Fatinikum, K. O., Gardette, J. L. and Lemaire, J., *Polym. Deg. Stab.*, **4**, 95 (1982).
5. Allen, N. S., Fatinikun, K. O. and Henman, T. J., *Polym. Degrad. Stab.*, **4**, 59 (1982).
6. Allen, N. S., Fatinikun, K. O. and Henman, T. J., *Eur. Polym. J.*, **19**, 551 (1983).
7. Carlsson, D. J., Chan, K. H., Garton, A. and Wiles, D. M., *Pure appl. Chem.*, **52**, 289 (1980).
8. Al-Malaika, S. and Scott, G., in ref. 1, Chapter 7, p. 283.

9. Al-Malaika, S. and Scott, G., *Eur. Polym. J.*, **16**, 709 (1980).
10. Allen, N. S. and Fatinikun, K. O., *Polym. Degrad. Stab.*, **3**, 327 (1980).
11. Garton, A., Carlsson, D. J. and Wiles, D. M., *Makromol. Chemie*, **181**, 1841 (1980).
12. Guillet, J. E., *Pure appl. Chem.*, **52**, 285 (1980).
13. Ginhac, J. M., Gardette, J. L., Arnaud, R. and Lemaire, J., *Makromol. Chemie*, **182**, 1017 (1981).
14. Ng, H. C. and Guillet, J. E., *Photochem. Photobiol.*, **28**, 571 (1978).
15. Knittel, T. W. and Kilp, T., *J. Polym. Sci., Polym. Chem. Ed.*, **21**, 3209 (1983).
16. Allen, N. S., *Polym. Degrad. Stab.*, in press.
17. Geuskens, G. and Kabama, M. S., *Polym. Degrad. Stab.*, **5**, 399 (1983).
18. Rabek, J. F. and Rånby, B., *Polym. Degrad. Stab.*, **5**, 65 (1983).
19. Hardy, W. B. In: *Developments in Polymer Photochemistry—3* (Ed. N. S. Allen), Elsevier Applied Science Publishers Ltd, London, Chapt. 8, p. 287, 1982.
20. Allen, N. S., *Polym. Photochem.*, **3**, 167 (1983) (and references cited therein).
21. Gupta, A., Scott, G. W. and Kilger, D., *Photodegradation and Stabilisation of Coatings*, (Ed. S. P. Pappas and F. H. Winslow), American Chemical Society Symposium Series No. 151, Chapt. 3, p. 27, 1981.
22. Allen, N. S., Gardette, J. L. and Lemaire, J., *Polym. Photochem.*, **3**, 251 (1983).
23. Allen, N. S., Mudher, M. and Green, P., *Polym. Degrad. Stab.*, **7**, 83 (1984).
24. Allen, N. S., Mudher, M. and Green, P., *Polym. Degrad. Stab.*, in press.
25. Allen, N. S. In: *Developments in Polymer Photochemistry—2* (Ed. N. S. Allen), Elsevier Applied Science Publishers Ltd, London, Chapt. 7, p. 239, 1982.
26. Sedlar, J., Marchal, J. and Petruj, J., *Polym. Photochem.*, **2**, 175 (1982) (and references cited therein).
27. Yang, Y. Y., Lucki, J., Rabek, J. F. and Rånby, B., *Polym. Photochem.*, **3**, 47 (1983).
28. Yang, Y. Y., Lucki, J., Rabek, J. F. and Rånby, B., *Polym. Photochem.*, **3**, 97 (1983).
29. Lucki, J., Rabek, J. F., Rånby, B. and Dai, G. S., *Polym. Photochem.*, **5**, 385 (1984).
30. Chakraborty, K. B. and Scott, G., *Polymer*, **21**, 252 (1980).
31. Allen, N. S., *Polym. Photochem.*, **1**, 243 (1981).
32. Allen, N. S., *Makromol. Chemie*, **181**, 2413 (1980).
33. Sedlar, J., Petruj, J., Pac, J. and Zahradnickova, A., *Eur. Polym. J.*, **16**, 659 (1980).
34. Son, P. N., *Polym. Degrad. Stab.*, **2**, 295 (1980).
35. Carlsson, D. J., Chan, K. H., Durmis, J. and Wiles, D. M., *J. Polym. Sci., Polym. Chem. Ed.*, **20**, 575 (1982).
36. Bagheri, R., Chakraborty, K. B. and Scott, G., *Polym. Degrad. Stab.*, **4**, 1 (1982).
37. Allen, N. S., Parkinson, A., Gardette, J. L. and Lemaire, J., *Polym. Degrad. Stab.*, **5**, 135 (1983).
38. Carlsson, D. J. and Wiles, D. M., *Polym. Degrad. Stab.*, **6**, 1 (1984).
39. Hodgeman, D. K. C., *J. Polym. Sci.*, **18**, 533 (1980).
40. Allen, N. S., Kotecha, J., Gardette, J. L. and Lemaire, J., *Polym. Degrad. Stab.*, to be published.

41. Zahradnickova, A., Petruj, J. and Sedlar, J., *Polym. Photochem.*, **3**, 295 (1983).
42. Felder, B., Schumacher, R. and Sitek, F., *Chemy Ind.*, **4**, 155 (1980).
43. Felder, B., Schumacher, R. and Sitek, F., *Helv. Chim. Acta*, **63**, 132 (1980).
44. Lucki, J., Rabek, J. F. and Rånby, B., *Polym. Photochem.*, **5**, 351 (1984).
45. Allen, N. S., Gardette, J. L. and Lemaire, J., *J. appl. Polym. Sci.*, **27**, 2761 (1982).
46. Allen, N. S. and Parkinson, A., *Polym. Degrad. Stab.*, **5**, 189 (1983).
47. Allen, N. S., Gardette, J. L. and Lemaire, J., *Dyes & Pigments*, **3**, 295 (1982).

14

Photo-antioxidants: A Review of Recent Developments

GERALD SCOTT

Department of Chemistry, University of Aston in Birmingham, UK

1. INTRODUCTION

During the 1970s, the view that carbonyl compounds were the primary initiators implicated in the photo-oxidation of most polymers became widely accepted and was invoked by photochemists to explain the photostabilising action of many nickel chelates. The position in 1975 was comprehensively reviewed by Rånby and Rabek.[1] The overall conclusion of this survey was that the action of UV stabilisers was either to absorb UV light or to quench the excited states of carbonyl or derived singlet oxygen.[2] Antioxidants were not considered to play a very important role in UV stabilisation, and their participation as synergists with UV absorbers merited only a few lines of comment in the above work.[3] In spite of the earlier pioneering work of Bateman and Gee and their co-workers[4,5] and later of Norrish and his co-workers[6] on the importance of hydroperoxides as photo-initiators in hydrocarbon polymers, the significance of these indigenous impurities in the photo-oxidation of polymers was unaccountably ignored by most investigators in the early 1970s.

The reason why antioxidants were not considered to be important in polymer photostabilisation appears to be associated with the fact that the well-known classes of chain-breaking antioxidants, the aromatic amines and hindered phenols, were generally ineffective UV stabilisers compared with the UV absorbers. However, even in the early 1960s, it was known[7] that some antioxidants acting by a peroxidolytic mechanism[8,9] were effective photostabilisers in their own right and that they synergised strongly with UV absorbers. Subsequent work has confirmed the important place of transition metal thiolate antioxidants in UV stabilisation technology and some recent uses of this class of photo-antioxidant will be discussed below.

The emphasis upon the role of triplet carbonyl and singlet oxygen in the study of the mechanism of polymer photodegradation was to a large extent a consequence of an ingenious theory put forward by Trozzolo and Winslow in 1968.[10] They proposed that carbonyl compounds, which are known to photolyse to give vinyl compounds by the Norrish II process (see Scheme 1), might also excite ground state dioxygen to the singlet state and that these two species might then react to give allylic hydroperoxide (Scheme 1). This postulate tacitly assumed that hydroperoxides are involved in photo-oxidation, an idea for which Carlsson and Wiles provided positive experimental evidence,[11] although these workers did not at this time consider them to be important during the early stages of photo-oxidation. The carbonyl sensitisation theory was considered to be a possible explanation of how hydroperoxides might come to be present in polymers. In practice this debate turned out to be entirely academic, since it is almost impossible to prevent the formation of hydroperoxides in commercial polymers except by using an efficient peroxide decomposer and these compounds are in practice good UV stabilisers.[12] With the development of sensitive methods for detecting and estimating hydroperoxides in polymers, it has been found possible to relate the initial rate of photo-oxidation of polyolefins[13–18] and PVC[17–19] to the concentration of hydroperoxides in conventionally fabricated films. Kinetic calculations by Carlsson and Wiles[20,21] have confirmed that hydroperoxides are always present in commercial polypropylene in sufficient amount to act as photosensitisers. Furthermore, since carbonyl compounds are the primary homolytic breakdown products of hydroperoxide, the carbonyl sensitisation theory appears now to have little practical relevance to the initial stage of polymer photo-oxidation. During the early stages of photo-oxidation, the kinetics of hydroperoxide formation are consistent with the

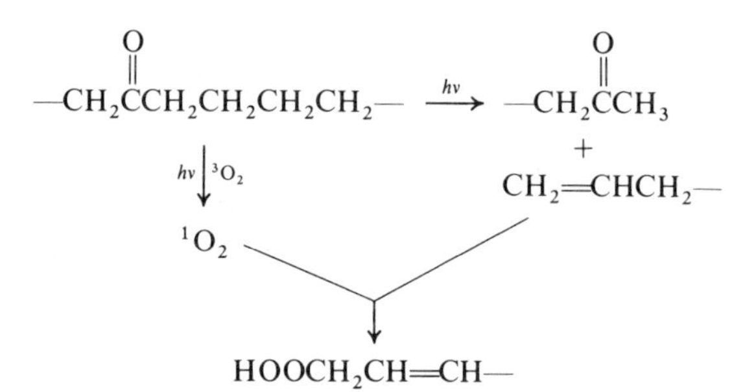

SCHEME 1. Carbonyl initiation mechanism for the photodegradation of polyethylene.

TABLE 1
Commercial UV Stabilisers: Photo-antioxidant Mechanisms

Technological classification	*Typical structures*	*Commercial name*	*Mechanism*
UV Absorbers	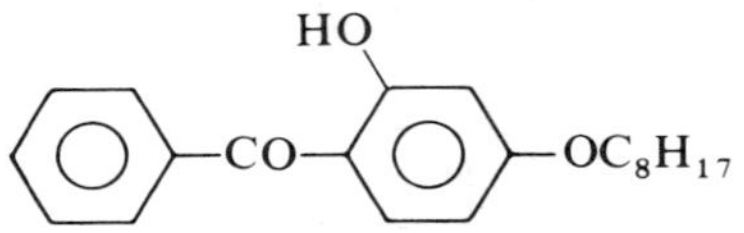 HO, CO, OC_8H_{17}	Cyasorb UV 531	UVA, CB-D
	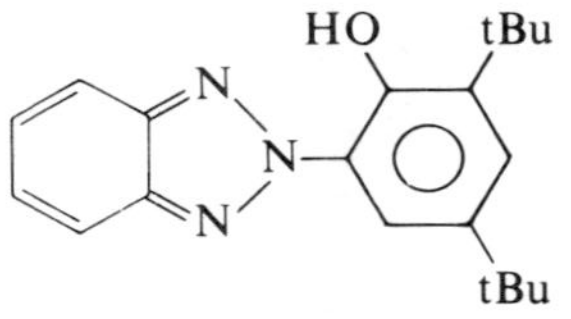 HO, tBu, N, N, N, tBu	Tinuvin 326	UVA, CB-D
	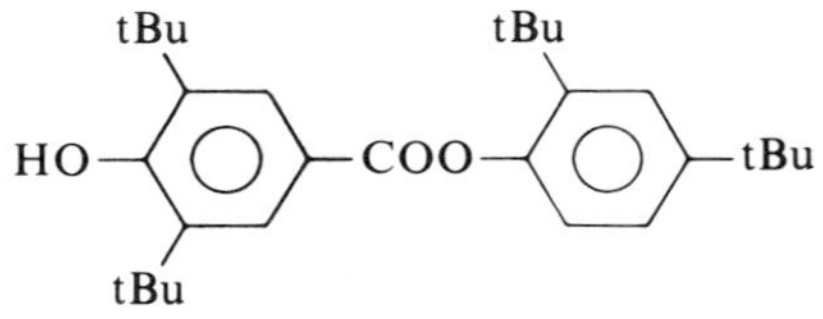 tBu, HO, tBu, COO, tBu, tBu	Tinuvin 120	CB-D, UVA
	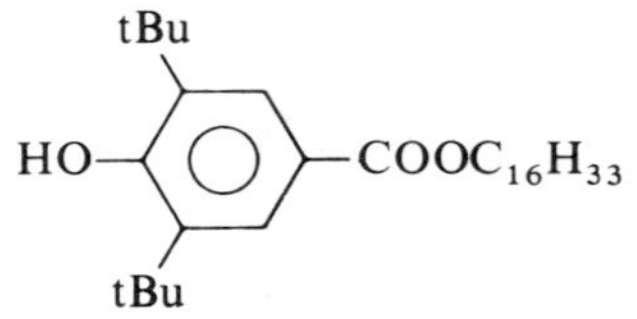 tBu, HO, tBu, $COOC_{16}H_{33}$	Cyasorb UV 2908	CB-D

(continued)

TABLE 1—*contd.*

Technological classification	*Typical structures*	*Commercial name*	*Mechanism*
Quenchers	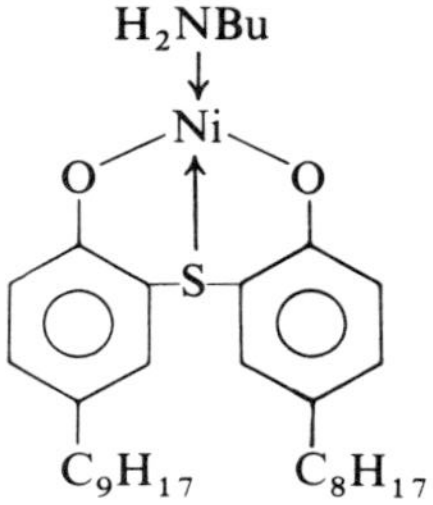 [Ni complex of 2,2'-thiobis(4-alkylphenol): $H_2NBu \rightarrow Ni$, O, O, S, C_9H_{17}, C_8H_{17}]	Cyasorb 1084	UVA, CB-D, PD-C
	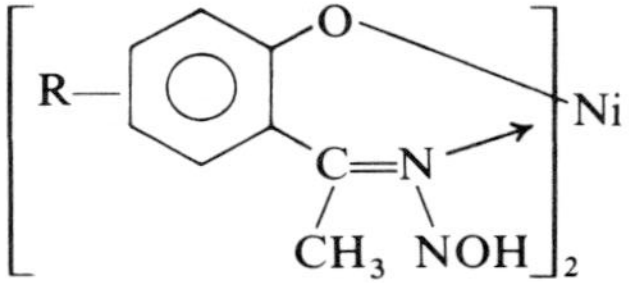 [R-phenoxy oxime Ni complex: $[R\text{—}C_6H_3(O)\text{—}C(CH_3)\text{=}N(NOH) \rightarrow]_2Ni$]	—	UVA, CB-D, PD-S
	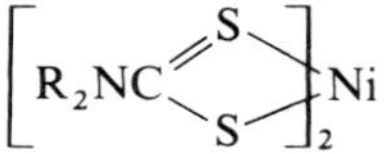 $[R_2NC(=S)S]_2Ni$ (NiDRC)		UVA, PD-C, CB-D
	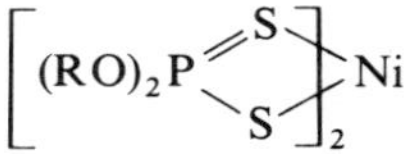 $[(RO)_2P(=S)S]_2Ni$ (NiDRP)	—	UVA, PD-C, CB-D

Hindered amine light stabilisers (HALS)	[—$(CH_2)_4COO$—(2,2,6,6-tetramethylpiperidin-4-yl, Me, Me, NH, Me, Me)]	Tinuvin 770	CB-D/CB-A
Synergists for UV absorbers	2,6-tBu-4-($CH_2CH_2COOC_{18}H_{37}$)-phenol (OH, tBu, tBu)	Irganox 1076	CB-D
	$(C_{12}H_{25}OCOCH_2CH_2)_2S$	DLTP	PD-C
	$R_2NC(=S)SSC(=S)NR_2$ (RDCS)	—	PD-C
	$(RO)_2P(=S)SSP(=S)(OR)_2$ (RDPS)	—	PD-C
	$(RO)_3P$	Various	PD-S
	$[R_2NC(=S)S]_2Zn$	ZnDRC	PD-C
	$[(RO)_2P(=S)S]_2Zn$	ZnDRP	PD-C

UVA UV Absorber
CB-A Chain-breaking electron acceptor
CB-D Chain-breaking electron donor
PD-S Stoichiometric peroxide decomposer
PD-C Catalytic peroxide decomposer
Q Excited-state quencher

idea that hydroperoxide is the only initiator.[15] However in the later stages, carbonyl photolysis appears to play an important role[22] and indeed, Guillet and his co-workers have utilised the Norrish II reaction (see Scheme 1) to deliberately sensitise polymers to light.[23]

In parallel with the above change in emphasis, it has also become increasingly evident that UV stabilisers do not act primarily by either absorbing UV light or by quenching excited states of indigenous molecules, although the first process certainly makes a contribution to their overall effect. Attempts to relate the effectiveness of photostabilisers to their quenching ability have generally been unsuccessful, and the effectiveness of most nickel complexes can now be explained in terms of more conventional antioxidant mechanisms (see Table 1).

2. PHOTO-ANTIOXIDANT MECHANISMS

Scheme 2 represents a reformulation of the original autoxidation mechanism of Bolland, Bateman and Gee. The broken lines indicate the

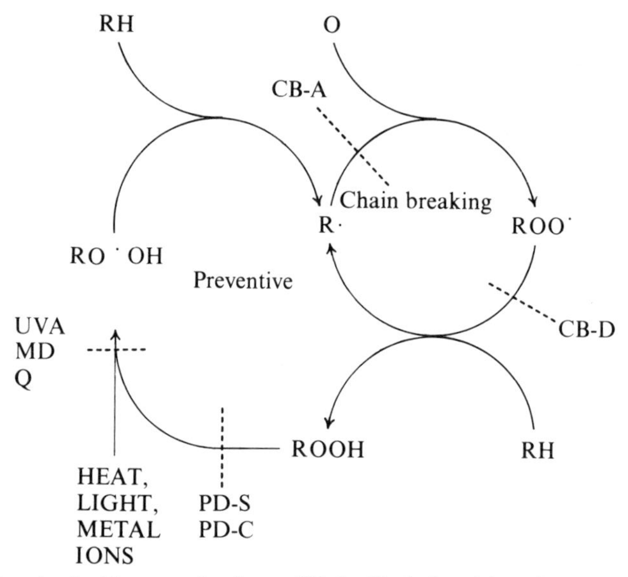

SCHEME 2. Antioxidant mechanisms. CB-A, Chain-breaking electron acceptor; CB-D, chain-breaking electron donor; PD-S, stoichiometric peroxide decomposer; PD-C, catalytic peroxide decomposer; UVA, UV absorber; MD, metal deactivator; Q, excited state quencher.

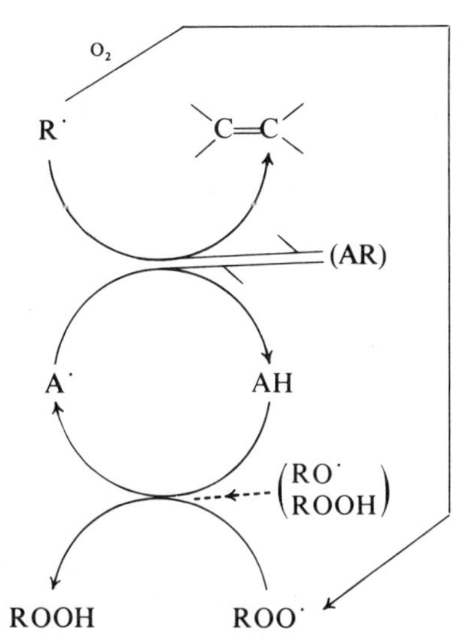

SCHEME 3. Generalised catalytic chain-breaking mechanism of antioxidant action.

points at which the individual steps in the process can be intercepted on the basis of current understanding of antioxidant mechanisms.[25,26] It can be seen that the alkyl radical is the common link between the two interphased cyclical processes. The chain-breaking (CB) antioxidants act by removing chain carrying radicals and the preventive antioxidants destroy the initiating species. The chain-breaking donor (CB-D) and the peroxidolytic (PD) mechanisms have been well documented in thermal oxidation of polymers,[27,28] and 'stable' radicals have recently been shown to be involved in a catalytic redox cycle in polyolefins in which the antioxidant is alternatively oxidised and reduced (see Scheme 3).[25–29] Of the preventive antioxidants, by far the most important are the catalytic peroxidolytic (PD-C) antioxidants, since they remove the main initiating species involved in the autoxidation cycle.

All the mechanisms outlined in Scheme 2 have been shown to operate under photo-oxidative conditions and Table 1 lists the most important classes of commercial UV stabilisers and their most probable mechanisms of action. It can be seen that most photo-antioxidants act by more than one mechanism.

Recent work has concentrated in two main areas.

(a) Chain-breaking antioxidants which can participate in the catalytic

CB-A/CB-D cycle (Scheme 3). The essential requirement is that A$^{\cdot}$ can compete with O_2 for the alkyl radical, R$^{\cdot}$, in the polymer.[28,29]

(b) Peroxidolytic antioxidants are strategically important because, not only do they protect the polymer under conditions of photo-oxidation, but they also provide a high thermal and processing stability.[30]

2.1. Photo-antioxidants Acting by a Catalytic (CB-A/CB-D) Mechanism

The oxygen pressure in autoxidation critically affects the mechanism and function of antioxidants. Under normal conditions, where oxygen solubility and rate of diffusion does not unduly reduce its partial pressure at the site of the reaction, alkylperoxyl radicals are present in large excess over alkyl radicals, and termination occurs almost exclusively by reactions of ROO$^{\cdot}$.[31] However, when the oxygen partial pressure is low, for example in a screw extruder,[32] or when there is diffusion limitation to the site of the reaction, alkyl radicals may then participate in the termination step, as outlined in Scheme 3. This has been shown to occur during mechano-oxidation of rubbers[33] and during photo-oxidation of polyolefins.[21,34–36]

Nitroxyl radicals are almost uniquely effective antioxidants under all conditions of restricted oxygen access. Their antioxidant mechanism has

TABLE 2
Photo-antioxidant Activity of a Hindered Piperidine and Related Oxidation Products in Polypropylene (all concentrations 6×10^{-4} mol/100 g)

Additive	*Time to embrittlement (h)*
$[-(CH_2)_4COOP-H]_2$ (Tinuvin 770)	750
$[-(CH_2)_4COOP-O^{\cdot}]_2$	960
$HOP-O^{\cdot}$	920
$HOP-OH$	1 040
Control (no additive)	90

P = 2,2,6,6-tetramethylpiperidin-4-yl (ring with two Me groups on each carbon adjacent to N—)

been shown to be described by Scheme 3, in which A$^{\cdot}$ is N—O$^{\cdot}$ and AH is N—OH. The extent of formation of the radical coupled product, AR, depends very strongly on the conditions. During the processing of polypropylene (high temperature mechano-oxidation), no O-alkyl hydroxylamines have been detected. During low temperature mechano-oxidation of rubbers (fatiguing),[37] on the other hand, macroalkyl hydroxylamines are formed reversibly. The same appears to be true during photo-oxidation.[21] Aliphatic nitroxyl radicals are more effective as UV stabilisers than the corresponding aromatic nitroxyls.[36] This appears to be associated with the photo-instability of the latter which generally absorb light in the actinic spectrum. The aliphatic hindered amines and their derived oxidation products in general do not.

Hindered piperidines, although effective UV stabilisers in polypropylene, are less effective than the derived nitroxyl/hydroxylamine combination (see Table 2).[35] Furthermore, the oxidation products are also much more effective as melt stabilisers,[35] since the amines are oxidised to nitroxyls by hydroperoxide in a radical generating process (reaction 1).

$$\rangle N{-}H + ROOH \longrightarrow [\rangle N^{\cdot}] + H_2O + RO^{\cdot} \xrightarrow{ROO^{\cdot}} \rangle N{-}O^{\cdot} + RO^{\cdot} \quad (1)$$

Other nitroxyl generators are much more effective mechano-antioxidants (melt stabilisers) than the hindered amines and some of them are also more effective as photo-antioxidants.[38,39] The hindered bis-nitrosamine derived from Tinuvin 770 is particularly effective (see Table 3). It has been shown that the nitroso compound dissociates at processing temperatures to give NO which in turn reacts with macroalkyl radicals.[39] The photo-antioxidant mechanism probably involves two different nitroxyl radical species (see Scheme 4), and supporting evidence has been found in the fact that nitroso-tert-alkanes (e.g. tBuNO, II) are also highly effective photo-antioxidants for polypropylene[38] (see Table 4). They are also effective processing stabilisers by virtue of their ability to trap macroalkyl radicals by the mechanism outlined in Scheme 4. Figure 1 shows the way in which nitroxyl radicals accumulate in the polymer during processing in the presence of C-nitroso compounds, and Fig. 2 follows the decay of nitroxyl

TABLE 3
Comparison of Tinuvin 770 with its Bis-nitroso Derivative as Mechano- and Photo-antioxidants (both additives at 10^{-3} mol/100 g)

Additive	*Induction period (IP) to MFI change (min)**	*Embrittlement time (h)†*
$[-(CH_2)_4COOP-H]_2$ (Tinuvin 770)	No IP	760
$[-(CH_2)_4COOP-N=O]_2$	>20	1 580

P as indicated in Table 2.
* At 180 °C in a closed mixer.
† In a sunlamp/actinic blue lamp.

during UV irradiation.[34] The latter behaviour is entirely analogous to the behaviour of the hindered piperidines.[31]

Phenolic nitrones, (I), are also effective mechano-antioxidants, but they are less effective than the nitroso-alkanes as photo-antioxidants. The evidence suggests that they function by being converted to the

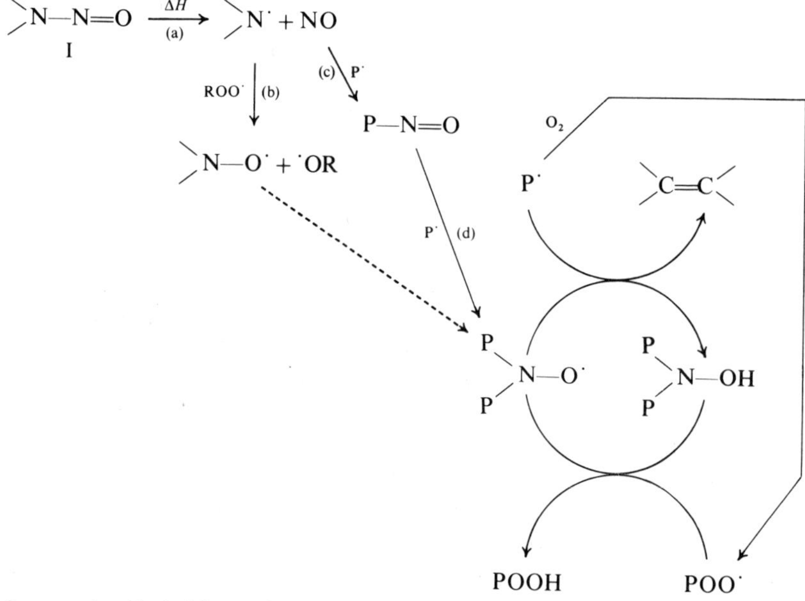

SCHEME 4. Probable mechano- and photo-antioxidant mechanism of bis-(N-nitroso-2,2,6,6-tetramethyl-4-piperidinyl) sebacate (I). P· is a sec. or tert. macroalkyl radical.

TABLE 4
Comparison of Tinuvin 770 with Nitroso-tert-alkanes as Photo-antioxidants in Polypropylene (concentrations, 10^{-3} mol/100 g)

Additive	*Processing time at 180°C (min)*	*UV embrittlement time (h)*
Tinuvin 770	10	760
tBuNO	10	770
tBuNO	15	875
tOctNO	10	690
tOctNO	15	650
Control (no additive)	10	85
Do.	15	75

corresponding phenoxyl/nitroxyl radical (II) during processing,[40,41] which then participates, together with the parent phenol, in the general redox antioxidant process shown in Scheme 3. The relatively poor photo-antioxidant activity of MHPBN(I) has been shown to be due to the photo-instability, not of the parent phenol itself, but of the derived radical, which appears to have an essentially quinonoid structure (IIb).[40]

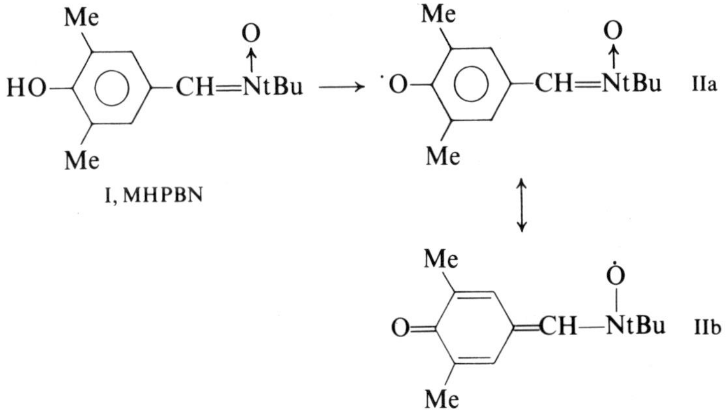

I, MHPBN → IIa ↔ IIb (2)

Galvinoxyl (G˙) is a very powerful melt stabiliser for polyolefins and it is partially converted during processing to the corresponding phenol (GH) by participation in the redox cycle shown in Scheme 3.[32] The G˙/GH combination is also a photo-antioxidant[43] (see Table 5) of similar activity to MHPBN, although both are less effective than the hindered

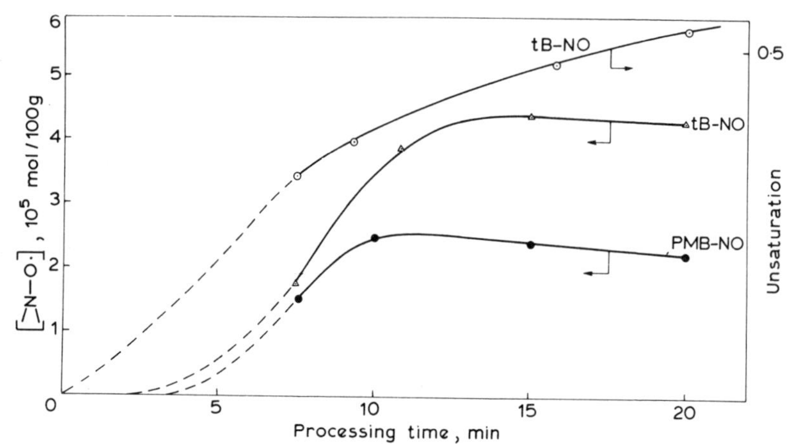

FIG. 1. Formation of nitroxyl radicals in polypropylene from tBuNO during processing at 200 °C (closed mixer).

piperidinoxyls (cf. Table 2). This is almost certainly due to the susceptibility to photolysis of the quinonoid structures of the 'stable' radicals.

$$\underset{\mathbf{G^{\cdot}}}{{}^{\cdot}O\text{–}C_6H_2(tBu)_2\text{–}CH{=}C_6H_2(tBu)_2{=}O} \xrightarrow{R^{\cdot}} \underset{\mathbf{GH}}{HO\text{–}C_6H_2(tBu)_2\text{–}CH{=}C_6H_2(tBu)_2{=}O} \tag{3}$$

It seems clear from this work and other research in progress[44] that the use of 'stable' radical precursors (spin-traps) is still at an early stage, and offers considerable promise for the future. We may expect to see other new

TABLE 5
Photo-antioxidant Activity of Galvinoxyl (G·) in Polypropylene (processed at 200 °C/10 min in a closed mixer)

Initial [$G^{\cdot}$], 10^4 *mol/100 g*	*Embrittlement time* (*h*)
2·37	171
4·75	220
9·50	286
Control	85

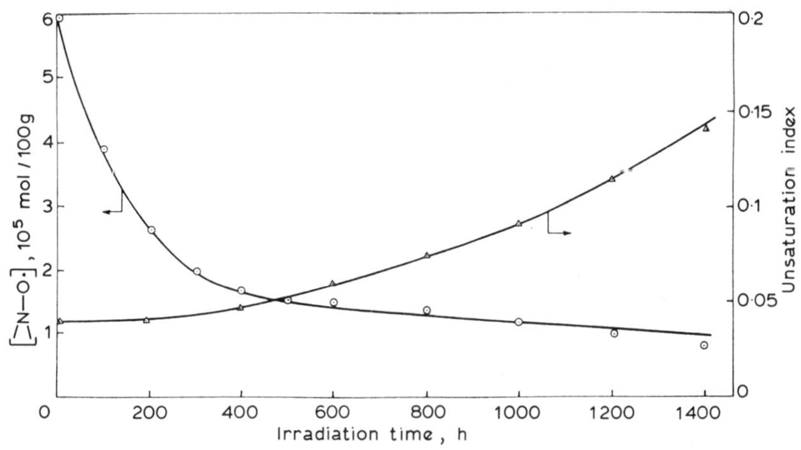

FIG. 2. Decay of nitroxyl radicals (from tBuNO) during UV irradiation of polypropylene.

and more effective general purpose antioxidant structures emerging as the CB-A/CB-D catalytic antioxidant mechanism is further clarified.

During the past ten years, many new commercial UV stabilisers have augmented the ranks of the UV absorbers and some representative examples are given in Table 1. Their function, like that of the earlier UV absorbers, cannot be accounted for in terms of UV absorption alone. Many (e.g. Tinuvin 326 and Cyasorb UV 2908) contain a conventional hindered phenolic CB-D function, but in addition they contain a 4-carbonyl group, which appears to increase their resistance to photolysis compared with the typical CB-D thermal antioxidants (e.g. Irganox 1076) which are normally used only in combination with a UV absorber (see Table 1). They appear therefore to combine the functions of UV screening and CB-D antioxidants.[27]

2.2. Photo-antioxidants Acting by a Peroxidolytic (PD-C) Mechanism

As was indicated earlier, peroxidolytic (PD) antioxidants were recognised to be mechanistically distinct from the chain-breaking antioxidants many years ago.[7] Reference to Scheme 2 shows that their action is complementary to that of the CB antioxidants and to the UV absorbers, and for this reason, they normally synergise with the former under thermal-oxidative conditions and with the latter under photo-oxidative conditions. Many peroxide decomposers which are effective thermal antioxidants are not by themselves highly effective photo-antioxidants. Thus the thiodipropionate esters (e.g. DLTP, Table 1), which are used almost universally in combination with CB-D antioxidants as heat stabilisers in polypropylene are weak light stabilisers, although they synergise with the UV absorbers.[7] Similarly, the zinc dialkyldithiocarbamates (Table 1, ZnDRC) and the zinc dialkyldithiophosphates (Table 1, ZnDRP) are powerful thermal antioxidants and melt stabilisers, but are again relatively ineffective as photo-antioxidants.[45–47] By contrast, the nickel complexes (Table 1, NiDRC and NiDRP) are highly effective UV stabilisers in their own right, and this has been shown to be a consequence of their much greater UV stability. Suitably substituted nickel dithiocarbamates[45] are now commercially available with similar photo-antioxidant activity to the CB antioxidants discussed above. A high level of polymer solubility is a primary criterion of photo-antioxidant activity for all the metal chelates.[45]

Dithiocarbamoyl and dithiophosphoryl disulphides (Table 1, RDCS and RDPS) are very effective thermal antioxidants and melt stabilisers for polyolefins, but, like the group II thiolates and the thiodipropionate esters, they are weak photo-antioxidants. However, they are converted to more effective photo-antioxidants under conditions of oxidative processing (see

TABLE 6
Effect of Processing on the Photo-antioxidant Activity of Dibutyldithiophosphoryl Disulphide

Additive	*Conc^n.*	*Treatment*	*Embrittlement time (h)*
DBDS	0·2	CM	300
DBDS	0·2	OM	375
DBDS	0·2	CM + TBH (0·1 g/100 g)	700
DBDS + HOBP	0·2 + 0·1	CM	1 300
DBDS + HOBP	0·2 + 0·1	OM	1 600
DBDS + HOBP	0·2 + 0·1	CM + TBH (0·1 g/100 g)	2 050
DBDS + HOBP + NiDP	0·2 + 0·1 + 0·1	CM	1 400
DBDS + HOBP + NiDP	0·2 + 0·1 + 0·1	CM + TBH (0·1 g/100 g)	5 000
Control	—	CM	90

DBDS = Dibutyl dithiophosphoryl disulphide.
HOBP = 2-Hydroxy-4-octoxy benzophenone (UV 531).
NiDBP = Nickel dibutyldithiophosphate.
TBH = Tert-butyl hydroperoxide.
CM = processed in a closed mixer for 10 min.
OM = processed in an open mixer for 10 min.

Table 6). The oxidation products so produced are very effective synergists with conventional UV absorbers.[49] Part of the reason for the effectiveness of the nickel thiolate chelates (NiDRC and NiDRP) is that they are converted to the corresponding disulphides and their further oxidation products during processing and under conditions of light exposure with the slow liberation of sulphur acids[30,50–53] (see Scheme 5). However, this is not the only reason why the nickel complexes are so effective. The other reason is that, unlike the disulphides, they are photostable and slowly liberate low molecular mass peroxide decomposers throughout the lifetime of the nickel complexes in the polymer.

UV absorbers synergise very effectively with the nickel dithiolates and even more so with their derived oxidation products[28,54] (see Table 6). They appear to protect the peroxidolytic antioxidants from photolysis in the same way as they protect the zinc dialkyl dithiocarbamates.[46,47] Quite remarkable synergistic effects can be achieved using oxidatively processed dithiophosphoryl disulphides in combination with the parent nickel complexes and a conventional UV absorber (Table 6).[28,54] This is an interesting illustration of the importance of the processing operation on the subsequent service performance of the polymer. The mechanism of this process is outlined in Scheme 5.

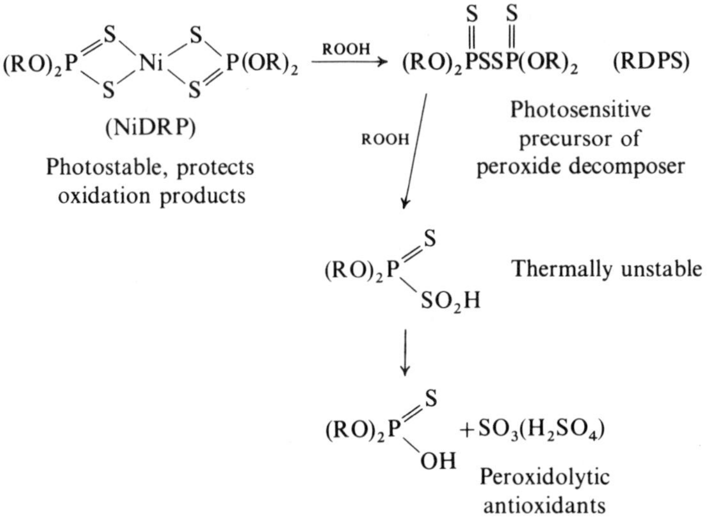

$$(RO)_2P(=S)S{-}Ni{-}S_2P(OR)_2 \xrightarrow{ROOH} (RO)_2P(=S)SSP(=S)(OR)_2 \quad (RDPS)$$

(NiDRP)
Photostable, protects oxidation products

Photosensitive precursor of peroxide decomposer

$$\downarrow ROOH$$

$$(RO)_2P(=S)SO_2H$$

Thermally unstable

$$\downarrow$$

$$(RO)_2P(=S)OH + SO_3(H_2SO_4)$$

Peroxidolytic antioxidants

SCHEME 5. Mechanism of photo-antioxidant synergism involving peroxidolytic antioxidants.

3. PHYSICAL ASPECTS OF PHOTO-ANTIOXIDANT BEHAVIOUR

The importance of photo-antioxidant solubility in the polymer has already been referred to and it appears that for maximum effect, such additives should be molecularly dispersed in the most photosensitive region of the polymer. In practice it is extremely difficult to manipulate the physicochemical partition of additives in heterophase polymers so that they are where they are most required. For example, in the case of rubber-modified plastics (HIPS, ABS, etc.) the most photosensitive region is the rubber phase which is the minor constituent of the matrix, but the stabilisers are distributed throughout the matrix. Consequently, photo-antioxidants are not very effective in polyblends.[55,56] In principle it would appear to be advantageous to anchor the stabiliser to the backbone of the elastomeric segment. Table 7 compares the effect of some conventional low molecular mass antioxidants and UV stabilisers and synergistic combinations with a polymer bound synergist containing similar antioxidant functions.[57] The actual components of the synergists are the CB-D/PD-C antioxidant, BHBM and the autosynergistic UVA/PD-C antioxidant, EBHPT, both of which are chemically attached to the polybutadiene segment (P) of the polyblend:

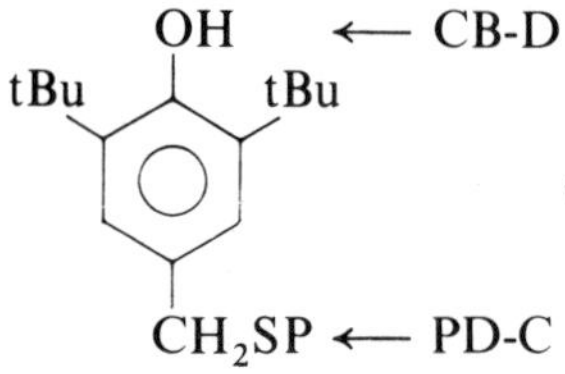

OH ← CB-D
tBu tBu
CH_2SP ← PD-C
BHBM

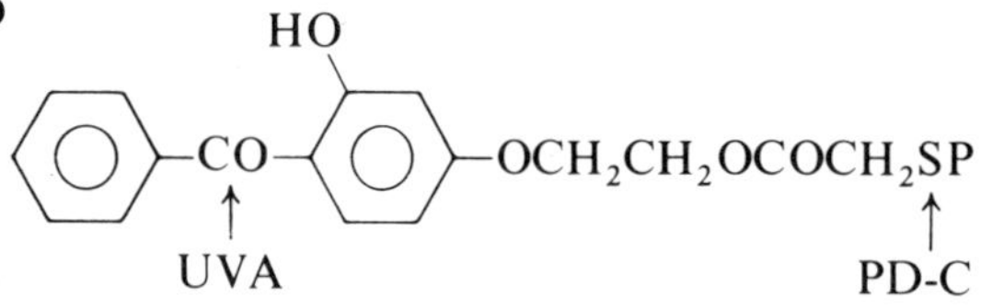

HO
—CO— —$OCH_2CH_2OCOCH_2SP$
↑ ↑
UVA PD-C
EBHPT

The remarkable photo-antioxidant activity of the polymer-bound synergist as compared with the equivalent conventional additives (UV531 + BHT + DLTP) is a reflection of the actual concentrations of the two systems in the rubber domain of the polyblend.[55]

4. FUTURE PERSPECTIVES

It is now accepted that UV stabilisers are essentially light stable antioxidants. Although much more research is required into the detailed chemistry by which photo-antioxidants exert their effect, the main

TABLE 7
Photo-antioxidant Activity of Bound Antioxidants and their Synergistic Additive Equivalents in ABS

Stabiliser	*Concentration (g/100 g)*	*Embrittlement time (h)*
Control	—	22
UV 531	1	35
BHT	1	34
DLTP	1	25
UV 531 + BHT	1 + 1	69
UV 531 + DLTP	1 + 1	77
UV 531 + BHT + DLTP	1 + 1 + 1	94
EBHPT	1	52
BHBM	1	50
EBHPT—B*	1	90
EBHPT—B* + BHBM—B*	1 + 1	380

* —B indicates that the antioxidant is chemically bound to the polymer.

mechanisms involved are now clear and provide a basis for further innovation.

The microenvironment of photo-antioxidants in polymers is a fruitful area for further research. Clearly, oxygen solubility and diffusion rate are of critical importance since they determine the efficiency of the catalytic cycle of Scheme 3. There is evidence to suggest that the CB antioxidants eventually disappear from the polymer by oxidation to non-active products.[32] Consequently there is a requirement for the rate of alkyl radical capture to be increased either by decreasing the rate of oxygen diffusion, or by increasing the oxidation potential of the CB-A antioxidant.

There appears to be no doubt that antioxidant solubility is critically important to photoantioxidant activity[45,46,59,60] and many effective antioxidant structures have probably been discarded in the past in industrial screening of UV stabilisers through not taking this into account. Billingham and his co-workers have put forward a coherent theoretical model to account for the loss of antioxidants from polymers in terms of migration and volatilisation rates.[61] However there is a need for more systematic experimental work to establish a relationship between additive structure and solubility and migration aptitude. Similarly the distribution of additives in multiphase polymers (including the crystalline and amorphous regions of semicrystalline polymers[62]) needs careful investigation in order to use photo-antioxidants to their maximum potential.

Carlsson and Wiles have suggested that hindered piperidines are particularly effective in polypropylene because of their ability to associate with hydroperoxides, the primary oxidation products of the polymer.[21] If further research shows this to be an important effect, then it has wide implications for all antioxidant systems and particularly the peroxide decomposers. Sulphoxides have been shown[63] to form hydrogen bonded complexes with hydroperoxides in solution, and this phenomenon could explain the exceptional activity of some sulphur compounds and their derived oxidation products as photoantioxidants.[30]

ACKNOWLEDGEMENTS

I am greatly indebted to my co-workers, Dr K. B. Chakraborty and Dr S. Al-Malaika for their contributions to the understanding of photo-antioxidant mechanisms.

REFERENCES

1. Rånby, B. and Rabek, J. F., *Photodegradation, Photooxidation and Photostabilisation of Polymers*, John Wiley, London, 1965.
2. Ref. 1, Chapter 10.
3. Ref. 1, p. 421.
4. Bateman, L. and Gee, G., *Proc. R. Soc., Ser. A*, **195**, 376, 391 (1948–49).
5. Martin, J. T. and Norrish, R. G. W., *Proc. R. Soc., Ser. A*, **220**, 322 (1953).
6. Norrish, R. G. W. and Searby, M. H., *Proc. R. Soc., Ser. A*, **237**, 464 (1956).
7. Scott, G., *Atmospheric Oxidation and Antioxidants*, Elsevier, London and New York, 1965, p. 216 et seq.
8. Ref. 7, p. 188 et seq.
9. Holdsworth, J. D., Scott, G. and Williams, D., *J. chem. Soc.*, 4692 (1964).
10. Trozzolo, A. M. and Winslow, F. H., *Macromolecules*, **1**, 98 (1968).
11. Carlsson, D. J. and Wiles, D. M., *Macromolecules*, **2**, 587, 597 (1969).
12. Scott, G., *Macromol. Chem.*, **8**, 319 (1972).
13. Amin, M. U., Scott, G. and Tillekeratne, L. M. K., *Europ. Polym. J.*, **11**, 85 (1975).
14. Scott, G., American Chemical Society Symposium Series, **25**, 340 (1976).
15. Chakraborty, K. B. and Scott, G., *Polymer*, **18**, 98 (1977); *Polym. Degrad. Stab.*, **1**, 37 (1979).
16. Chakraborty, K. B. and Scott, G., *Europ. Polym. J.*, **15**, 731 (1977).
17. Scott, G., *Developments in Polymer Degradation* (Ed. N. Grassie), Applied Science Publishers, London, 1977, p. 205.
18. Scott, G., *Adv. in Chem. Ser.*, **169**, 30 (1978).
19. Cooray, B. B. and Scott, G., *Polym. Degrad. Stab.*, **3**, 127 (1980/1).

20. Carlsson, D. J. and Wiles, D. M., *J. Macromol. Sci., Rev. Macromol. Chem.*, **C14**, 65 (1976).
21. Carlsson, D. J., Garton, A. and Wiles, D. M., *Developments in Polymer Stabilisation—1*, Applied Science Publishers, London, 1979, p. 219.
22. Chew, C. H., Gan, L. M. and Scott, G., *Europ. Polym. J.*, **13**, 361 (1977).
23. Guillet, J. E., *Polymers and Ecological Problems* (Ed. J. Guillet), Plenum Press, New York, 1973, p. 1.
24. Scott, G., *Pure appl. Chem.*, **52**, 365 (1980).
25. Al-Malaika, S. and Scott, G., *Degradation and Stabilisation of Polyolefins* (Ed. N. S. Allen), Applied Science Publishers, London, 1983, p. 283.
26. Grassie, N. and Scott, G., *Degradation and Stabilisation of Polymers*, Cambridge University Press, 1985.
27. Ref. 25, p. 247.
28. Scott, G., *Brit. Polym. J.*, **15**, 208 (1984).
29. Scott, G., *Developments in Polymer Stabilisation—7* (Ed. G. Scott), Elsevier Applied Science Publishers, 1984, Chap. 2.
30. Al-Malaika, S., Chakraborty, K. B. and Scott, G., *Developments in Polymer Stabilisation—6* (Ed. G. Scott), Applied Science Publishers, London, 1983, p. 73.
31. Bateman, L. and Morris, A. L., *Trans. Faraday Soc.*, **49**, 1026 (1953).
32. Bagheri, R., Chakraborty, K. B. and Scott, G., *Chemy Ind.*, 865 (1980); *Polym. Degrad. Stab.*, **5**, 145 (1983).
33. Katbab, A. A. and Scott, G., *Europ. Polym. J.*, **17**, 599 (1981).
34. Chakraborty, K. B. and Scott, G., *Chemy Ind.*, 239 (1978); *Polymer*, **21**, 252 (1980).
35. Bagheri, R., Chakraborty, K. B. and Scott, G., *Polym. Degrad. Stab.*, **4**, 1 (1982).
36. Shlyapintokh, V. Ya. and Ivanov, V. B., *Developments in Polymer Stabilisation—5* (Ed. G. Scott), Applied Science Publishers, London, 1982, p. 41.
37. Nethsinghe, L. P. and Scott, G., *Europ. Polym. J.*, **20**, 213 (1984).
38. Chakraborty, K. B., Scott, G. and Yaghmour, H., *J. app. Polym. Sci.*, in press.
39. Chakraborty, K. B., Scott, G. and Yaghmour, H., *Polym. Degrad. Stab.*, in press.
40. Chakraborty, K. B., Scott, G. and Yaghmour, H., *J. app. Sci.*, in press.
41. Nethsinghe, L. P. and Scott, G., *Rubb. Chem. Tech.*, in press.
42. Gugumus, F., *Developments in Polymer Stabilisation—1* (Ed. G. Scott), Applied Science Publishers, London, 1979, p. 261.
43. Chakraborty, K. B. and Scott, G., *J. Polym. Sci., Letters*, in press.
44. American Chemical Society, Polymer Preprints, **25**, St. Louis, April 1984, pp. 32, 36, 40.
45. Chakraborty, K. B., Scott, G. and Poyner, W. R., *Plast. Rubb. Proc. Appl.*, **3**, 59 (1983).
46. Chakraborty, K. B., Scott, G. and Poyner, W. R., *Polym. Degrad. Stab.*, **8**, 1 (1984).
47. Chakraborty, K. B. and Scott, G., *Europ. Polym. J.*, **13**, 1007 (1977).
48. Chakraborty, K. B. and Scott, G., *Polym. Degrad. Stab.*, **1**, 37 (1979).
49. Scott, G. and Al-Malaika, S., *Brit. Pat.*, 2,117,779A (1983).

50. Al-Malaika, S. and Scott, G., *Polymer Communications*, **24**, 24 (1983).
51. Al-Malaika, S. and Scott, G., *Europ. Polym. J.*, **19**, 241 (1983).
52. Al-Malaika, S. and Scott, G., *Europ. Polym. J.*, **19**, 235 (1983).
53. Al-Malaika, S. and Scott, G., *Polym. Degrad. Stab.*, **5**, 415 (1983).
54. Al-Malaika, S., *Brit. Polym. J.*, in press.
55. Scott, G., American Chemical Society, Polymer Preprints, **25**, St. Louis, April 1984, p. 62.
56. Fernando, W. S. E. and Scott, G., *Europ. Polym. J.*, **16**, 971 (1980).
57. Ghaemy, M. and Scott, G., *Polym. Degrad. Stab.*, **3**, 405 (1980–81).
58. Ref. 7, p. 164.
59. Gugumus, F. L., *Org. Coatings and App. Polym. Sci. Proc.*, ACS 183rd National Meeting, Las Vegas, March 25–April 2 (1982), p. 449.
60. Al-Malaika, S., Chakraborty, K. B., Scott, G. and Tao, Z. B., *Polym. Degrad. Stab.*, in press.
61. Billingham, N. C. and Calvert, P. D., *Developments in Polymer Stabilisation—3* (Ed. G. Scott), Applied Science Publishers, London, 1980, p. 139.
62. Billingham, N. C. and Calvert, P. D., *Dev. Polym. Char.*, **3**, 229 (1982).
63. Armstrong, C., Plant, M. A. and Scott, G., *Europ. Polym. J.*, **11**, 161 (1975).

15

New Polymerizable 2(2-Hydroxyphenyl)2H-Benzotriazole Ultraviolet Absorbers: 2[2,4-Dihydroxy-5-Vinyl(Isopropenyl)Phenyl]1,3-2H-Dibenzotriazole*

SHOUKUAN FU,[1] AMITAVA GUPTA,[2] ANN CHRISTINE ALBERTSSON[3] and OTTO VOGL[1] †

[1] *Polytechnic Institute of New York, New York, USA*

[2] *Jet Propulsion Laboratory, California Institute of Technology, Pasadena, USA*

[3] *Royal Institute of Technology, Stockholm, Sweden*

1. INTRODUCTION

Polymeric materials, when used outdoors or indoors under light rich in ultraviolet radiation, must be stabilized by ultraviolet stabilizers. Several types of ultraviolet stabilizers exist which include ultraviolet screens, quenchers, and compounds (HALS) which regulate the photodecomposition of hydroperoxides. The most frequently used ultraviolet stabilizers are ultraviolet screens, molecules which absorb preferentially the damaging ultraviolet radiation; the photoexcited stabilizer molecule is capable of returning to its ground state by giving up its energy in the form of harmless vibrational energy.

For many years simple low molecular weight organic compounds have been used as stabilizers in plastics materials. These compounds, although normally quite effective, when used in long-term applications or when the surface-to-volume ratio is high, such as in films, coatings, and fibers have several disadvantages: limited compatibility, volatility, leaching, to mention a few.

In recent years efforts have been made to prepare stabilizers with better compatibility and higher molecular weight. In the last decade, examples of polymerizable and polymeric stabilizers of several types have been

* Functional Polymers, Part XXXIX. Part XXXVIII: A.C. Albertsson, L. G. Donaruma, and O. Vogl, Proceedings, NY Academy, in press.

† To whom all inquiries should be addressed.

developed: vinylsalicylates, vinyl(and acryloxy)-2-hydroxybenzophenones, vinyl-α-cyano-β-phenylcinnamates, and most recently vinyl (and acryloxy)2(2-hydroxyphenyl)2H-benzotriazoles.

In our continued search for more effective polymerizable ultraviolet stabilizers of the 2(2-hydroxyphenyl)2H-benzotriazole type, we have synthesized styrene-type 2(2-hydroxyphenyl)2H-benzotriazole monomers[1-3] and acrylate-type 2(2-hydroxyphenyl)2H-benzotriazole monomers.[4-6] In the course of these investigations, compounds with more than one benzotriazole group in the molecule[7,8] and compounds with 2(2-hydroxyphenyl)2H-benzotriazole and 2-hydroxybenzophenone units in the same molecule[9] have been synthesized. We have reported recently on monomers (acrylates and methacrylates) of ultraviolet absorbers with more than one benzotriazole group in the same molecule.[10] The reactivities of acrylates and methacrylate derivatives of the ultraviolet absorber were found to be somewhat different from those of styrene derivatives. In addition, acrylic esters where the functional group is in the ester function are always suspect for hydrolysis or other deterioration.

We have therefore made it our objective to synthesize styrene derivatives (compounds in which the vinyl or isopropenyl group is directly attached to the phenyl ring of the dihydroxyphenyl) of 2(2,4-hydroxyphenyl)1,3-2H-dibenzotriazoles, to attempt their homopolymerization, and to copolymerize these monomers with such resonance-stabilized monomers as styrene, methyl methacrylate, or n-butyl acrylate.

2. EXPERIMENTAL PART

2.1. Materials

o-Nitroaniline (Eastman Kodak Co.) was used as received. 2,4-Dihydroxyacetophenone (Aldrich Chemical Co.) was recrystallized from a dilute solution of hydrochloric acid (water/conc. hydrochloric acid, 12/1). Iodomethane (Aldrich Chemical Co.) was used without purification. Picric acid (Aldrich Chemical Co.) was dried in small quantity in vacuum for one day.

Azobisisobutyronitrile (AIBN) (Aldrich Chemical Co.) was recrystallized three times from absolute methanol and dried for one day at 0·05 mm Hg at room temperature. Styrene (St) (Aldrich Chemical Co.), methyl methacrylate (MMA) (Aldrich Chemical Co.), and n-butyl acrylate (BA) (Fisher Scientific Co.) were distilled under reduced pressure before use.

Magnesium turnings (Fisher Scientific Co.), sodium borohydride (Fisher

Scientific Co.), and zinc powder (Fisher Scientific Co.) were used as received. Potassium hydrogen sulfate (Fisher Scientific Co.) and zinc chloride (Fisher Scientific Co.) were freshly fused before use.

Anhydrous diethyl ether (American Scientific Products) was used from freshly opened cans. Tetrahydrofuran (THF) (Aldrich Chemical Co.) was heated to reflux over calcium hydride for 3 h, distilled, and kept over molecular sieves.

Solvents such as chloroform, toluene, methanol, dichloromethane, and anhydrous ethylene glycol dimethyl ether (DME) were used as received. Deutero chloroform (99·8 % D) and dimethyl sulfoxide-D_6 (99·9 % D) were obtained from Aldrich Chemical Co.

N,N′-Dimethylacetamide (DMAc), dichloromethane, when used as solvent for polymerization, were reagent grade (Aldrich Chemical Co.) and were used directly from freshly opened bottles.

2.2. Measurements

Infrared spectra were recorded on a Perkin-Elmer Spectrophotometer, Model 727. Solid samples were measured in the form of potassium bromide pellets.

^{1}H NMR spectra were recorded on a Varian A-60 spectrometer and ^{13}C NMR spectra on a Varian CTF-20 spectrometer with complete proton decoupling; TMS was used as the internal standard. The compounds were measured in deuterated $CDCl_3$ or DMSO-D_6 in 15 % or saturated solutions. Conditions for acquiring the spectra were as follows.

Ultraviolet absorptions were measured in chloroform (Spectro-grade, Fisher Scientific Co.) solution with a Beckman M VI Spectrometer in a double-beam servo mode (1·0 cm optical path length). The maximum absorbances and corresponding wavelengths were determined by dialing in the wavelength and recording the absorbance value presented on the digital display.

Melting points were determined on a MELT-TEMP capillary melting point apparatus at a heating rate of 2 °C/min and are uncorrected.

Microanalyses were carried out at the Microanalytical Laboratory Office of Research Services, University of Massachusetts, Amherst, Massachusetts.

2.3. Procedures

2.3.1. Preparation of 3,5[Di(2H-benzotriazole-2-yl)]2,4-dihydroxyacetophenone (DBAP)

The procedure for the preparation of DBAP was essentially the same as

TABLE 1
Ultraviolet Spectral Data[a] for DBDH-5HE,[b] DBDH-5V,[c] DBDH-5P,[d] Homopolymer and Copolymer of DBDH-5V and Copolymer of DBDH-5P

Compound	λ_{max} (*nm*)	ε (*l/mol cm* $\times 10^{-4}$)	λ_{max} (*nm*)	ε (*l/mol cm* $\times 10^{-4}$)	λ_{max} (*nm*)	ε (*l/mol cm* $\times 10^{-4}$)
DBDH-5HE	332	3·10			248	0·97
DBDH-5V	332	3·28	276 (sh)	1·95	248	1·77
Poly-DBDH-5V	340	2·60	290 (sh)	1·73	248	1·04
St-*co*-DBDH-5V	335	3·00	260	1·58		
MMA-*co*-DBDH-5V	335	3·34			248	1·12
DBDH-5P	333	3·26			246	1·89
St-*co*-DBDH-5P	335	2·68	260	1·68	248	1·09
MMA-*co*-DBDH-5P	333	3·45				

[a] Absorption spectra determined in solutions of chloroform. Concentration: 2×10^{-5} mol/l.
[b] DBDH-5HE 2[2,4-Dihydroxy-5(1-hydroxyethyl)phenyl]1,3-2H-dibenzotriazole.
[c] DBDH-5V 2(2,4-Dihydroxy-5-vinylphenyl)1,3-2H-dibenzotriazole.
[d] DBDH-5P 2(2,4-Dihydroxy-5-isopropenylphenyl)-1,3-2H-dibenzotriazole.

that described previously.[1] However, during the final step, the reductive cyclization with zinc powder, some of the product was found to be tightly bound to the zinc powder residue and was difficult to extract with dilute alkali solution. This residue was therefore treated with hydrochloric acid (1:1) to dissolve the zinc powder; the remaining precipitate was then extracted with benzene. This procedure increased the yield of DBAP to 50%, as compared to the previously reported yield of 32%.

2.3.2. Monomer Preparations

2[2,4-Dihydroxy-5(1-hydroxyethyl)phenyl]1,3-2H-dibenzotriazole (DBDH-5HE): DBAP (1·55 g, 4 mmol) was dissolved in chloroform (50 ml) and DME (35 ml) in a 250 ml round bottom flask, and the solution was cooled under a nitrogen atmosphere to −5 °C with stirring. A solution of zinc borohydride [$Zn(BH_4)_2$] in DME (16 ml of 0·5 M solution, 8 mmol) was then added via a syringe to this cold solution slowly over a period of 8 h. (Zinc borohydride was prepared by the reaction of sodium borohydride with freshly fused zinc chloride in DME at 0–5 °C.) The reaction was followed by TLC; after one day, no starting material remained. Dilute hydrochloric acid (40 ml of 1:8) was added to decompose the excess zinc borohydride, and the mixture was stirred for an additional hour. The organic layer was then separated and the aqueous layer extracted twice with chloroform. The combined organic layers were washed twice with water and dried over magnesium sulfate. The filtered solution was concentrated until about 50 ml of solution were left. On cooling, yellowish crystals were obtained (1·40 g, 90%). Recrystallization from chloroform gave white crystals, m.p. 200–202 °C.

The ultraviolet absorption data are presented in Table 1 and the spectrum in Fig. 1. IR (KBr): 3420 cm^{-1} (O—H stretching). ^{13}C NMR chemical shift data are presented in Table 2 and the spectrum in Fig. 1.

Elemental Analysis: Calculated for $C_{20}H_{16}N_6O_3$: C, 61·85%; H, 4·12%; N, 21·65%. Found: C, 61·66%; H, 4·06%; N, 21·43%.

2(2,4-Dihydroxy-5-vinylphenyl)1,3-2H-dibenzotriazole (DBDH-5V): DBDH-5HE (3·48 g, 9 mmol) was dissolved in toluene (150 ml) in a 250 ml three-neck flask equipped with a condenser and freshly fused potassium bisulfate (1·74 g); picric acid (135 mg) was also added. The reaction mixture was heated to reflux for one day with stirring under a nitrogen atmosphere. After the reaction was judged to be complete (TLC), the mixture was cooled, filtered, and the toluene evaporated under reduced pressure. The crude product was dissolved in dichloromethane (80 ml) and the solution washed twice with 5% aqueous sodium bicarbonate solution. The water

TABLE 2

^{13}C NMR Chemical Shift Data for 2[2,4-Dihydroxy-5(1-hydroxyethyl)-phenyl]1,3-2H-dibenzotriazole (DBDH-5HE) and 2(2,4-Dihydroxy-5-isopropenylphenyl)1,3-2H-dibenzotriazole (DBDH-5P)

Assignment	*Chemical shift*	*DBDH-5HE in ppm*	*DBDH-5P in ppm*
Phenoxy group	a	146·0	145·0
	b	118·1	115·6
	c	152·6	149·6
	d	125·5	124·2
	e	120·4	125·3
	f	118·7	120·6
Benzotriazole group	1(1′)	142·5/144·5	142·1/143·7
	2(2′)	117·4/118·3	117·5/117·9
	3(3′)	127·6/126·9	128·4/127·4
Substituent	g	66·2	—
	g′	—	141·7
	h	24·3	23·2
	i	—	116·7

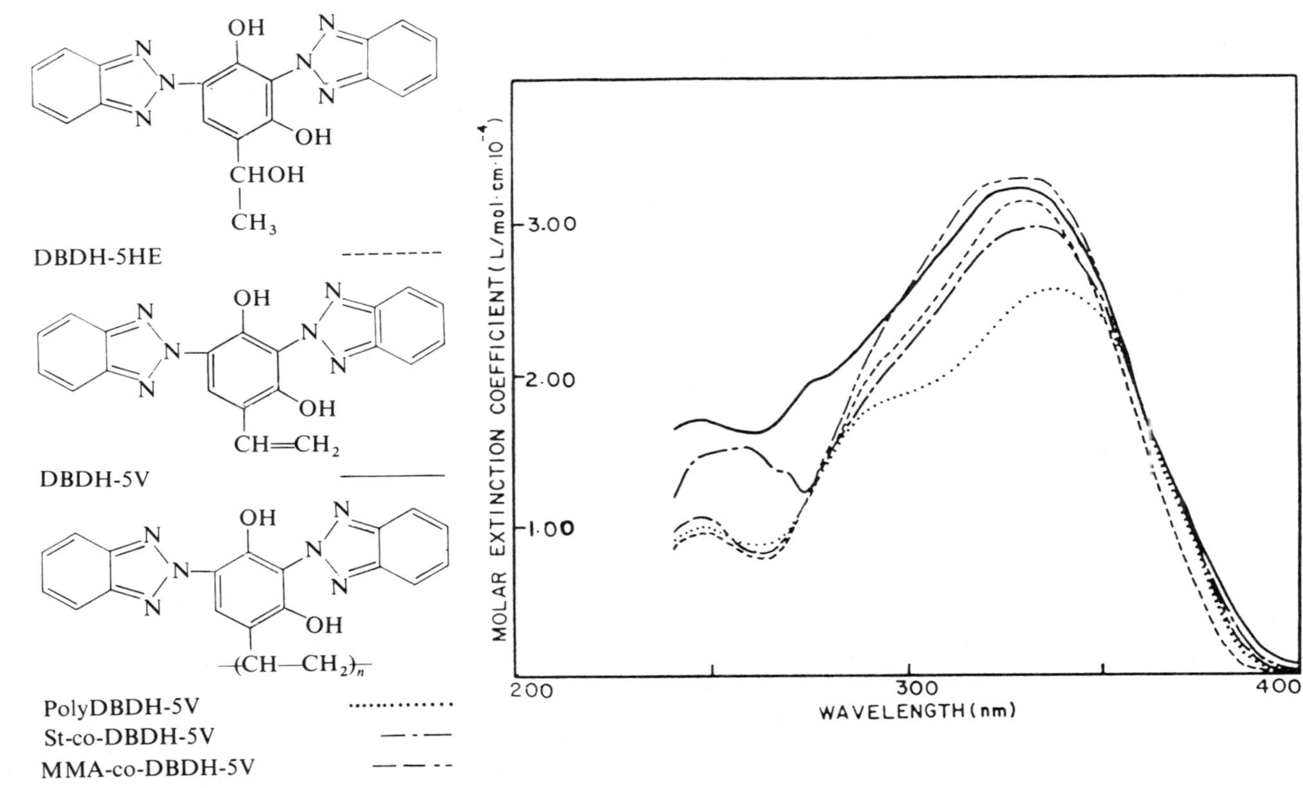

FIG. 1. Ultraviolet spectra of DBDH-5HE, DBDH-5V, and homopolymer and copolymers of DBDH-5V.

layer was extracted twice with dichloromethane and the combined organic layers washed with dilute hydrochloric acid and twice with water before drying the solution over anhydrous magnesium sulfate. After filtration, the solution was concentrated and the product which precipitated was collected (2·8 g, 84 %). Recrystallization from dichloromethane/n-hexane (1:1) gave yellowish crystals, m.p. 194–195 °C.

The ultraviolet absorption data are presented in Table 1 and the spectrum in Fig. 1. IR (KBr): 890 cm^{-1} ($\rangle$C=CH$_2$, C—H bending). ^{13}C NMR chemical shift data are presented in Table 2.

Analysis: Calculated for $C_{20}H_{14}N_6O_2$: C, 64·86 %; H, 3·78 %, N, 22·70 %. Found: C, 65·02 %; H, 3·84 %; N, 22·52 %.

2(2,4-Dihydroxy, 5-isopropenylphenyl)1,3-2H-dibenzotriazole (DBDH-5P): In a 1 l, three-neck flask equipped with a pressure-equalizing dropping funnel, a reflux condenser, and a stirring bar were placed dry magnesium turnings (1·44 g, 62·5 mmol). The reaction apparatus was dried by heating it under dry nitrogen. Anhydrous diethyl ether (50 ml) was then added; iodomethane (8·53 g, 62·5 mmol) dissolved in anhydrous diethyl ether (7 ml) was added dropwise to the stirred mixture. The addition was made over a period of 30 min; the solution was stirred for an additional 2 h.

DBAP (2·5 g, 6·5 mmol) was dissolved in dry THF (250 ml) and anhydrous diethyl ether (50 ml), and the solution was added to the Grignard reagent such that the ethereal solution was kept at a gentle reflux. The reaction was stirred for an additional 4-h period and then treated with an aqueous solution of ammonium chloride (25 g) and concentrated sulfuric acid (5 ml) in water (90 ml) for 1 h. The organic layer was separated and the residue was extracted twice with diethyl ether (2 × 50 ml). The combined solutions were washed three times with water and dried over anhydrous magnesium sulfate. The filtered solution was evaporated and gave 1·6 g (62 %) of slightly pink crystals.

The ^{1}H NMR spectrum showed it to be a mixture of 2[2,4-dihydroxy-5(2′-hydroxy-2′-propylphenyl)]1,3-2H-dibenzotriazole (DBDH-5PR) (δ 1·63 ppm, $-\overset{\displaystyle OH}{\overset{|}{C}}-(C\underline{H}_3)_2$) and DBDH-5P ($\delta$ 2·12 ppm, $=C-C\underline{H}_3$). (The ratio of these two compounds may be different in different batches of dehydration reactions.)

It was found that vacuum drying at 110 °C dehydrated DBDH-5PR readily and quantitatively to DBDH-5P, m.p. 174–176 °C; recrystallization was done from benzene.

The ultraviolet absorption data are presented in Table 1 and the

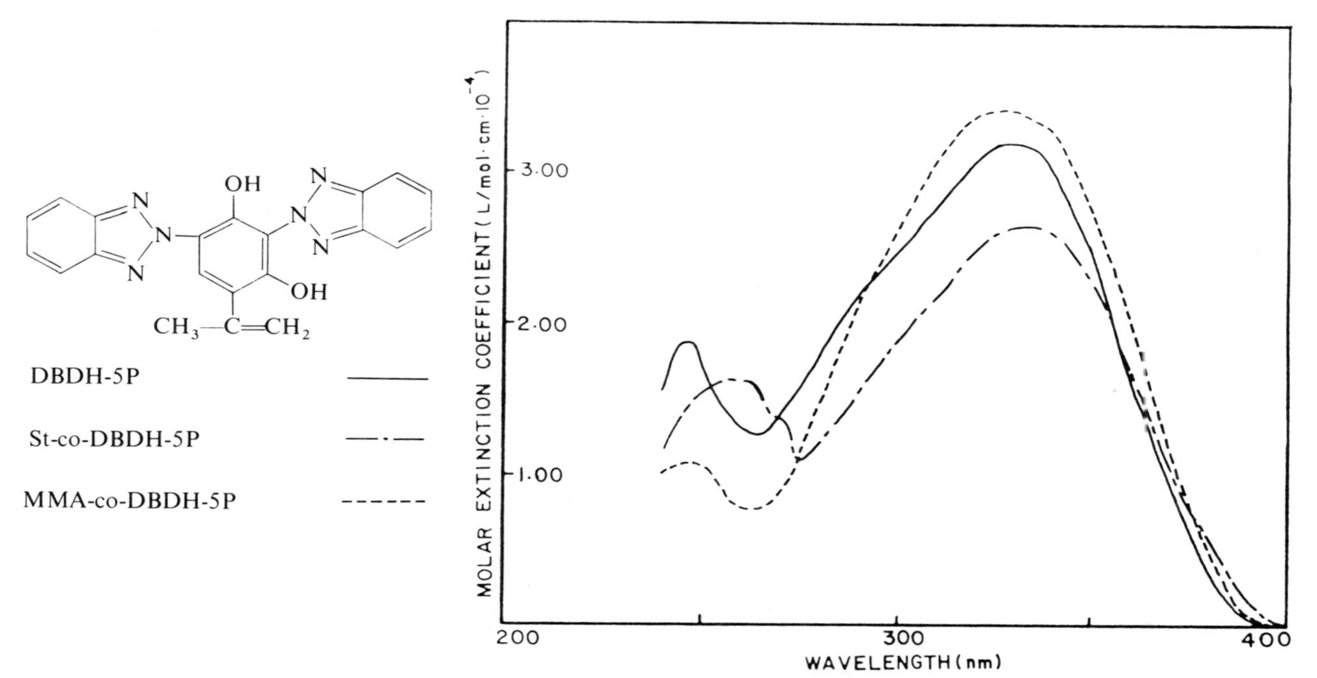

FIG. 2. Ultraviolet spectra of DBDH-5P and its copolymers.

TABLE 3

Homopolymerization of DBDH-5V and its Copolymerization with Styrene, Methyl Methacrylate, and n-Butyl Acrylate. Polymerization Conditions: Temperature: 50 °C; Time: 3 Days; Initiator: AIBN, 0·3 mol per cent; Sealed Tube (Pressure 0·05 mm Hg)

Monomers					*Total amount of monomers*		*Solvents*		*Polymerization yield*		*Polymer composition (mol % DBDH-5V unit)*	η_{inh} [a] *(dl/g)*
M_1 (DBDH-5V)		*M_2*										
g	*mmol*	*Type*	*g*	*mmol*	*g*	*mmol*	*Type*	*ml*	*g*	*%*		
0·56	1·5	—	—	—	0·56	1·5	DMAc CH_2Cl_2	2 1	0·30	54	100	0·10
0·04	0·1	St	1·03	9·9	1·07	10	DMAc CH_2Cl_2	1 1	0·60	54	1	0·25
0·11	0·3	St	1·01	9·7	1·12	10	DMAc CH_2Cl_2	1 1	0·58	52	3	0·24
0·04	0·1	MMA	0·99	9·9	1·03	10	DMAc CH_2Cl_2	1 1	0·94	91	0·6	0·69
0·11	0·3	MMA	0·97	9·7	1·08	10	DMAc CH_2Cl_2	1 1	0·93	86	1·5	0·58
0·26	0·7	MMA	0·63	6·3	0·89	7	DMAc CH_2Cl_2	2 1	0·77	87	6·5	0·50
0·04	0·1	BA	1·27	9·9	1·31	10	DMAc CH_2Cl_2	3 2	1·10	84	0·7	0·44
0·11	0·3	BA	1·24	9·7	1·35	10	DMAc CH_2Cl_2	3 2	1·10	82	1·9	0·43

[a] Approximately 0·5 g/dl, in chloroform, 30 ° ± 0·1 °C.

spectrum in Fig. 2. IR (KBr): 890 cm^{-1} ($>$C=CH$_2$, C—H bending). ^{13}C NMR chemical shift data are presented in Table 2 and the spectrum in Fig. 1c.

Analysis: Calculated for $C_{21}H_{16}N_6O_2$: C, 65·62%, H, 4·17%; N, 21·87%. Found: C, 65·56%; H, 4·05%; N, 21·75%.

2.3.3. Polymerizations

Homopolymerization of 2(2,4-dihydroxy-5-vinylphenyl)1,3-2H-dibenzotriazole (DBDH-5V): A 10 ml polymerization tube was charged with 0·5 ml of dichloromethane solution of AIBN (0·75 mg, 0·0045 mmol), DBDH-5V (0·56 g, 1·5 mmol), DMAc (2 ml), and dichloromethane (0·5 ml). After three freeze–thaw cycles at 0·05 mm Hg pressure to degas the homogeneous polymerization mixture, the tube was sealed and allowed to react for 3 days at 50 °C; the tube was opened and the solution was added dropwise to methanol (200 ml) to precipitate. The suspension was allowed to settle for 2 h; it was filtered and dried at 0·05 mm Hg. The solid polymer was dissolved in chloroform and precipitated again into methanol. After drying for 1 day at 60 °C and 0·05 mm Hg, poly-DBDH-5V was obtained (0·30 g, 54% yield) with an inherent viscosity of 0·10 dl/g (see Table 3).

The ultraviolet absorption data are presented in Table 1.

Copolymerization of 2(2,4-dihydroxy-5-vinylphenyl)1,3-2H-dibenzotriazole (DBDH-5V) with styrene (St), methyl methacrylate (MMA), and n-butyl acrylate (BA), respectively: DBDH-5V was copolymerized with styrene (St), methyl methacrylate (MMA), or n-butyl acrylate (BA) in different ratios, respectively.

The procedure was essentially the same as that described for the homopolymerization. The amount of material used for the polymerization and the results of all polymerizations, including polymerization yield, polymer composition, and inherent viscosity of polymers, are presented in Table 3.

The ultraviolet absorption data are presented in Table 1 and the spectrum in Fig. 1.

Homopolymerization attempt of 2(2,4-dihydroxy-5-isopropenylphenyl)-1,3-2H-dibenzotriazole (DBDH-5P): Attempts were made to homopolymerize DBDH-5P, the procedure being essentially the same as for the homopolymerization of DBDH-5V. The amount of materials used for this attempt is presented in Table 4. After allowing the reaction mixture to polymerize for 3 days at 50 °C, the tube was opened and the solution was added dropwise to methanol (200 ml). No precipitation was observed initially, but upon cooling in the refrigerator, a white precipitate formed

TABLE 4
Copolymerization of DBDH-5P with Styrene or Methyl Methacrylate or n-Butyl Acrylate. Polymerization Conditions: Temperature: 50 °C; Time: 3 Days; Initiator: AIBN, 0·3 mol per cent; Sealed Tube (Pressure 0·05 mm Hg)

Monomers					*Total amount of monomers*		*Solvents*		*Polymerization yield*		*Polymer composition (mol% DBDH-5P unit)*	η_{inh} [a] *(dl/g)*
M_1 (DBDH-5P)		*M_2*										
g	*mmol*	*Type*	*g*	*mmol*	*g*	*mmol*	*Type*	*ml*	*g*	*%*		
0·39	1	—	—	—	0·39	1	DMAc CH_2Cl_2	4 2	[b]		—	—
0·04	0·1	St	1·03	9·9	1·07	10	DMAc CH_2Cl_2	1 1	0·60	56	0·5	0·36
0·12	0·3	St	1·01	9·7	1·13	10	DMAc CH_2Cl_2	3 2	0·44	39	0·7	0·11
0·27	0·7	St	0·66	6·3	0·93	7	DMAc CH_2Cl_2	3 2	0·29	31	2·2	0·10
0·04	0·1	MMA	0·99	9·9	1·03	10	DMAc CH_2Cl_2	1 1	0·95	92	0·9	0·67
0·12	0·3	MMA	0·97	9·7	1·09	10	DMAc CH_2Cl_2	3 2	0·93	85	2·1	0·17
0·27	0·7	MMA	0·63	6·3	0·90	7	DMAc CH_2Cl_2	3 2	0·51	56	5·7	0·09
0·04	0·1	BA	1·27	9·9	1·31	10	DMAc CH_2Cl_2	1 1	1·1	84	1·1	1·31
0·12	0·3	BA	1·24	9·7	1·36	10	DMAc CH_2Cl_2	3·5 3·5	0·92	67	3·1	0·16
0·27	0·7	BA	0·81	6·3	1·08	7	DMAc CH_2Cl_2	3·5 3·5	0·39	36	13·6	0·08

[a] Approximately 0·5 g/dl, in toluene, 30° ± 0·1 °C.

which after filtration was identified as monomer DBDH-5P, m.p. 172–174 °C.

Copolymerization of 2(2,4-dihydroxy-5-isopropenylphenyl)1,3-2H-dibenzotriazole (DBDH-5P) with styrene (St), methyl methacrylate (MMA), or n-butylacrylate (BA), respectively: The copolymerization of DBDH-5P with styrene (St), methyl methacrylate (MMA), or n-butylacrylate (BA) in different ratios, respectively, was carried out. The procedure was essentially the same as that described for the homopolymerization of DBDH-5V. The amount of materials used for the polymerization and the results of all polymerizations, including polymerization yield, polymer composition, and inherent viscosity of polymers, are presented in Table 4.

The ultraviolet absorption data are presented in Table 1 and the spectrum in Fig. 2.

3. RESULTS AND DISCUSSION

Previously synthesized 3,5[di(2H)-benzotriazole-2-yl]2,4-dihydroxyacetophenone, also called 2(2,4-dihydroxy-5-acetylphenyl)1,3-2H-dibenzotriazole (DBAP), was transformed successfully in 4 steps to 2(2,4-dihydroxy-5-vinyl-phenyl)1,3-2H-dibenzotriazole (DBDH-5V) and 2(2,4-dihydroxy-5-isopropenylphenyl)1,3-2H-dibenzotriazole (DBDH-5P), respectively (eqns (1)–(4)).

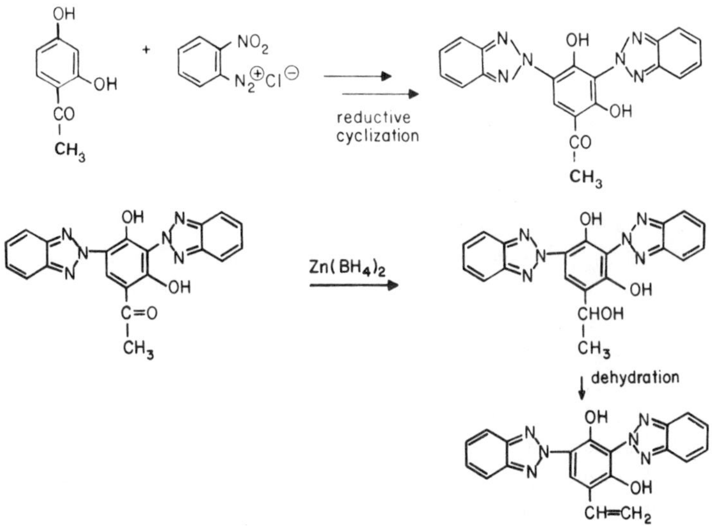

reductive cyclization (1)

$Zn(BH_4)_2$ (2)

dehydration

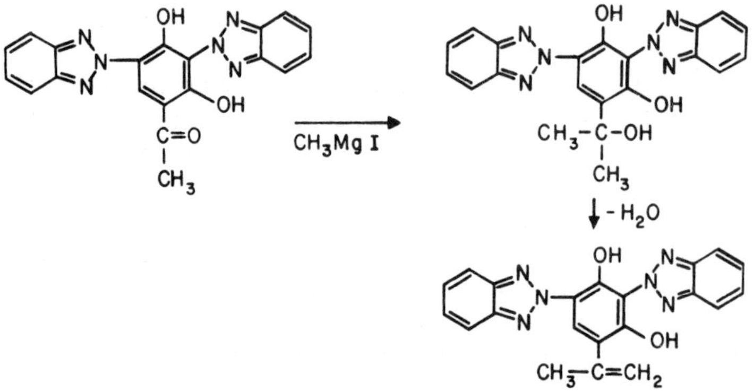

$\xrightarrow{CH_3MgI}$ $\xrightarrow{-H_2O}$ (3)

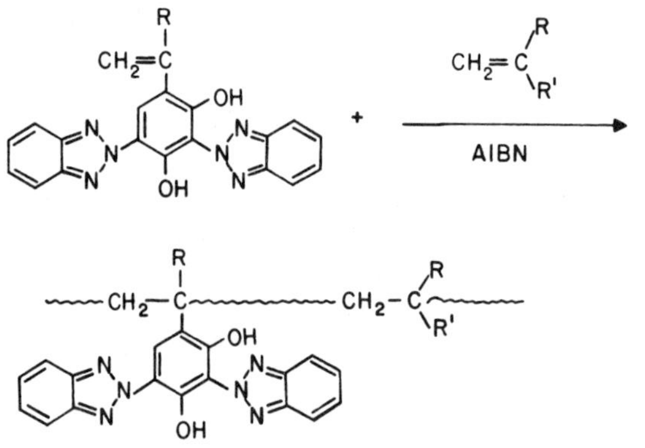

$+\ CH_2{=}C(R)(R')\ \xrightarrow{AIBN}$ (4)

R = CH_3 , H

R′ = C_6H_5 , $COOCH_3$, $COOC_4H_9$

For the synthesis of DBDH-5V, DBAP was reduced with zinc borohydride in over 90% yield to 2[2,4-dihydroxy-5(1-hydroxyethyl)phenyl]1,3-2H-dibenzotriazole (DBDH-5HE). The use of zinc borohydride as reducing agent greatly facilitated the synthesis of DBDH-5V. In the past, when sodium borohydride had been used as reducing agent of the carbonyl group of the acetyl function of benzotriazole derivatives, free phenolic hydroxyl groups had to be acetylated or otherwise protected; otherwise insoluble salts of the phenol group with the sodium borohydride were formed, which precipitated and did not undergo further reduction,[1] or the reduction of the carbonyl group proceeded further to the methylene group.[11] Furthermore, zinc borohydride is readily prepared from sodium borohydride and zinc chloride in DME.[12] The reduction of DBAP with zinc borohydride proceeded in homogeneous solution. The dehydration of

DBDH-5HE was accomplished with potassium bisulfate in boiling toluene in the presence of a small amount of picric acid to prevent DBDH-5V as it is formed from polymerizing by radicals formed adventitiously during the dehydration reaction.

For the synthesis of DBDH 5P, DBAP was allowed to react with methyl magnesium iodide and gave a mixture of the tertiary carbinol DBDH-5PR and DBDH-5P. Dehydration of the DBDH-5PR part of the mixture was accomplished by simply heating the product to 110 °C under reduced pressure. It is remarkable that this dehydration of the tertiary carbinol to DBDH-5P was so readily and quantitatively accomplished without even a small amount of polymerization, and this step could ultimately be very useful. This point will be discussed in more detail in the section on polymerization of DBDH-5V and DBDH-5P.

DBDH-5V and DBDH-5P were characterized by their elemental analysis, infrared, ^{1}H and ^{13}C NMR spectra. DBDH-5V showed the typical infrared absorption of a compound of this type, particularly the vinyl group at 890 cm^{-1}. The ^{1}H NMR spectrum of DBDH-5V showed the normal chemical shift values expected for the parent compound, 2(2,4-hydroxyphenyl)1,3-2H-dibenzotriazole (DBDH), but in addition, the methine proton at 6·8 ppm and the methylene proton at 5·3 ppm. DBDH-5P was similar to DBDH-5V in its spectral behavior, but showed in addition the chemical shift of the methyl group at 2·1 ppm.

The ^{13}C NMR chemical shift values are tabulated in Table 2.

The numeric values of the ultraviolet data of DBDH-5HE, DBDH-5V, and DBDH-5P are shown in Table 1 and the actual spectra, measured in chloroform, in Figs 1 and 2. DBDH-5HE has two maxima: one at λ_{max} of 332 nm and an extinction coefficient of $3{\cdot}10 \times 10^{4}$ l/mol cm and a small maximum at λ_{max} of 248 nm with an extinction coefficient of 0·97 l/mol cm. DBDH-5V had a λ_{max} of 332 nm and an ε of $3{\cdot}28 \times 10^{4}$ l/mol cm with a shoulder at 276 nm (ε of $1{\cdot}95 \times 10^{4}$ l/mol cm) and another maximum at a λ of 248 nm (ε of $1{\cdot}77 \times 10^{4}$ l/mol cm). DBDH-5P had its main maximum at 333 nm with an extinction coefficient of $3{\cdot}26 \times 10^{4}$ l/mol cm; a small maximum was observed at 246 nm (ε of $1{\cdot}89 \times 10^{4}$ l/mol cm).

DBDH-5V was homopolymerized with AIBN as the initiator in DMAc solution. DBDH-5P (as an α-methylstyrene derivative), similar to 2H5P, did not undergo homopolymerization; the monomer was recovered after attempted polymerizations in nearly quantitative yield. DBDH-5V and DBDH-5P copolymerized readily with St, MMA, and BA. The copolymers were characterized by elemental analysis, infrared, ^{1}H and ^{13}C NMR spectroscopy and by their ultraviolet spectra. The copolymer composition was primarily checked by nitrogen analysis and the ultraviolet spectra. The polymerization conditions and the results of DBDH-5V homo- and copolymerizations are summarized in Table 3 and those for DBDH-5P in Table 4.

The polymerizations were carried out in 10–30% solutions in DMAc/dichloromethane mixtures for reasons of solubility of monomers and polymers (copolymers); polymers from moderate to reasonable molecular weight were obtained in yields from 50% to a nearly quantitative yield.

The homopolymerization of DBDH-5V, carried out in 17% solution, gave a polymer of an inherent viscosity of 0·10 dl/g in over 50% yield. The copolymers with styrene were also obtained in 50–60% yield and an η_{inh} of 0·25 dl/g (1–3 mol% composition of DBDH-5V in the copolymer). With MMA as the comonomer, the copolymers with DBDH-5V were obtained in about 90% yield and an η_{inh} of 0·50–0·70 dl/g. BA as the comonomer gave copolymers of DBDH-5V with an η_{inh} of about 0·45 dl/g.

It is significant that the molecular weights of all copolymers, as judged by their inherent viscosity, were lower at higher comonomer contents of DBDH-5V. In the case of MMA/DBDH-5V copolymers, the η_{inh} decreased from the polymer with a DBDH-5V content of 0·6 mol% of 0·69 dl/g to an η_{inh} of 0·50 dl/g for a 6·5 mol% of DBDH-5V content. Similar but less pronounced trends could be seen for St or BA copolymers of DBDH-5V.

Although it might be argued simply that DBDH-5V had a persistent impurity in our polymer grade sample, we would like to think that this decrease of the molecular weight (as judged by the inherent viscosity) is actually caused by the monomer itself. Although in our work on the polymerization and copolymerization of vinyl and acrylic monomers that also have phenolic hydroxyl groups in the molecule, we have fairly convincingly shown that the phenolic hydroxy group does not act as an efficient inhibitor, we were aware that this statement was an over-simplification. High molecular weight homo- and copolymers of vinyl and acrylic monomers with phenolic hydroxy groups can indeed by made when the formation of any oxygen radicals (including the use of azo initiators rather than peroxides as initiators) is avoided or minimized (vacuum technique for the polymerization with several freeze/thaw cycles), in our case of the 2(2,4-dihydroxyphenyl)1,3-2H-dibenzotriazoles which have the polymerizable vinyl or isopropenyl group in 5 position of the benzene ring with an ortho- *and* para- hydroxyl group in the same benzene ring. The circumstances are such that transfer of the hydrogen atom from the hydroxyl group, especially from the ortho-hydroxyl group, to the growing polymeric carbon radical followed by an easy resonance stabilization of the resulting phenoxy radical is not unreasonable.

DBDH-5P did not homopolymerize; we explain this phenomenon strictly by the fact that DBDH-5P is an α-methylstyrene derivative; α-methylstyrene itself has a ceiling temperature of 61 °C, and a more highly substituted α-methylstyrene derivative, such as DBDH-5P, most likely has a lower ceiling temperature. Otherwise, the copolymerization behavior of DBDH-5P is similar to that of DBDH-5V; if anything, it is even more characteristic. St copolymers were obtained in 30–55 % yield with yield and η_{inh} dropping with increasing comonomer (DBDH-5P) content. The η_{inh} decreased from 0·36 dl/g for a 0·5 mol % to 0·10 dl/g for a 2·2 mol % composition. The MMA copolymers showed a similar trend with a high yield of 92 % and an inherent viscosity of 0·67 dl/g for a copolymer with 0·9 mol % DBDH-5P to a 56 % yield and an η_{inh} of 0·1 dl/g for a 5·7 mol % DBDH-5P containing MMA copolymer. For BA as the comonomer, the numbers are 84 % for an η_{inh} of 1·3 dl/g and 1·1 mol % DBDH-5P containing BA copolymer to a 36 % yield with an η_{inh} of 0·08 dl/g and a 13·6 mol % DBDH-5P-containing composition.

The homopolymer of DBDH-5V showed a λ_{max} of 340 nm with an ε of $2{\cdot}6 \times 10^4$ l/mol cm and a shoulder at 290 nm. The copolymers of DBDH-5V and DBDH-5P showed λ_{max} of 333–335 nm and extinction coefficients of $3{\cdot}0$–$3{\cdot}5 \times 10^4$ l/mol cm. The slight variations are not very significant and are within our experimental error limits.

In conclusion, we have prepared DBDH-5V and DBDH-5P, the first styrene-type polymerizable stabilizers with two benzotriazole units in the monomer. DBDH-5V and DBDH-5P could be incorporated into copolymers of St, MMA, and BA; we studied only the range of incorporation of up to 10 mol %. Only DBDH-5V homopolymerized.

ACKNOWLEDGEMENTS

This work was supported by Grant No. 956413 from the Jet Propulsion Laboratory, California Institute of Technology, Pasadena, California. S. K. Fu would like to thank Fudan University, Shanghai, P.R.C., for granting him a leave of absence to work at the Polytechnic Institute of New York, Brooklyn, New York. We would like to thank S. J. Li for the first preparation of the starting material DBAP and initial but unsuccessful attempts for the synthesis of DBDH-5V and DBDH-5P. We would like to express our appreciation to Peter M. Gomez for many suggestions concerned with this work. We would also like to express our appreciation to W. Bassett, Jr, for measuring and calculating the NMR spectra and to E. Cary for her assistance in the preparation of this manuscript.

REFERENCES

1. Yoshida, S. and Vogl, O., *Makromol. Chem.*, **183**, 259 (1982).
2. Nir, Z., Vogl, O. and Gupta, A., *J. Polym. Sci., Polym. Chem. Ed.*, **20**, 2735 (1982).
3. Elbs, V., *J. Prakt. Chem.*, **108**, 209 (1924).
4. Li, S. J., Albertsson, A. C., Gupta, A., Bassett, W. Jr and Vogl, O., *Monatsh. Chem.*, **115**, 853 (1984).
5. Mannens, M. G., Hove, J. J., Aarschot, W. J. and Priem, J. J., U.S. Pat. 3,813,255 (1974); Germ. Pat. 2,128,005 (1971); CA: **76**, 119908m (1972).
6. Xi, F., Bassett, W. Jr and Vogl, O., *Polym. Bull.*, **11**(4), 329 (1984).
7. Li, S. J., Gupta, A. and Vogl, O., *Monatsh. Chem.*, **114**, 937 (1983).
8. Xi, F., Bassett, W. Jr and Vogl, O., *Makromol. Chem.*, in press.
9. Li, S. J., Bassett, W. Jr, Gupta, A. and Vogl, O., *J. Macromol. Sci.-Chem.*, **A20**, 309 (1983).
10. Li, S. J., Gupta, A., Albertsson, A. C., Bassett, W. Jr and Vogl, O., *Polym. Bull.*, in press.
11. Bell, K. H., *Aust. J. Chem.*, **22**, 601 (1969).
12. Crabble, P., Garcia, G. A. and Rius, C., *J. chem. Soc., Perkin I*, 810 (1973).

16

Polymers in Solar Energy

J. F. RABEK

Department of Polymer Technology,
The Royal Institute of Technology, Stockholm, Sweden

POTENTIAL APPLICATIONS OF POLYMERS IN SOLAR TECHNOLOGIES[1-5] AND SOLAR ENERGY CONVERSION[5-7]

(i) Passive Solar Heating Systems[4]

Polymers (plastics) can control incident solar radiation in a passive solar heating system. The buildings themselves are designed and modified with polymers so they retain more heat in winter and absorb less in summer. Plastics application include foam insulation, thermal windows, and weather stripping. Movable wall panels or flaps made of polymers serve to direct the heat where desired. A concrete wall painted a dark colour with polymeric emulsions serves as both collector and storage. A large south facing array made of material less thermally conductive than glass, sheets of polyacrylates, polycarbonate or glass-reinforced thermoset polyester on a building, retains more heat from solar energy. Polyester films, well UV stabilised with doped dyes and covered with a thin coat of aluminium, used as glass windows, provide passive cooling.[8]

(ii) Solar Collectors[4,8]

Polymers (plastics) placed in the sun must withstand UV radiation, thermal cycling (day/night), and the resulting expansion and contraction. The simplest solar collector (Fig. 1) consists of an absorber plate, usually black, which may be steel, aluminium, copper, or plastic. The material used determines efficiency and transferability of sun radiation to the heat but also cost and lifetime of a collector. Glazing the plate with glass or polymers reduces radiation and convection losses. The glazing material should transmit solar radiation and not release radiation from absorber. Higher operating solar collectors employ two glazings. The outer glazing

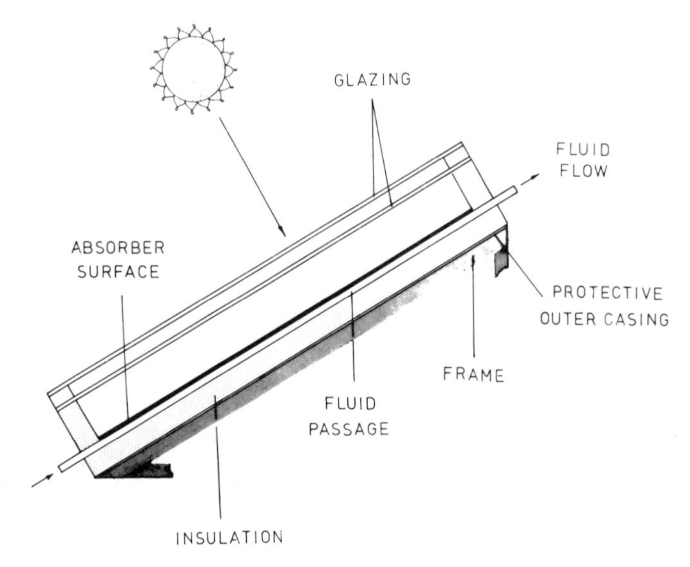

FIG. 1. A flat plate solar collector panel. Reproduced from ref. 2 by permission of Ann Arbor Science Publishers, Ann Arbor, Michigan.

must withstand the environmental corrosion while the inner glazing must be temperature resistant. None of the polymeric materials are completely satisfactory either as an outer or inner glazing. The high temperature inside the collector eliminates most polymers other than fluorocarbon polymers. Glass is the most common outer glazing, because most polymers have low environmental durability. Other solar collectors such as backing and heat-transfer fluid pathways can be made from extruded polycarbonate or designed from polyacrylates and polyethersulphone. In low-temperature application solar collectors (e.g. for swimming pool heating) plastics are widely employed because of their high resistance to corrosive pool water chemicals. Polymers presently being used include polypropylene, ethylene–propylene copolymer, EPDM and poly(vinyl chloride). Absorber plates are produced of polypropylene with carbon black filler, which acts as a heat absorber and UV screen. Such open solar collectors where the absorber has no glazing are widely used and their normal operating temperature is 20–30 °C with maximum 40 °C. In solar collectors for space and water heating, higher operating temperatures are involved, and most absorber plates are made of metal (aluminium), and plastics are used to glaze these collectors (mirrors). Aluminised acrylics have performed well for up to five years in the tested heliostats. As protective covers metallised

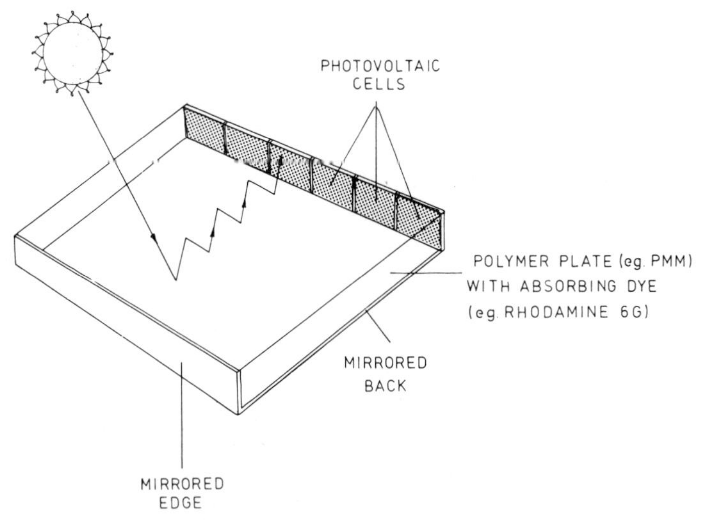

FIG. 2. A luminescent solar concentrator. Reproduced from ref. 8 by permission of the American Chemical Society, Washington, DC.

polymeric mirrors are used with bubbles made of thin, air-supported transparent polyesters films. Biaxially oriented poly(vinylidene fluoride) is an excellent material for this purpose having good optical properties and weatherability.

Polymers (plastics) employed in sun collectors must not only withstand UV radiation but have suitable mechanical properties and durability to maintain operation under the combined effects of wind, hail, atmospheric corrosion, pollutants, for several years (5–20 years).

(iii) Luminescence Solar Concentrators[9–11]

In a planar solar concentrator (Fig. 2) a transparent polymer (e.g. poly(methyl methacrylate) is impregnated with guest luminescent absorbers (e.g. organic dye molecules) having strong absorption bands in the visible and UV regions of the spectrum, and also having an efficient quantum yield of emission. Solar photons entering the upper face of the plate are absorbed, and luminescent photons are then emitted. A large fraction of these luminescent photons are trapped by total internal reflection (e.g. 74% of an isotropic emission will be trapped in a poly(methyl methacrylate) plate with an index of refraction of 1·49). Successive reflections transport the luminescent photons to the edge of the plate where they can enter an edge-mounted array of photovoltaic cells.

(iv) Encapsulating Photovoltaic Cells with Plastics[12-15]

Photovoltaic solar cells are semiconductor devices that convert sunlight directly to d.c. electricity. Exposure to sunlight causes electrons to flow from one layer of the cell to another. This electric current can be tapped by attaching wires to the layers. The core of the encapsulation system (Fig. 3) is the pottant, which embeds the solar cells and related electrical conductors. Polyacrylates, polysilicones and polyfluorocarbons are employed as pottants. Functional requirements for a polymer pottant are: high transparency in the range of solar cell response, rigid support for the brittle solar cells, electrical insulation to isolate module voltage and to prevent leakage of electric current, minimal temperature build-up, mechanical resistance against wind, ice, and other objects, resistance to oxidation, hydrolysis, and other degradation, 20-year tolerance of daily thermal cycles, and cost-effective and module fabrication processes. Several polymers can be considered as pottant material, such as ethylene–vinyl acetate copolymers, ethylene–propylene rubber, and plasticised poly(vinyl chloride). Self-contained UV protection can be

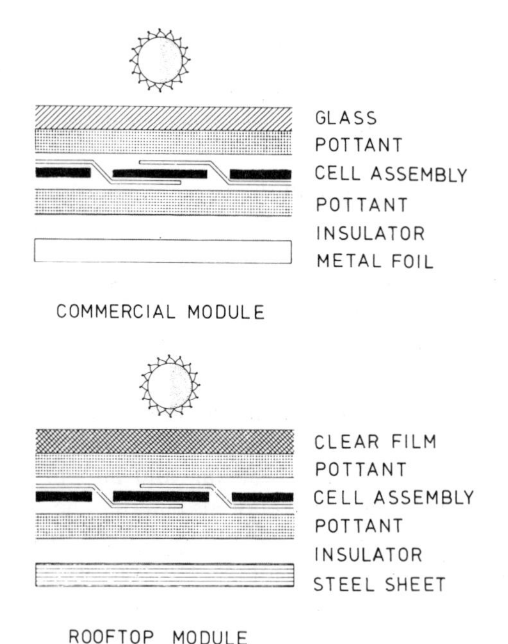

FIG. 3. Photovoltaic cell components. Reproduced from ref. 14 by permission of the American Chemical Society, Washington, DC.

provided by application of UV stabilisers or UV-screening hardtop cover films, such as UV-stabilised polyacrylates, poly(vinyl fluoride) or polyesters.

(v) Other Applications in Solar Technologies[4,5]

In addition to such conventional application as adhesives, coatings, moisture barriers, electrical and thermal insulation, and structural frames, polymers are used as optical components such as Fresnel lenses, light conductors in solar systems. Glass-reinforced plastics have been employed for the frame construction. Strong, high-temperature-resistant frames can be produced using liquid epoxy resins to impregnate continuous-strand glass mat. Using the reaction injection moulding (RIM) process it is possible to produce urethane foam frame. This cellular material provides good impact strength. Sealants in solar components are usually ethylene–propylenediene modified rubber (EPDM) or silicone. Most insulators used in collectors are polyurethane foams.

Polymers employed in solar technologies must maintain optical, chemical and mechanical properties during prolonged exposure to solar ultra-violet radiation. All polymers deteriorate under exposure to outdoor weathering and solar radiation but at greatly varying rates. All polymer photodegradation processes which involve chain scission and/or crosslinking occur by free radical mechanism by the following steps:

(i) photoinitiation step:
 absorption of ultra-violet radiation and visible light by internal and/or external chromophoric groups present in polymer;
 energy transfer of electronic excitation to some bonds;
 dissociation of some bonds into free radicals;

(ii) propagation step:
 subsequent reactions of free radicals in a chain process like abstraction of hydrogen atoms, reactions with molecular oxygen, disproportionation reactions, rearrangement reactions, etc.

(iii) termination step:
 subsequent reactions of free radicals between each other.

Chain scission reactions may occur during the photo-initiation, propagation and termination steps, whereas crosslinking is a result of the termination step. At elevated temperatures (up to 100 °C and even higher), the photodegradation process becomes even more complex. In the case of poly(vinyl chloride)[16] we have observed spontaneous dehydrochlorination

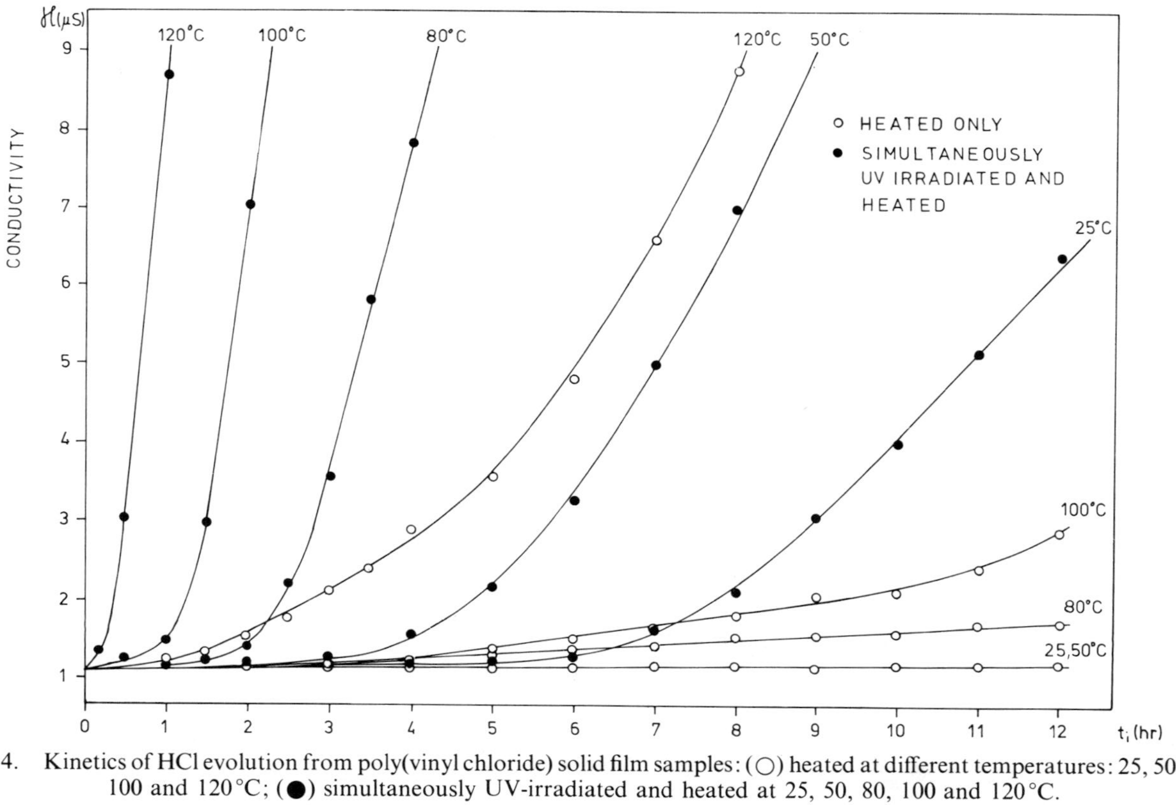

FIG. 4. Kinetics of HCl evolution from poly(vinyl chloride) solid film samples: (○) heated at different temperatures: 25, 50, 80, 100 and 120 °C; (●) simultaneously UV-irradiated and heated at 25, 50, 80, 100 and 120 °C.

when samples were UV irradiated at temperatures higher than 50 °C (Fig. 4). Photothermal degradation of polypropylene, poly(methyl methacrylate), poly(methyl acrylate) and their copolymers and blends have been also a subject of intensive investigation.[17–21] Chain scission and/or crosslinking are both effects of photodegradation and thermal degradation of all polymers and they exhibit different effects on mechanical properties of polymeric (plastic) materials (Fig. 5).[22] Simultaneously, during long exposure to sunlight radiation, low molecular weight materials, such as plasticisers, processing aid additives, stabilisers and antioxidants diffuse out of the polymer matrix. Loss of UV stabilisers, antioxidants and antiozonants causes the polymer to become more susceptible to the ensuing degradation. The less volatile degradation products can coat the absorber plate and the glazing, thus decreasing the efficiency of the collector. The cumulative changes may result in yellowing, loss of transparency, change in

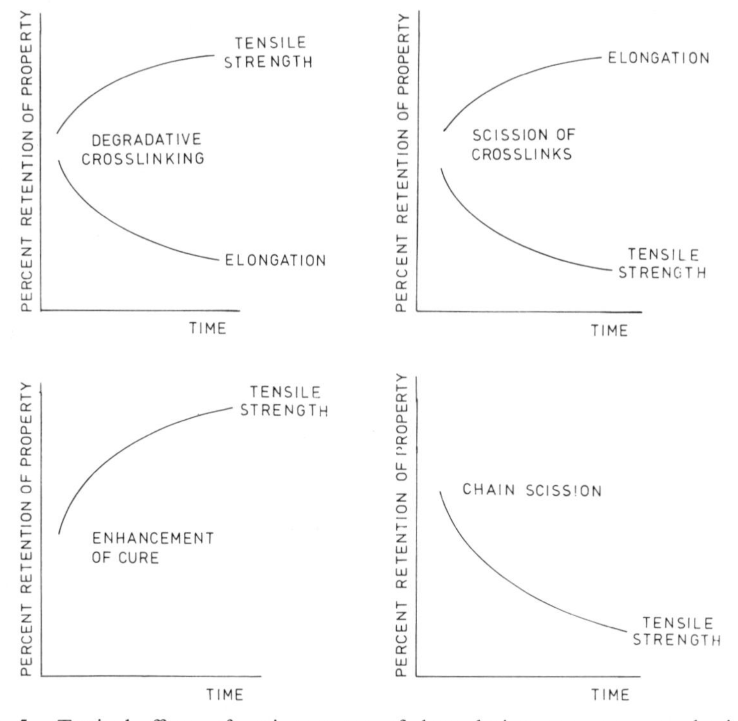

FIG. 5. Typical effects of various types of degradation processes on physical properties. Reproduced from ref. 22 by permission of the American Chemical Society, Washington, DC.

refractive index, and deterioration of surface properties (e.g. crazing). Surface phenomena play a significant role in the long-term application of polymers in solar technologies. The accumulation of airborne particulates and aerosols on the optical surfaces of optical elements causes unwanted absorption and scattering which lower operating efficiencies more than 30%. Examples of surface problems affecting mirrors include abrasion, dust adhesion, and cleaning procedures. Regular cleaning is required to maintain high optical performance of all solar collectors. Loss of adhesion of photothermally degraded polymers causes permeation problems. Oxygen, moisture, air pollutants (SO_2, NO_x, O_3) can penetrate polymer film and attack underlying reflector metallisation, conductors, or other functional elements.

Environmental corrosion of polymers[23] in solar technology application is a very complex problem which includes: photodegradation, thermal degradation, photothermal degradation, weathering, bio(soil) degradation, change of physical (optical) and mechanical properties, dimensional stability, permeability, etc. Effective methods for the detection of early stages of degradation are specially required. The latest methods employed are luminescence matrix spectroscopy,[24] chemiluminescence, Fourier-Transform[25] Infrared Spectroscopy, photoacoustic techniques.[26] Improved analytical and test methods are still needed to provide correlation between different environmental corrosion parameters and real-time behaviour. Performance prediction modelling is a new method for evaluating polymeric materials performance.[27]

Photodegradation of polymers can be stabilised by ultraviolet screens, absorbers, quenchers, radical scavengers and antioxidants.[28] Ultraviolet stabilisers can be incorporated by simple addition or in form of immobilised parts of polymers (polymeric absorbers). Several efficient UV stabilisers exhibit, in the presence of other additives (e.g. antioxidants), synergetic or antagonistic effects. Testing of all combinations of polymer additives for their synergetic/antagonistic effects requires an unacceptable number of experiments. The number of such combinations studied can be limited by better understanding of mechanism of synergism. Computer-controlled experimentation now makes it possible. It has been estimated that almost two thirds of all residential and commercial energy consumption in Sweden in the year 2000 will be obtained from solar energy conversion, wind, geothermal sources, etc. The wealth of this idea is a promoting source for us to study the problems of photodegradation and photostabilisation of polymers which would be practically employed in solar technologies.

ROLE OF POLYMERS IN DIRECT SOLAR ENERGY CONVERSION PROCESSES

(i) Photoreduction of Water[6,7]

Water photolysis is the simplest among the chemical conversion systems of solar energy, and can be brought about by a number of reactions, e.g. two-electron photoreduction:

$$2H^+ + 2e \dashrightarrow H_2 \tag{1}$$

and four-electron photo-oxidation:

$$2H_2O \dashrightarrow O_2 + 4H^+ + 4e \tag{2}$$

Both processes can be practically realised by employing photochemical redox systems which rely on electron-transfer reactions between the excited state of light-absorbing species (sensitiser (S)) and another substance which is either reduced (Scheme 1) or oxidised (Scheme 2):

$$S + h\nu \dashrightarrow S^* \tag{3}$$

$$S^* + A \dashrightarrow S^+ + A^- \tag{4}$$

$$S^+ + D \dashrightarrow S + D^+ \tag{5}$$

$$A^- + H_2O \xrightarrow{\text{catal.}} A + OH^- + \tfrac{1}{2}H_2 \tag{6}$$

SCHEME 1.

or

$$S^+ + h\nu \dashrightarrow S^{+*} \tag{7}$$

$$S^{+*} + D \dashrightarrow S + D^+ \tag{8}$$

$$S + A \dashrightarrow S^+ + A^- \tag{9}$$

$$A^- + H_2O \xrightarrow{\text{catal.}} A + OH^- + \tfrac{1}{2}H_2 \tag{10}$$

SCHEME 2.

Some components of systems for the photoreduction of water are shown in Table 1. Many of these components can be replaced by polymeric materials, which improve of such systems efficiency.

The most investigated system (Scheme 1) consist of $Ru^{2+}(bpy)_3$, EDTA, MV^{2+} and colloidal platinum (cf. Table 1 for abbreviation).[6,7] The ruthenium complex (sensitiser, S) employed in such system has a strong charge-transfer absorption with maximum at 452 nm (near the peak of solar spectrum, which is ca. 500 nm), it luminesces strongly and its excited

TABLE 1
Basic Compounds Used as Components of Systems for the Photoreduction of Water in Direct Solar Energy Conversion Processes

Photosensitisers:
Tris(2,2′-bipyridyl)ruthenium (II) complex (abbreviated as $Ru(bpy)_3^{2+}$)

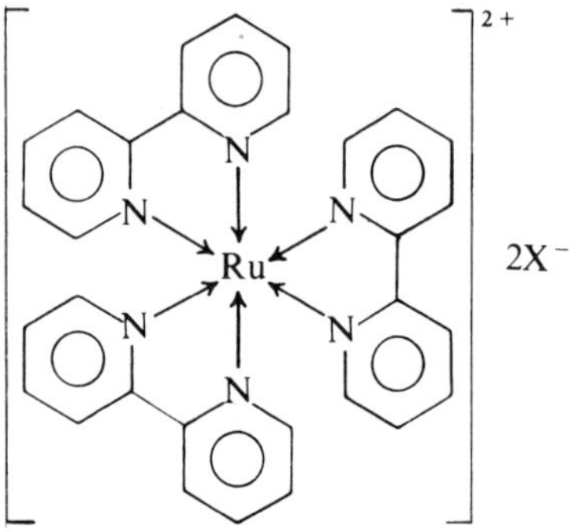

$2X^-$

Proflavin

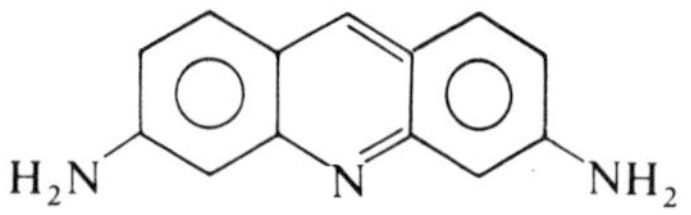

Acceptor (electron mediator, electron relay):
Methyl viologen (1,1′-dimethyl-4,4′-dipyridinium chloride) (abbreviated as MV^{2+})

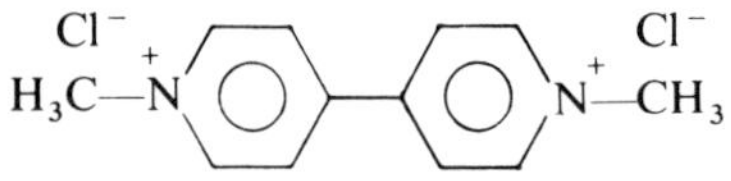

Donor:
Ethylenediamine tetra-acetic acid disodium salt (abbreviated as EDTA)

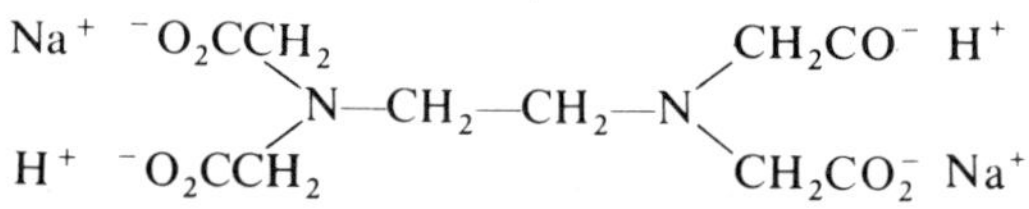

$Na^+ \; ^-O_2CCH_2$, $H^+ \; ^-O_2CCH_2$ >N—CH_2—CH_2—N< $CH_2CO^- \; H^+$, $CH_2CO_2^- \; Na^+$

Catalysator:
Colloidal platinum (abbreviated as Pt)
Ruthenium dioxide (RuO_2)

state is relatively long-lived (ca. 0·6 μs) with excitation energy of 2 eV (1·23 eV is required for each of four electron transfers in the four-electron photo-oxidation system). The first step is excitation of the ruthenium complex by light:

$$Ru(bpy)_3^{2+} + h\nu \dashrightarrow Ru(bpy)_3^{2+*} \quad (11)$$

An excited molecule of ruthenium complex reacts further with an acceptor molecule (methyl viologen) to yield the ion MV^+:

$$Ru(bpy)_3^{2+*} + MV^{2+} \dashrightarrow Ru(bpy)_3^{3+} + MV^+ \quad (12)$$

In the absence of donor the reverse of above reaction, to give ground state $Ru(bpy)_3^{2+}$, occurs rapidly, but in the presence of EDTA (ethylene-diaminetetra-acetic acid) it is prevented:

$$Ru(bpy)_3^{3+} + EDTA \dashrightarrow Ru(bpy)_3^{2+} + EDTA^+ \quad (13)$$

The reverse reaction does not occur because $EDTA^+$ is a weak oxidant. The methyl viologen ion MV^+ is thermodynamically capable of reducing water, but the reaction does not occur spontaneously. It must be catalysed by colloidal platinum:

$$2MV^+ + 2H^+ \xrightarrow{Pt} 2MV^{2+} + H_2 \quad (14)$$

The reaction is catalytic with respect to the ruthenium sensitiser and MV^{2+}. The role of the platinum catalyst in these reactions is probably that of a microelectrode.

A similar mechanism occurs with proflavin (P) as sensitiser, EDTA and colloidal platinum (Scheme 2):

$$P + h\nu \dashrightarrow P^* \quad (15)$$

$$P^* + EDTA \dashrightarrow P^- + EDTA^+ \quad (16)$$

$$2P^- + 2H^+ \xrightarrow{Pt} 2P + H_2 \quad (17)$$

Employing two catalysts, colloidal platinum for reduction and RuO_2 powder for oxidation, it is possible to carry out combined reduction and oxidation of water simultaneously, according to the reactions:

$$Ru(bpy)_3^{2+*} + MV^{2+} \dashrightarrow Ru(bpy)_3^{3+} + MV^+ \quad (18)$$

$$2MV^+ + 2H^+ \xrightarrow{Pt} 2MV^{2+} + H_2 \quad (19)$$

$$4Ru(bpy)_3^{3+} + 4OH^- \xrightarrow{RuO_2} 4Ru(bpy)_3^{2+} + O_2 + 2H_2O \quad (20)$$

The ion MV^+ is known to react rapidly with oxygen, but in this system MV^+ is apparently oxidised competitively by water on the colloidal platinum.

One of the most important problems in a photochemical solar energy conversion system is to prevent energy consuming back electron transfer which inevitably occurs in a homogeneous reaction system. One approach to solve this problem is the utilisation of a microheterogeneous reaction medium such as micelle, or liposome.[6,7] Another direction is the use of a macroheterogeneous reaction system such as solid–solution interphase, solid–solid interphase, or solution–solution interphase. In order to construct such a heterogeneous reaction system, polymer carriers are useful to immobilise the reaction components.[7] Several pendant tris(bipyridyl) ruthenium (II) complexes have been used as photosensitisers:[29–31]

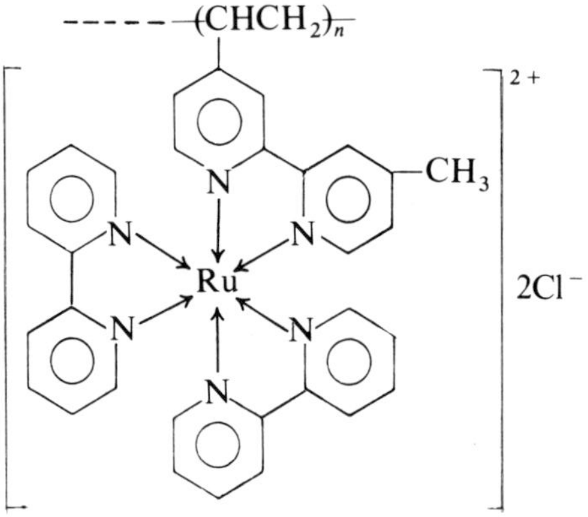

Poly(vinyl-Ru(bpy)$_3^{2+}$)

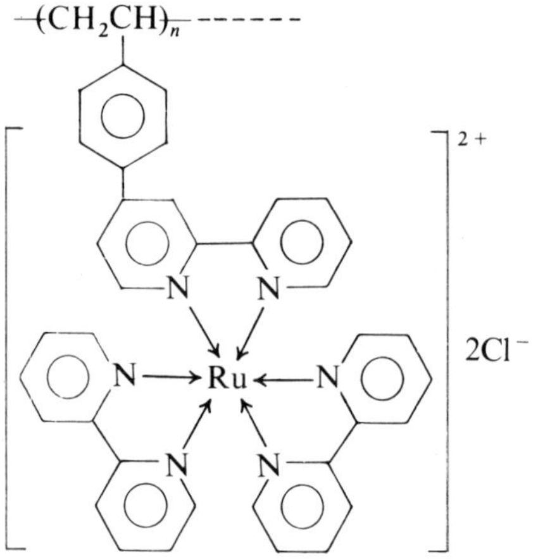

Poly(styryl-Ru(bpy)$_3^{2+}$)

Poly(styryl-Ru(bpy)$_3^{2+}$) complex has also been covalently linked to viologen:[32,33]

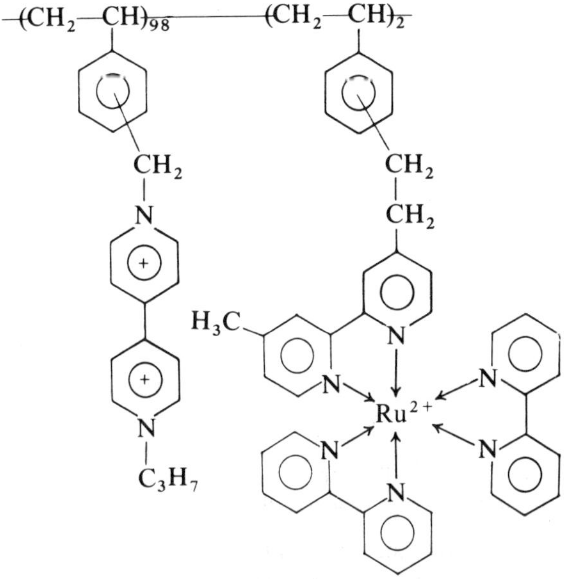

Such polymeric structures show high charge separation efficiency in MV$^+$ in the presence of EDTA as reducing agent.

A solid phase photochemical reaction in which the polymer matrix plays an important role in constructing heterogeneous conversion systems has been reported. The irradiation of cellulose paper after absorbing Ru(bpy)$_3^{2+}$, MV^{2+} and EDTA induces rapid formation of MV$^+$ in the solid phase.[33] The reducing power of the MV$^+$ in a solid phase can be transferred to liquid phase. The irradiation of the mixture of water-insoluble polystyrene–Ru(bpy)$^{2+}$ complex containing MV^{2+} and EDTA induces MV$^+$ formation in the liquid phase.[34]

Metal colloid and sol supported on a polymer matrix have been reported to be effective for photolysis of water. Platinum colloids are stabilised by water-soluble polymers such as poly(vinyl alcohol)[35] or poly(vinyl pyrrolidone).[36]

A completely polymeric unit which accomplishes solar induced water splitting to hydrogen and/or oxygen has not yet been devised.

(ii) Polymeric Photogalvanic Cells[5,7]

When a photochemical process in solution gives a photoresponse at the

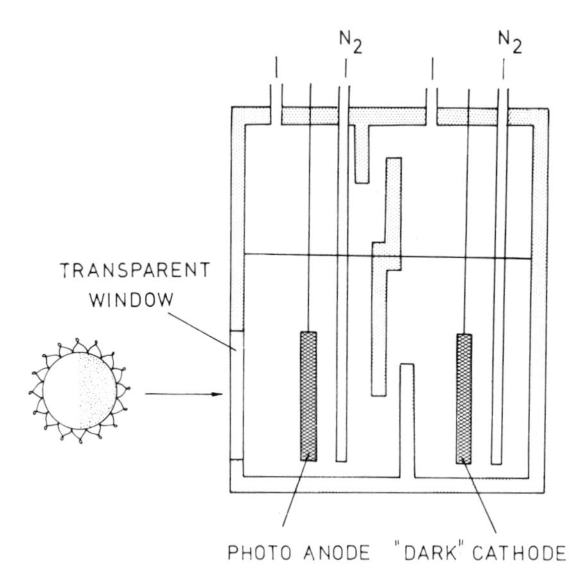

FIG. 6. Photochemical cell composed of light and dark chambers. Reproduced from ref. 7, by permission of Springer-Verlag, Heidelberg.

electrode, the system can form a photogalvanic cell. The photo-induced redox reaction is typical for photogalvanic cells. The most well known photoredox system is thionine (abbreviated to TH^+) and ferrous ion. The photoexcited TH^+ is reduced by Fe^{2+} to give TH_2^+ and Fe^{3+}:[37]

H$_2$N–(thionine ring: N, S)=$\overset{+}{N}H_2$

$$\underset{\text{(violet)}}{TH^+ + Fe^{2+} + H^+} \underset{\text{dark}}{\overset{\text{light}}{\rightleftharpoons}} \underset{\text{(colourless)}}{TH_2^{2+} + Fe^{3+}} \qquad (21)$$

This reversible photoredox system gives no photopotential when irradiated in a photochemical cell composed of light and dark chambers[38] (Fig. 6). A photogalvanic cell with polycation polymer pendant thionine (TH^+) has been constructed:[39]

$-(O-CH-CH_2)-$ with pendant CH_2–HN–(thionine ring: N, S)=$\overset{+}{N}H_2$ Cl^- ; $-(O-CH-CH_2)-$ with pendant CH_2–$^+N(CH_2)_3$ Cl^-

Several other polymeric photoredox systems which can be used for construction of photogalvanic cells have been developed.[40]

(iii) Polymeric Photovoltaic Materials

Thin films of organic photoconductive polymers can be employed for construction of photovoltaic devices. Photoconductivity is defined as a significant increase (by a factor $\geq 10^3$ to be of practical utility) in the conductivity under illumination. This increase is attributed to an increase in the number of charge carriers (electron and holes) as a direct result of electronic excitation. Photoconduction depends on:

photogeneration of charge carriers which have been pursued by excitation of the spectral response by addition of small molecules 'dopants' (spectral sensitisers) or increase in efficiency by chemical modification;

transport of charge carriers which have been pursued by 'doping' of polymers with small molecules. Transport of charge may then occur via either cationic (holes) or anionic (electrons) states.

The most promising polymers which can be employed for the photoconducting materials are semiconductive polymers such as poly(acetylene),[41-44] poly(1,6-heptadiyne)[45,46] and doubly conjugated poly(acrylonitrile):[47]

$-\!(CH{=}CH)_n\!-$

Poly(acetylene) Poly(1,6-heptadiyne) Poly(acrylonitrile)

Poly(acetylene) is the most extensively investigated 'metallic polymer' and can exist in three structural forms:[44]

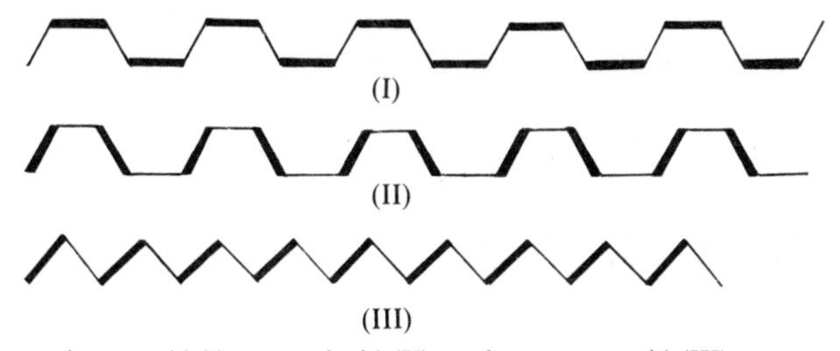

cis-transoid (I), trans-cisoid (II), and trans-transoid (III)

Poly(acetylenes) have a number of unpaired spins which are not localised on a single carbon atom, which are called solitons.[44] Such solitons are not fixed spatially and can migrate:

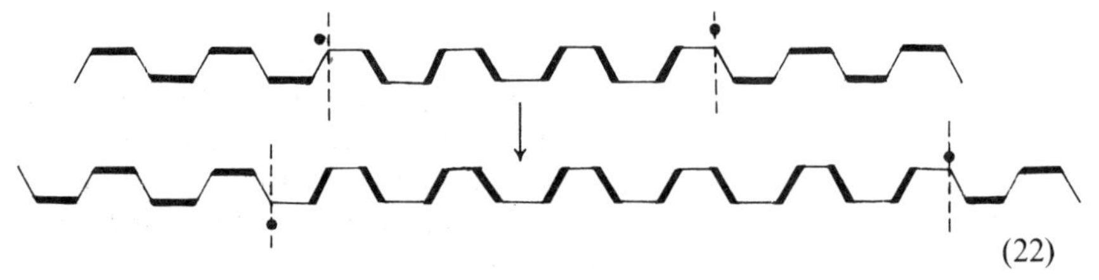

(22)

In the case of poly(acetylene), doping techniques include exposure of the films to the vapours of dopants such as I_2, HBr, AsF_5 and $SbCl_5$ or to solutions of dopants such as I_2, sodium naphthalide and NO_2SbF_6. Electrochemical oxidation of poly(acetylene) is a useful doping technique. In this manner the counter ion is incorporated in the electrolyte solution, e.g. ClO_4^-, I^- and BF_4^-. The photoconductivity can be controlled by controlling the dopant concentration. Doping with a donor (e.g. iodine or AsF_5) or an acceptor (alkaline metal) gives p- or n-junctions respectively.

The p-type dopant such as iodine can oxidise a neutral soliton directly, converting it to a delocalised carbenium ion:[44]

$+ I_2 \longrightarrow$ $I_x^{\ominus}$ $\oplus$ (23)

In this process a neutral soliton is converted to a positively charged spinless soliton.

The n-type dopant such as alkali metal reduces a neutral soliton:

$+ Na \longrightarrow$ (24)

$\ominus$ $Na^{\oplus}$

In this process a neutral soliton is converted to a negatively charged spinless soliton.

Alternatively iodine can oxidise the poly(acetylene) chain directly to produce a cation radical:

$+ I_2 \longrightarrow$ (25)

$I_x^{\ominus}$ $\oplus$

whereas alkali metal can reduce the poly(acetylene) chain to radical anion:

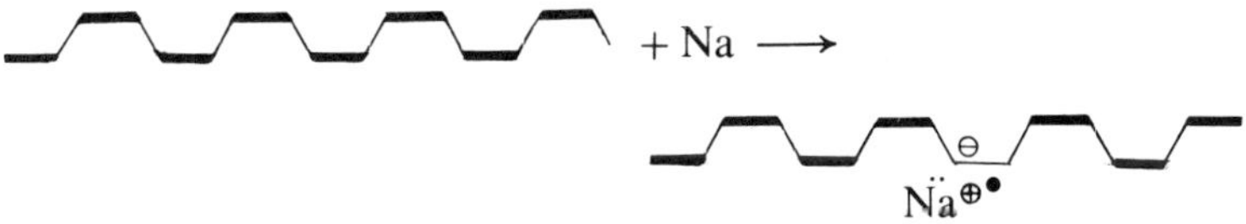

$$\text{poly(acetylene)} + \text{Na} \longrightarrow \text{poly(acetylene)}^{\ominus}\ \ddot{\text{Na}}^{\oplus\bullet} \qquad (26)$$

Such doped poly(acetylene) film containing spectral sensitisers has been employed for the design of a new type of solar cell[48,49] (Fig. 7).

Photovoltaic cells can also be designed using other polymeric materials such as poly(vinyl carbazole) or poly(vinylidenefluoride) doped with phthalocyanine[50] or polyester doped with merocyanine dye and coated with indium–tin oxide.[51]

Specially interesting are polymeric phthalocyanines in which metallo-macrocycles have face-to-face orientation:[52]

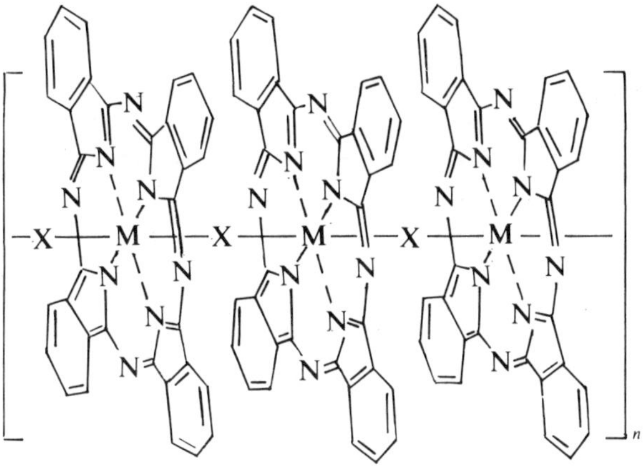

X = O or ORO

R: CH_2, $(CH_2)_2$, $(CH_2)_4$, $(CH_2)_6$, $(CH_2)_{12}$, $-C_6H_4-$, $-C_6H_4-C_6H_4-$, $-C_6H_4-C(CH_3)_2-C_6H_4-$

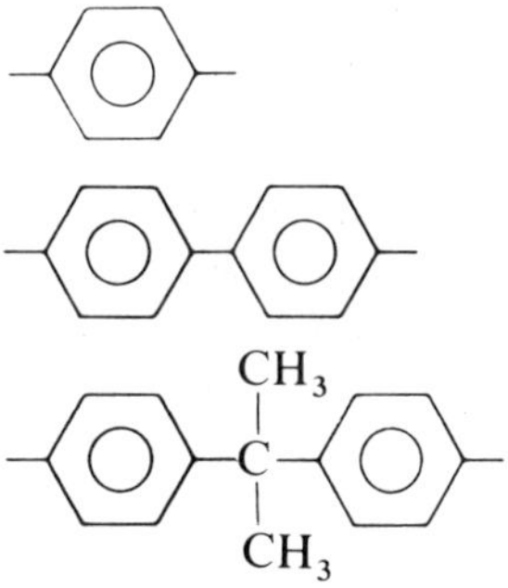

Strong photoactivity is associated with a molecular arrangement in the solid state corresponding to a staggered parallel plane dimers.

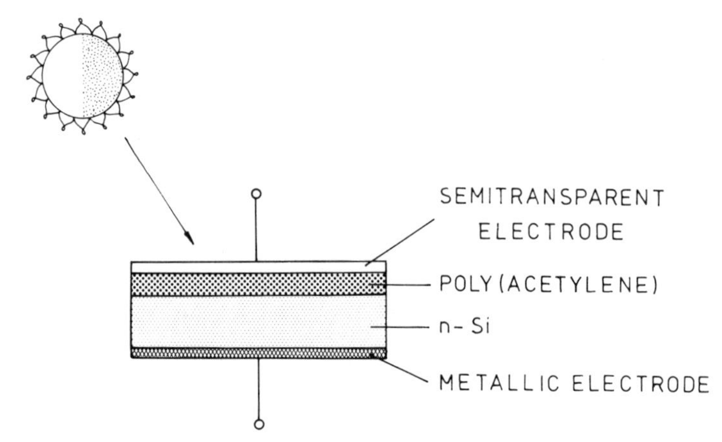

FIG. 7. A solar cell. Reproduced from ref. 7 by permission of Springer-Verlag, Heidelberg.

(iv) Solar Energy Storage

Solar energy can be stored by organic substances and polymers which can rapidly isomerise to an energy-rich unstable isomer under irradiation.[53] The most well-known system to store solar energy is the photoisomerisation of norbornadiene (NBD) to quadricyclane (QC):[54]

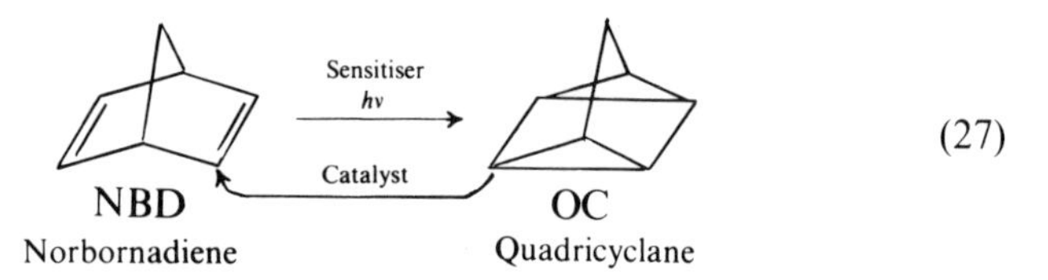

(27)

There are the following advantages of such a NBD–QC solar energy storage system:

norbornadiene (NBD) is readily available and comparatively inexpensive;
quadricyclane (QC) has a high storage capacity of 21 kcal/mol;
both norbornadiene and quadricyclane are liquids;
conditions exist for which both steps in the cyclical process can be virtually quantitative;

and also several disadvantages, such as:

norbornadiene has no absorption in the visible region and sensitisers such as aromatic ketones are required for the norbornadiene isomerisation by visible light;

the reverse reaction of quadricyclane to norbornadiene, which releases heat, must be catalysed, e.g. by Co(II)tetraphenylporphyrine;
both synthesiser and catalyst must be immobilised on polymers, ketones in the form of polyketones (e.g. poly-4-(N,N-dimethylamino)benzophenone)[55,56] and Co(II) tetraphenylporphyrine adsorbed on silica gel or anchored to polystyrene:[57]

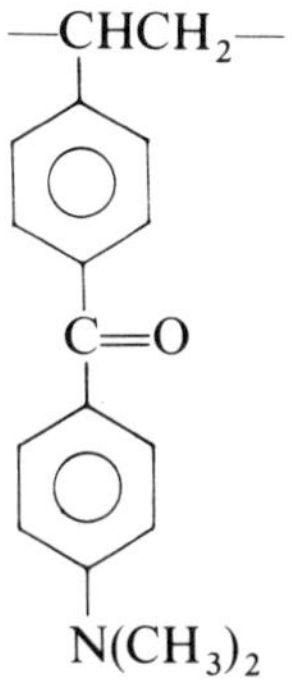

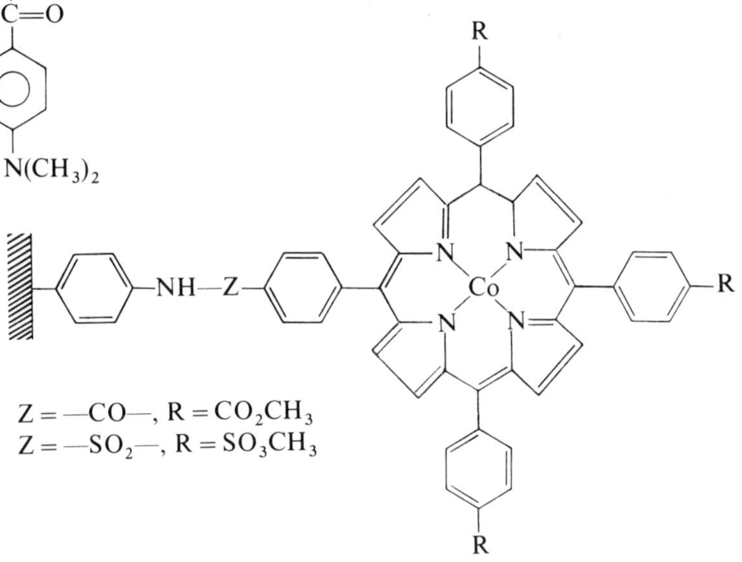

$Z = -CO-$, $R = CO_2CH_3$
$Z = -SO_2-$, $R = SO_3CH_3$

Low molecular sensitisers and catalysts may produce undesirable photoadducts with either norbornadiene or quadricyclane. Usually, the sensitising activity of the polymeric sensitiser or catalyst remains almost unchanged through immobilisation, but sometimes decreases depending on their structure. Activity decrease can be observed after several times with the recyclic NBD-CQ system.

A flow type model of solar energy storage device utilising polymer pendant sensitiser and catalyst is shown in Fig. 8.

Particularly important in direct solar energy conversion processes is the long-term stability of polymer pendant elements of such systems and of the photosensitisers, in particular the effect on its useful properties, when exposed for prolonged periods to solar radiation.

Polymeric photosensitisers employed in the solar energy storage systems, mainly poly(vinyl ketones) are very susceptible towards photolysis, which occurs by the Norrish type I and type II mechanisms.[58,59] The occurrence of Norrish types I and II reactions leads to modified photochemical behaviour of the polymer, and this influences the energy transfer characteristic of the polymer. When oxygen is present, the reactions become much more complicated on account of the concurrent photo-oxidation reactions, because atmospheric oxygen reacts readily with the initially formed radicals. The extent of degradation and photo-oxidation, however, is dependent on the position of the carbonyl group in the polymer, i.e. whether incorporated into the main chain or attached to a side group:

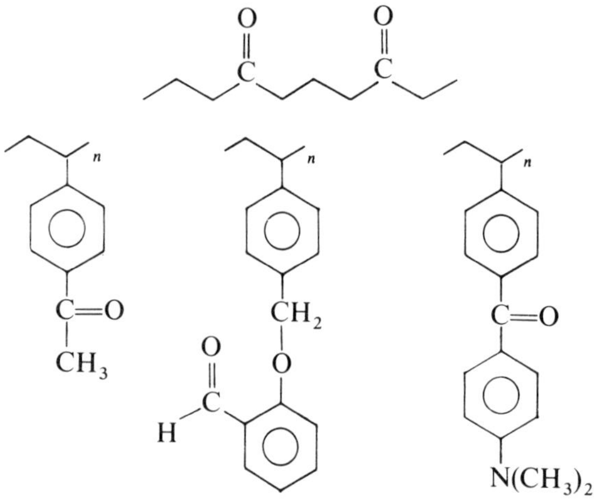

Energy transfer studies[60] have proved to be useful in the investigation of many problems related to the photodegradation and photostabilisation of polymers and has importance in the solution of several problems of solar energy conversion by polymers. Various factors affect the extent and rate of energy transfer between the excited donor and acceptor, whether they are in separate molecules (intermolecular energy transfer) or in different parts of the same molecule (intramolecular energy transfer):

the distance between them;
their orientation relative to each other;

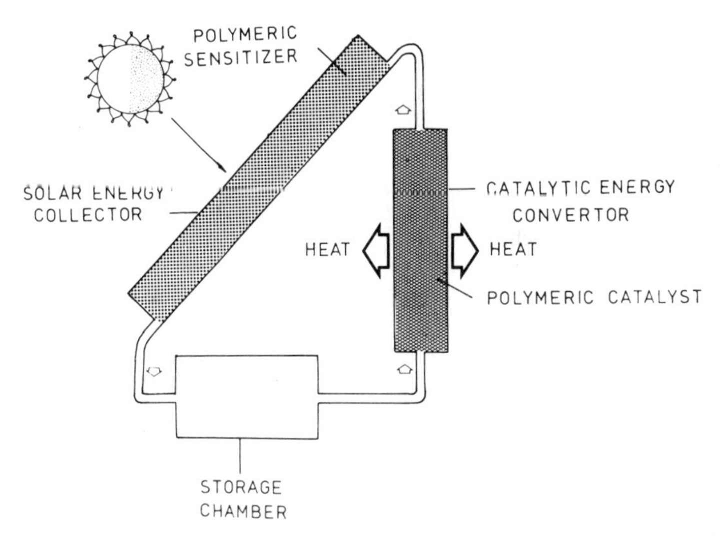

FIG. 8. Solar energy storage device. Reproduced from ref. 55 by permission of Pergamon Press, Oxford.

their spectroscopic properties;
the optical properties of the medium;
the effect of molecular collisions on the motion of the donor and acceptor in the period during which the donor is excited.

Studies of the photophysics of chromophores bound to polymers have revealed that these systems are often far more complex than corresponding small-molecule systems. In view of this complexity, the main goal in the practical application of polymers in solar energy should be a better understanding of energy transfer processes in polymeric systems.

ACKNOWLEDGEMENT

The investigations on photodegradation, photo-oxidation and photostabilisation of polymers and plastics which can be employed in solar technologies have been continuously, since 1971, carried out in the Department of Polymer Technology, the Royal Institute of Technology, and supported by the Swedish National Board for Technical Developments (STU), which the author gratefully acknowledges.

REFERENCES*

1. Sayigh, A. A. M. (Ed.), *Solar Energy Engineering*, Academic Press, New York, 1977.
2. Cheremmisinoff, P. N. and Regino, I. C., *Principles and Applications of Solar Energy*, Ann Arbor Science Publishers Inc., Ann Arbor, 1978.
3. Williams, R. H. (Ed.), *Towards a Solar Civilization*, The MIT Press, Cambridge, Massachusetts, 1978.
4. Geddes, K. A. and Deanin, R. A., *CHEMTECH*, Dec., 736 (1982).
5. Gebelein, C. G., Williams, D. J. and Deanin, R. D., (Eds), *Polymers in Solar Energy Utilization*, American Chemical Society Symposium Series, No. 220, ACS, Washington, D.C., 1983.
6. Connolly, J. S. (Ed.), *Photochemical Conversion and Storage of Solar Energy*, Academic Press, New York, 1981.
7. Kaneko, N. and Yamada, Y., *Adv. Polym. Sci.*, **55**, 1 (1984).
8. Carrol, W. F. and Schissel, P. In: American Chemical Society Symp. Ser. No. 220, p. 1, ACS, Washington, D.C., 1983.
9. Batchelder, J. S., Zewail, A. H. and Cole, T., *Appl. Optics*, **18**, 3090 (1979).
10. Batchelder, J. S., Zewail, A. H. and Cole, T., *Appl. Optics*, **20**, 3733 (1981).
11. Zewail, A. H. and Batchelder, J. S. In: American Chemical Society Symp. Ser. No. 220, p. 331, ACS, Washington, D.C., 1983.
12. Addeo, A., Bonadies, V., Carfagna, C., Guerne, G. and Moschetti, A., *Solar Energy*, **30**, 421 (1983).
13. Cuddihy, E. F., Coulbert, C. D., Willis, P., Baum, B., Garcia, A. and Minning, C. In: American Chemical Society Symp. Ser. No. 220, p. 355, ACS, Washington, D.C., 1983.
14. Lewis, K. J. In: American Chemical Society Symp. Ser. No. 220, p. 387, ACS, Washington, D.C., 1983.
15. Lewis, K. J. and Megerle, C. A. In: American Chemical Society Symp. Ser. No. 220, p. 387, ACS, Washington, D.C., 1983.
16. Rabek, J. F., Skowronski, T. and Rånby, B., in press.
17. Grassie, N., Scotney, A., Jenkins, R. and Davis, T. I., *Chem. Zvesti*, **26**, 208 (1972).
18. Grassie, N. and Leeming, W. B. H., *Europ. Polym. J.*, **11**, 809 (1975).
19. Grassie, N. and Leeming, W. B. H., *Europ. Polym. J.*, **11**, 819 (1975).
20. Grassie, N. and Davidson, A. J., *Polym. Degrad. Stab.*, **3**, 25, 45 (1980–81).
21. Grassie, N. and Holmes, A. S., *Polym. Degrad. Stab.*, **3**, 145 (1980–81).
22. Mendelson, M. A., Navish, F. W., Jr, Luck, R. M. and Yeoman, F. A. In: American Chemical Society Symp. Ser. No. 220, p. 39, ACS, Washington, D.C., 1983.
23. Rånby, B. and Rabek, J. F. In: American Chemical Society Symp. Ser. No. 229, p. 291, ACS, Washington, D.C., 1983.
24. Rabek, J. F., *Experimental Methods in Photochemistry and Photophysics*, John Wiley, Chichester, 1982.

* Only some of representative references are given in this chapter. The number of published papers, reports and patents in this field exceed at present several hundreds.

25. Rabek, J. F., *Experimental Methods in Polymer Chemistry*, John Wiley, Chichester, 1980.
26. Liang, R. H., Coulter, D. R., Dao, C. and Gupta, A. In: American Chemical Society Symp. Ser. No. 220, p. 265, ACS, Washington, D.C., 1983.
27. Guillet, J. E., Somersall, A. C. and Gordon, J. W. In: American Chemical Society Symp. Ser. No. 220, p. 217, ACS, Washington, D.C., 1983.
28. Rånby, B. and Rabek, J. F., *Photodegradation, Photooxidation and Photostabilization of Polymers*, John Wiley, London, 1975.
29. Kaneko, M., Yamada, A. and Kurimura, Y., *Inorg. Chem. Acta*, **45**, L73 (1980).
30. Kaneko, M., Nemoto, S., Yamada, A. and Kurimura, Y., *Inorg. Chem. Acta*, **44**, L289 (1980).
31. Kaneko, M., Yamada, A., Tsuchida, E. and Kurimura, Y., *J. Polym. Sci.*, **20**, 593 (1982).
32. Matsuo, T., Sakamoto, T., Takura, K., Sakurai, K. and Ohsako, T., *J. phys. Chem.*, **85**, 1277 (1981).
33. Kaneko, M., Motoyashi, J. and Yamada, A., *Nature*, **285**, 468 (1980).
34. Kaneko, M., Ochiai, M., Kinoshita, K., Jr and Yamada, Y., *J. Polym. Sci., A1*, **20**, 1011 (1982).
35. Kiwi, J. and Grätzel, M., *Nature*, **281**, 657 (1979).
36. Toshima, N., Kuriyama, M., Yamada, Y. and Hirai, H., *Chem. Lett.*, **1981**, 793.
37. Rabinowich, E., *J. chem. Phys.*, **8**, 551, 560 (1940).
38. Kaneko, M. and Yamada, A., *Rep. Ind. Phys. Chem. Res.*, **52**, 210 (1976).
39. Shigehara, K., Sano, H. and Tsuchida, E., *Makromol. Chem.*, **179**, 1531 (1978).
40. Kaneko, M. and Yamada, Y., *J. Phys. Chem.*, **81**, 1213 (1977).
41. Ito, T., Shirakawa, H. and Ikeda, S., *J. Polym. Sci., A1*, **12**, 11 (1974).
42. Chiang, C. K., Fincher, C. R., Jr, Park, Y. W., Heager, A., Shirakawa, J., Louis, E. J., Gau, S. C. and MacDiarmid, A. C., *Phys. Rev. Lett.*, **39**, 1098 (1977).
43. Shirakawa, H., Ito, I. and Ikeda, S., *Makromol. Chem.*, **179**, 1565 (1978).
44. Chien, J. C. W., *J. Polym. Sci., B*, **19**, 249 (1981).
45. Gibson, H. W., Bailey, F. C., Epstein, A. J., Rommelman, H. and Pochan, J. M., *J. Chem. Soc.* (*Chem. Commun.*), **1980**, 426.
46. Pochan, J. M., Pochan, D. F. and Gibson, H. W., *Polymer*, **22**, 1367 (1981).
47. Metz, P. D., Teoh, H., Vanderhart, D. L. and Wiche, W. G. In: American Chemical Society Symp. Ser. No. 220, p. 421, ACS, Washington, D.C., 1983.
48. Kaneko, M. and Yamada, A., *Photochem. Photobiol.*, **33**, 793 (1981).
49. Shirakawa, H. and Ikeda, S., *Kobunski*, **28**, 369 (1979).
50. Miniami, N. and Sasaki, K., *Proc. Autumn Meet. Chem. Soc. Japan*, **1981**, 1502.
51. Moriizumi, T. and Kudo, K., *Appl. Phys. Lett.*, **38**, 85 (1981).
52. Branson, B., Duff, J., Hsiao, C. K. and Loutfy, R. O. In: American Chemical Society Symp. Ser. No. 220, p. 437, ACS, Washington, D.C., 1983.
53. Jones, G., Chiang, S. H. and Xuan, P. P., *J. Photochem.*, **10**, 1 (1979).
54. Schwendiman, D. P. and Kutel, C., *J. Am. chem. Soc.*, **93**, 5677 (1977).
55. Hautala, R., Little, J. and Sweet, E., *Solar Energy*, **19**, 503 (1977).

56. Hautala, R. R. and Little, J., *Adv. Chem. Ser.*, **184**, 1 (1980).
57. King, R. B. and Sweet, E. M., *J. Org. Chem.*, **44**, 385 (1979).
58. David, C., Demarteau, W. and Geuskens, G., *Polymer*, **8**, 497 (1967).
59. Weir, N. A., *J. Polym. Sci., A1*, **17**, 3723 (1979).
60. Owen, E. D. In: *Developments in Polymer Photochemistry* (Ed. N. S. Allen) Vol. 1, Applied Science Publishers, London, 1980, p. 1.

17

Photocatalytic H_2 Production from Aqueous Polymer Solution by a Suspension of TiO_2/Pt

S. NISHIMOTO, B. OHTANI and V. T. KAGIYA

Department of Hydrocarbon Chemistry, Kyoto University, Japan

1. INTRODUCTION: 'SEMICONDUCTOR PHOTOCATALYSIS'

One of the current interests in photochemistry is the photocatalytic reaction on semiconductors, i.e. redox-including chemical reaction on the semiconductor surface induced by photo-irradiation.[1-3] The primary step of the photocatalytic reaction involves the absorption of light energy by a semiconductor which is characterised by a band structure, i.e. a filled valence band separated by an energy gap from a vacant conduction band (Fig. 1). Light of energy greater or equal to the band gap promotes excitation of an electron from the valence band to the conduction band, leaving the valence band electron-deficient, thus generating a positive hole. These active species, electron and positive hole, can reduce and oxidise substrates, respectively. The activity of photo-irradiated semiconductors for chemical reactions depends not only on the formation of these active species but also the subsequent charge transfer through the semiconductor–solution interface.[4-5]

Among semiconductor materials n-type titanium dioxide (TiO_2) has been extensively used as a stable photocatalyst suspended in an aqueous solution. According to the electrochemical studies on an irradiated TiO_2 anode,[6] the energies of the electron and the positive hole are evaluated to be sufficient for the reduction and oxidation of water into hydrogen and oxygen,[7] respectively (Fig. 2).

Noticeable attempts have been made to utilise the photogenerated active species in separation without their mutual recombination. One of the methods toward efficient charge separation is modification of the TiO_2 surface by partial coverage with metal and/or metal oxide, such as platinum[8] and/or ruthenium dioxide,[9-10] according to the idea of

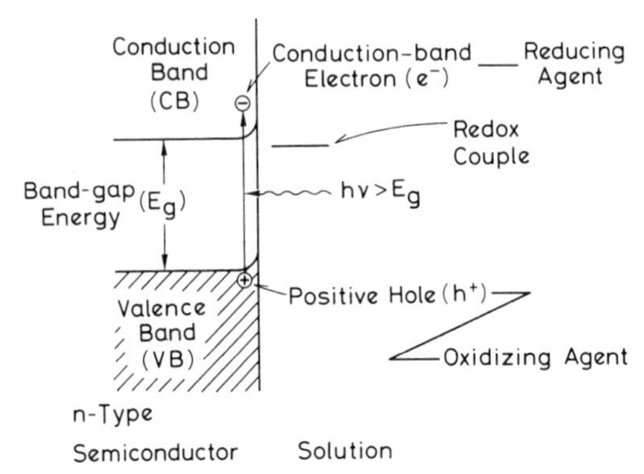

FIG. 1. Photo-excitation of n-type semiconductor with energy of $h\nu > E_g$ to produce a pair of conduction-band electron and positive hole.

coupling semiconductor anode and metal cathode which have been used separately in a photoelectrochemical cell.[11] In the modified, typically platinised, TiO_2 particles the photo-excited electron could effectively reduce proton to hydrogen on the metal surface[12] to enhance total photocatalytic activity (Fig. 3). Such platinised TiO_2 (TiO_2/Pt) powders have been successfully prepared by *in situ* photoreduction of hexachloroplatinum ion[13-14] or simply by mixing with platinum black.[15] The photocatalyst thus prepared has ability sufficient to decompose gaseous water into oxygen and hydrogen[16] or for the oxidation of organic compounds, such as primary alcohols,[17-18] amines,[19] or sugar,[20-21] with the simultaneous formation of hydrogen.

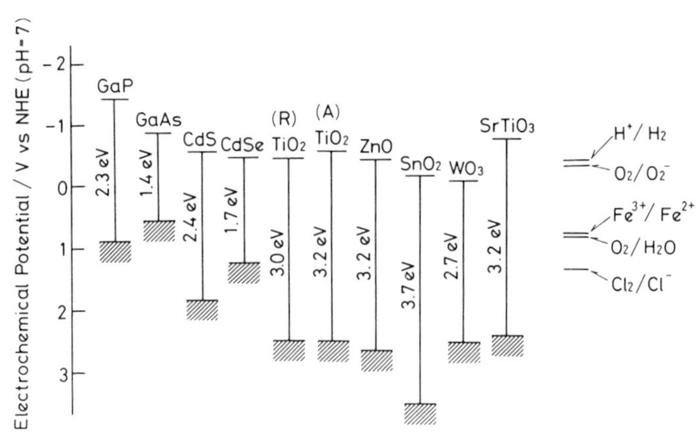

FIG. 2. Potential of conduction and valence bands of various semiconductors; (A) and (R) show anatase and rutile (TiO_2), respectively.

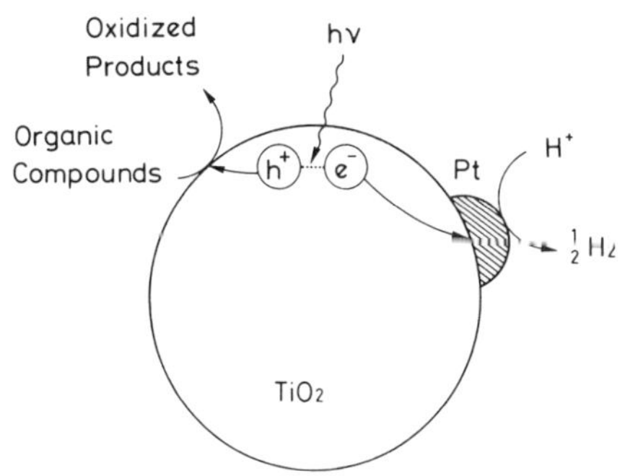

FIG. 3. Oxidation of organic compounds along with H_2 formation by platinised TiO_2 particle.

Thus, the metallised TiO_2 catalyst has been favourably applied to photo-induced reactions of several organic compounds. In these systems, however, the fate of organic substrates other than simple alcohols or acids has not yet been well characterised, particularly in the primary stage of the reaction,[22-23] because attention has been paid almost exclusively to the production of hydrogen from water,[9,24] but not to the oxidation products in aqueous solution. Consequently, semiconductor photocatalysis needs to be characterised in further detail.

In this respect, this report is concerned with reaction characteristics of a series of polymeric materials, especially poly(vinyl alcohol) (PVA), and low molecular-weight model compounds.

2. TiO_2/Pt PHOTOCATALYTIC REACTION OF POLYMERS IN AQUEOUS SUSPENSION

The TiO_2 photocatalytic reaction of polymers is summarised in Table 1. Kawai and Sakata have reported the photocatalytic H_2 production from aqueous solution (or suspension) of polymers in the presence of suspended TiO_2/Pt catalyst under deaerated conditions.[20] They showed that the TiO_2/Pt catalyst prepared by photocatalytic reduction of hexachloro-platinum ion was effective for the photocatalytic H_2 formation, and that water-insoluble polymers such as polyethylene and poly(vinyl chloride) could also be oxidised to CO_2 as a final oxidised product along with the liberation of H_2 in the gas phase. In strongly alkaline suspensions (5 mol dm^{-3} NaOH), the amount of H_2 liberated from these polymers is larger than that in distilled water; CO_2 could not be detected in the gas phase due to the dissolution in the alkaline solution.

TABLE 1
Hydrogen Evolution from the Mixture of Aqueous Suspension of TiO_2/Pt and Polymers

Polymer	*$H_2/\mu mol$*			*Ref.*
	in H_2O	*in NaOH aq.*	*in H_2SO_4 aq.*	
Polyethylene	17	93	—	20[a]
Poly(vinyl alcohol)	45	86	—	20
Poly(vinyl chloride)	20	45	—	20
Polytetrafluoroethylene (teflon)	0·4	3·5	—	20
Poly(vinyl alcohol)	6	30	23	[b]
Poly(methyl vinyl ether)	1	20	1	[b]
Poly(acrylic acid)	7	22	9	[b]
Polyacrylamide	6	36	2	[b]
Poly(ethylene oxide)	6	27	9	[b]
Polyethyleneimine	45	45	34	[b]

[a] Polymer (150 mg), TiO_2/5 wt % Pt photocatalyst (photoelectrochemically deposited), and solvent (30 cm^3) were placed in a pyrex bulb and irradiated by a 500-W xenon lamp for 10 h under deaerated conditions. NaOH aq: 5 N.
[b] (This work) aqueous polymer (410 μmol monomer unit except for the case of polyethyleneimine (700 μmol)) solution (3·0 cm^3) containing TiO_2/5 wt % Pt (30 mg) was irradiated under Ar for 20 h by a 400-W high-pressure mercury arc. NaOH aq: 6 mol dm^{-3}, H_2SO_4 aq: 3 mol dm^{-3}.

TABLE 2
Hydrogen Evolution and Change in Intrinsic Viscosity $[\eta]$ in the Photoreaction of Aqueous Polymer Solutions by TiO_2/Pt (5 wt %)[a]

Polymer	*X*	*$H_2/\mu mol$*	$[\eta]_{t=0}$	$[\eta]_{t=20}$
$(\text{—}CH_2CH\text{—})_n$ with side group X	—OH[b]	123·2	0·27	0·22
	—COOH	7·0	3·10	2·20
	$\text{—}CONH_2$	6·5	28·68	16·84
$(\text{—}CH_2CH_2X\text{—})_n$	—O—	15·3	0·17	0·08
	—NH—	99·4	0·92	0·08

[a] Polymer 410 μmol monomer unit, TiO_2/Pt 30 mg, and water 3·0 cm^3 were placed in a test tube and irradiated under Ar for 20 h.
[b] Poly(vinyl alcohol) 0·90 g (21 mmol monomer unit), TiO_2/Pt 600 mg, and water 600 cm^3 were placed in an immersion-type reaction vessel equipped with a 200-W high pressure mercury arc and irradiated under Ar for 20 h.

Table 1 also shows the results of the photocatalytic reactions (high-pressure mercury arc, $\lambda_{ex} > 300$ nm) of water-soluble polymers by the TiO_2/Pt photocatalyst suspended in 3 mol dm^{-3} H_2SO_4, 6 mol dm^{-3} NaOH, and distilled water. In the typical experiment, anatase TiO_2 powder (Merck) was mixed with 5 wt % of platinum black to prepare the photocatalyst. Regardless of the types of polymers, the amount of H_2 evolved from 6 mol dm^{-3} NaOH solution was 3–40 times as large as that in H_2O, except the case of poly(ethylene-imine) (PEI). This observation is similar to the result reported by Kawai and Sakata as discussed in the following section. Among the gaseous products other than H_2, CO_2 was obtained from poly(acrylic acid) (PAA), and NH_3 from polyacrylamide and PEI.

Table 2 shows that intrinsic viscosity ($[\eta]$) of the polymer solutions decreases as the H_2 evolution proceeds, indicating the occurrence of the main-chain scission of polymers by the photocatalytic reaction.

3. TiO_2 PHOTOCATALYTIC REACTION OF POLY(VINYL ALCOHOL) IN AQUEOUS SOLUTION[25]

Photo-irradiation ($\lambda_{ex} > 300$ nm) of TiO_2/Pt-suspended aqueous PVA solution led to the formation of H_2 in the gas phase of the reaction system as shown in Table 1. Figure 4 shows the dependence of the H_2 formation in the PVA system on the amount of Pt loading. Negligible amounts of H_2 could be obtained by unplatinised TiO_2 powder. The H_2 yield over the irradiation period of 20 h increased drastically upon increasing Pt up to 1·0 wt %, and was practically constant over the Pt content of 5–20 wt %. Thus, a small amount of the Pt loading is effective for H_2 formation in the present system. Furthermore, the reaction could be observed neither in the dark nor with Pt alone, indicating that the reaction is initiated by photo-absorption ($h\nu <$ ca. 390 nm)[26] of anatase TiO_2 to form electron–hole pairs as primary active species.

$$TiO_2 + h\nu \longrightarrow e^- + h^+ \qquad (1)$$

The electron thus formed would reduce a proton to produce H_2 at the Pt side of the photocatalyst.[22]

$$e^- + H^+ \rightleftharpoons Pt{-}H \longrightarrow \tfrac{1}{2}H_2 \qquad (2)$$

Figure 4 also shows the pH-dependence of the H_2 formation by TiO_2/Pt (5 wt %). The H_2 yield (Y_{H_2}) by the irradiation for 20 h was almost constant

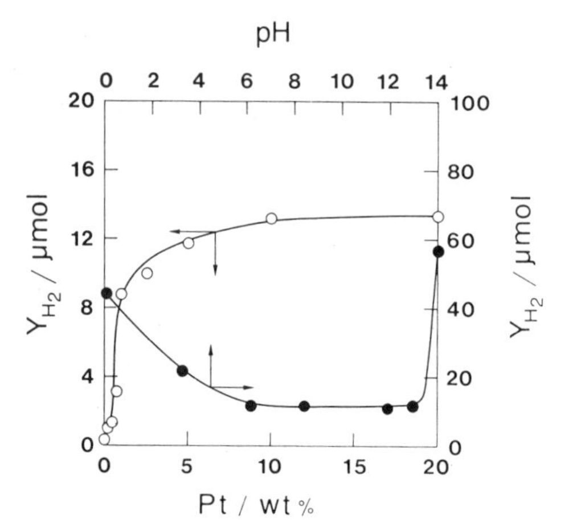

FIG. 4. Influence of Pt loading (○) and pH of the reaction mixture (●) on the photocatalytic H_2 formation in aqueous TiO_2/Pt (30 mg) suspension (3·0 cm^3) containing poly(vinyl alcohol) (410 μmol monomer unit). The suspensions were irradiated for 20 h by a 500-W high-pressure mercury arc.

(12–20 μmol) in the pH range of 3–13, while drastically increased in either lower (3 M H_2SO_4) or higher (6 M NaOH) pH regions. It is noted that PVA was much less soluble in alkaline solution (pH > 11). Similar enhancement of the TiO_2/Pt activity in the higher pH region (5 M NaOH) has been reported by Kawai and Sakata[20] as described in the preceding section. Such a pH-dependence of the reaction rate is attributable to modification of the TiO_2 powder by strong base treatment[27] to yield a surface OH group, which is expected to become an 'oxidation site' as discussed below.

$$\begin{array}{c} \diagdown \\ -\mathrm{Ti} \\ \diagup \quad \diagdown \\ \mathrm{O} \qquad \mathrm{O} \\ \diagdown \quad \diagup \\ -\mathrm{Ti} \\ \diagup \end{array} + H_2O \xrightarrow{\mathrm{NaOH}} \begin{array}{c} \diagdown \\ -\mathrm{Ti}-\mathrm{OH} \\ \diagup \\ \mathrm{O} \\ \diagdown \\ -\mathrm{Ti}-\mathrm{OH} \\ \diagup \end{array} \qquad (3)$$

Along with the H_2 formation, the intrinsic viscosity ([η]) of PVA solution decreased as a result of the main-chain scission. By the use of the viscosity–molecular weight relationship,[28–29] the average number of the main-chain scission was evaluated as 32 μmol at 190 μmol of H_2 liberation in the preparative photolysis for 150 h. In this photolysis, 18 μmol of CO_2 was also detected in the form of $BaCO_3$.

In a similar manner to the PVA system, photocatalytic activities of the

TABLE 3
Photocatalytic Reaction of Poly(vinyl alcohol) and Model Compounds by TiO_2 and TiO_2 Composites[a]

Catalyst	*$H_2/\mu mol$*		
	Poly(vinyl alcohol)	*2,4-pentane-diol*	*2-propanol*
TiO_2	0·4		1·1
TiO_2 + Pt black (5%)	8·7	16	81
$TiO_2 + RuO_2$ (10%)	4·6		12
TiO_2 + Pt black (5%) + RuO_2 (10%)	7·8		85

[a] Substrate (amount of OH group in the starting material; 410, 820, and 500 μmol for poly(vinyl alcohol)), 2,4-pentanediol, and 2-propanol, respectively), TiO_2/Pt (5 wt %, 30 mg), and distilled water (3·0 cm^3) were placed in a test tube and irradiated by a 400-W high-pressure mercury arc under Ar at room temperature for 20 h.

TiO_2/Pt catalyst for the reaction o 2-propanol and 2,4-pentanediol (PDO) as model compounds were studied in aqueous system (Table 3). In both cases H_2 was liberated in the gas phase together with formation of an almost equimolar amount of the corresponding carbonyl compound, acetone or 2-hydroxy-4-pentanone (HPO).[17,30–32]

$$(CH_3)_2CHOH \xrightarrow[h\nu > 300\,nm]{TiO_2/Pt} (CH_3)_2CO + H_2 \quad (4)$$

$$CH_3CH(OH)CH_2CH(OH)CH_3 \xrightarrow[h\nu > 300\,nm]{TiO_2/Pt} CH_3COCH_2CH(OH)CH_3 + H_2 \quad (5)$$

The amount of H_2 liberated over the 20-h irradiation decreased with the increase in molecular weight (Table 3); approximately 10-fold larger amount of H_2 was obtained using 2-propanol compared with the PVA solution. These facts are possibly accounted for by the adsorptive ability of the hydrophilic TiO_2 surface[33] for the alcohols.

The effect of co-catalyst Pt was evident in both cases; the activity of TiO_2 (anatase) alone was negligible but was increased drastically by the platinisation. The RuO_2 loading was similarly effective for the photocatalytic H_2 formation, but the enhancement ratio (ca. 10 times for both PVA and 2-propanol systems) was considerably smaller than that by Pt (20–70 times). It has been suggested that RuO_2 catalyses the H_2 generation with the photo-excited electron of TiO_2 (reaction 2).[34–35] The H_2 yield by

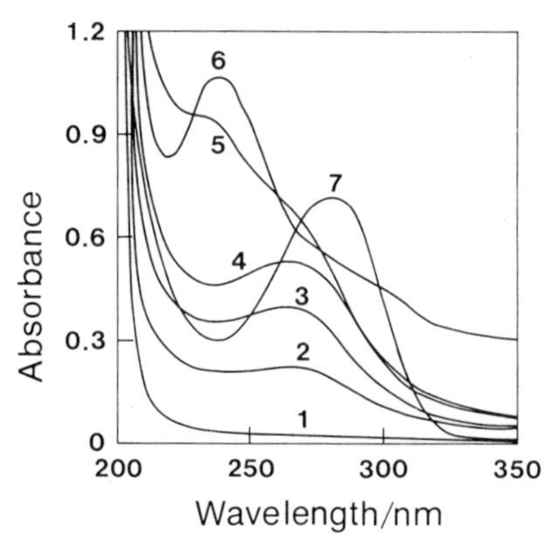

FIG. 5. Ultra-violet absorption spectra of aqueous poly(vinyl alcohol) solution (48 μmol monomer unit in 3·0 cm^3 of water) photo-irradiated ($\lambda_{ex} > 300$ nm) with TiO_2/Pt catalyst (30 mg) by a 400-W high-pressure mercury arc: 1, 0 h; 2, 5 h; 3, 24 h; 4, 37 h; 5, 75 h; 6, freeze-dried sample of poly(vinyl alcohol) irradiated for 150 h; 7, authentic sample of 4-hydroxy-2-pentanone.

TiO_2 loaded with both Pt and RuO_2 was practically the same as that with Pt alone. These facts suggest that the enhancement of photocatalytic activity by Pt and RuO_2 is achieved in a similar way; other factors, e.g. decrease of the incident photons by coverage with a large amount of deposits (Pt and/or RuO_2), decreases the activity. Another important point determining the photocatalytic activity is the crystal form of TiO_2; anatase TiO_2 shows sufficient activity for H_2 formation but the rutile analogue could only produce negligible amounts of H_2 even when loaded with Pt, in both PVA and 2-propanol. This behaviour can be attributed to the lower energy of the conduction-band electron (see Fig. 2 of the previous section) which reduces H^+ to H_2.

The formation of a carbonyl group on the PVA chain was suggested by comparison of the UV spectra of photo-irradiated aqueous solutions of PVA and the related model compounds (Fig. 5). The characteristic absorption band of PVA solution at $\lambda_{max} \simeq 270$ nm, which increased upon increasing the irradiation period, is assigned to β-ketol structure by reference to the spectrum of 2-hydroxy-4-pentanone (HPO) ($\lambda_{max} \simeq 280$ nm).

$$\text{—CH}_2\text{CH(OH)CH}_2\text{CH(OH)—} \xrightarrow[h\nu > 300\,\text{nm}]{TiO_2/Pt} \text{—CH}_2\text{COCH}_2\text{CH(OH)—} + \text{H}_2 \qquad (6)$$

Another absorption appeared at $\lambda_{max} \simeq 230$ nm by further irradiation (> 50 h). Since a dehydrated derivative of HPO, 3-penten-2-one, shows a similar absorption spectrum,[36] the shorter wavelength absorption band is possibly due to the formation of α,β-unsaturated ketone by dehydration of the β-ketol.

$$\text{—CH}_2\text{COCH}_2\text{CH(OH)—} \longrightarrow \text{—CH}_2\text{COCH=CH—} + \text{H}_2\text{O} \quad (7)$$

However, the proportion of α,β-unsaturated ketone must be small compared with the β-ketol, because the molar extinction coefficients (ε) of such α,β-unsaturated ketones are larger ($\log \varepsilon \simeq 4$) than those of the β-ketols ($\log \varepsilon = 1$–2).[36]

Since the two absorption bands were observed after freeze drying of the photo-irradiated PVA (see also Fig. 5), the PVA chain is suggested to bear the ketones. The number of ketones on the PVA chain was evaluated conveniently by the use of the reported molar extinction coefficient of HPO ($63\ M^{-1}\ cm^{-1}$),[36] being of the same order as the liberated H_2 (10 and 5·0 μmol, respectively, by 24-h irradiation).

A possible mechanism of the photocatalytic reaction including the formation of polymer radical is outlined as follows. The positive hole, which is produced together with electron on the irradiated TiO_2 (eqn 1), acts as an oxidising agent strong enough to oxidise a surface hydroxyl group to a hydroxyl radical adsorbed on the TiO_2 surface. Jaeger and Bard have demonstrated the formation of hydroxyl radical by ESR study on the TiO_2/Pt suspension.[37] The hydroxyl radical presumably abstracts hydrogen from the polymer chain to form a α-hydroxy polymer radical.[38]

$$\text{—CH}_2\text{—}\underset{\text{OH}}{\underset{|}{\text{CH}}}\text{—CH}_2\text{—}\underset{\text{OH}}{\underset{|}{\text{CH}}}\text{—} + {}^{\cdot}\text{OH} \longrightarrow \text{—CH}_2\text{—}\underset{\text{OH}}{\underset{|}{\dot{\text{C}}}}\text{—CH}_2\text{—}\underset{\text{OH}}{\underset{|}{\text{CH}}}\text{—} + \text{H}_2\text{O} \quad (8)$$

The ketone formation on the polymer chain could occur by subsequent oxidation of the polymer radical. Disproportionation of the polymer radicals to produce the ketone would be minor. It is likely that the α,β-unsaturated ketone is derived from dehydration of the ketol (eqn (7)). On the other hand, the main-chain cleavage is accounted for by the β-scission of the polymer radical with simultaneous formation of a ketone (terminal acetyl group).

$$\text{—CH}_2\text{—}\underset{\text{OH}}{\underset{|}{\dot{\text{C}}}}\text{—CH}_2\text{—}\underset{\text{OH}}{\underset{|}{\text{CH}}}\text{—} \xrightarrow{\beta\text{-scission}} \text{—CH}_2\text{—}\underset{\text{OH}}{\underset{|}{\text{C}}}\text{=CH}_2 + \underset{\text{OH}}{\underset{|}{{}^{\cdot}\text{CH}}}\text{—}$$

$$\updownarrow$$

$$\text{—CH}_2\text{—CO—CH}_3 \quad (9)$$

Kawai and Sakata reported the photocatalytic CO_2 formation by TiO_2/Pt suspension from various organic compounds as a result of complete oxidation of the backbone.[20] However, it seems more likely that the CO_2 formation in the present preparative photolysis is due to the decomposition[39,40] of a small amount of acetate ion eliminated catalytically in the dark or photocatalytically from the residual acetyl groups on the PVA, because maximally 200 μmol of acetyl groups are estimated to remain in 1·8 g of PVA.

4. SUMMARY

(1) Photocatalytic H_2 production was observed in the presence of the TiO_2/Pt photocatalyst suspended in the aqueous solution of polymers.

(2) Pt loaded on the TiO_2 surface facilitated the reduction of H^+ with the photo-excited electron in the TiO_2.

(3) The amount of H_2 formation was enhanced in alkaline conditions.

(4) PVA chain was oxidised to carbonyl derivatives or cleaved by way of an α-hydroxy polymer radical. The formation of the polymer radical was attributed to hydrogen abstraction by $^{\cdot}OH$ which is generated presumably by the oxidation of surface OH^- due to the positive hole.

REFERENCES

1. Bard, A. J., *J. Photochem.*, **10** (1), 59–75 (1979).
2. Bard, A. J., *Science*, **207** (4227), 139–44 (1980).
3. Bard, A. J., *J. Phys. Chem.*, **86** (2), 172–77 (1982).
4. Nozik, A. J., *Ann. Rev. Phys. Chem.*, **29**, 189–222 (1978).
5. Wrighton, M. S., *Acc. Chem. Res.*, **12** (9), 303–10 (1979).
6. Dutoit, E. C., Cardon, F. and Gomes, W. P., *Ber. Bunsenges. Phys. Chem.*, **80** (6), 475–81 (1976).
7. (a) Nishimoto, S., Ohtani, B., Kajiwara, H. and Kagiya, T., *J. Chem. Soc., Faraday Trans. 1*, **79** (11), 2685–94 (1983). (b) Ohtani, B., Nishimoto, S. and Kagiya, T., *Shokubai* (*Catalyst*), **25** (5), 383–5 (1983).
8. Lehn, J.-M., Sauvage, J.-P. and Ziessel, R., *Nouv. J. Chim.*, **4** (6), 355–8 (1980).
9. Kawai, T. and Sakata, T., *Nature*, **286** (5772), 474–6 (1980).
10. Kalyanasundaram, K., Borgarello, E. and Grätzel, M., *Helv. Chim. Acta*, **64** (1), 362–6 (1981).
11. (a) Fujishima, A. and Honda, K., *Bull. Chem. Soc. Japan*, **44** (4), 1148–50 (1971). (b) Fujishima, A. and Honda, K., *Nature, London*, **238** (5358), 37–8 (1972).
12. Nakabayashi, S., Fujishima, A. and Honda, K., *Chem. Phys. Lett.*, **102** (5), 464–5 (1983).

13. Kraeutler, B. and Bard, A. J., *J. Am. chem. Soc.*, **100** (13), 4317–18 (1978).
14. Dunn, W. W. and Bard, A. J., *Nouv. J. Chim.*, **5** (12), 651–5 (1981).
15. Kawai, T. and Sakata, T., *J. chem. Soc., Chem. Commun.* (23), 1047–8 (1979).
16. Sato, S. and White, J. M., *J. Catal.*, **69** (1), 128–39 (1980).
17. Pichat, P., Herrmann, J.-M., Disdier, J., Courbon, H. and Mozzanega, M. N., *Nouv. J. Chim.*, **5** (12), 627–36 (1981).
18. Kawai, T. and Sakata, T., *J. chem. Soc., Chem. Commun.* (15), 694–95 (1980).
19. Nishimoto, S., Ohtani, B., Yoshikawa, T. and Kagiya, T., *J. Am. chem. Soc.*, **105** (24), 7180–2 (1983).
20. Kawai, T. and Sakata, T., *Chem. Lett.* (1), 81–4 (1981).
21. St. John, M.-R., Furgala, A. J. and Sammells, A. F., *J. Phys. Chem.*, **87** (5), 801–5 (1983).
22. Kraeutler, B. and Bard, A. J., *J. Am. chem. Soc.*, **100** (19), 5985–92 (1978).
23. Yoneyama, H., Takao, Y., Tamura, H. and Bard, A. J., *J. Phys. Chem.*, **87** (5), 1417–22 (1983).
24. Sakata, T. and Kawai, T., *Nouv. J. Chim.*, **5** (5), 279–81 (1981).
25. Nishimoto, S., Ohtani, B., Shirai, H. and Kagiya, T., *J. Polym. Sci., Polym. Lett. Ed.*, in press.
26. Pak, V. N. and Ventov, N. G., *Russ. J. Phys. Chem.* (*Engl. Transl.*), **49** (10), 1489–90 (1975).
27. Okazaki, S. and Kanto, T., *Nippon Kagaku Kaishi* (*J. chem. Soc. Japan*) (3), 404–9 (1976).
28. Hayashi, K. and Otsu, T., *Makromol. Chem.*, **127**, 54–65 (1969).
29. Flory, P. J. and Leutner, F. S., *J. Polym. Sci.*, **3** (6), 880–90 (1948).
30. Nishimoto, S., Ohtani, B., Sakamoto, A. and Kagiya, T., *Nippon Kagaku Kaishi* (*J. chem. Soc. Japan*) (2), 246–52 (1984).
31. Nishimoto, S., Ohtani, B., Kajiwara, H. and Kagiya, T., *J. chem. Soc., Faraday Trans. 1*, **80** (1984).
32. Teratani, S., Nakamichi, J., Taya, K. and Tanaka, K., *Bull. chem. Soc. Japan*, **55** (6), 1688–90 (1982).
33. Boehm, H. P., *Disc. Faraday Soc.*, **52**, 264–75 (1971).
34. Sakata, T., Kawai, T. and Hashimoto, K., *Denki Kagaku*, **51** (1), 79–80 (1980).
35. Amouyal, E., Keller, P. and Maradpour, A., *J. chem. Soc., Chem. Commun.* (21), 1019–20 (1980).
36. Clarke, J. T. and Blout, E. R., *J. Polym. Sci.*, **1** (5), 419–28 (1946).
37. Jaeger, C. D. and Bard, A. J., *J. Phys. Chem.*, **83** (24), 3146–52 (1979).
38. Bahnemann and co-workers reported the formation of α-hydroxyl polymer radical via hydrogen abstraction by hydroxyl radical ($^{\cdot}OH$) generated by γ-radiolysis of water in a homogeneous system; Bahnemann, D., Henglein, A., Lilie, J. and Spanhal, L., *J. Phys. Chem.*, **88** (4), 709–11 (1984).
39. Kraeutler, B. and Bard, A. J., *J. Am. chem. Soc.*, **100** (7), 2239–40 (1978).
40. Izumi, I., Dunn, W. W., Wilbourn, K. O., Fan, F. F. and Bard, A. J., *J. Phys. Chem.*, **84** (24), 3207–10 (1980).

18

Photochemical Synthesis of New Biomedical Polymers

HAJIME MIYAMA

Department of Materials Science and Technology, Technological University of Nagaoka, Niigata, Japan

1. INTRODUCTION

Recently, synthesis of new biolomedical polymers, especially antithrombogenic polymers, has been urgently required for medical use as artificial internal organs. Many studies on antithrombogenic polymers have been carried out by changing the composition and microstructure of the polymer surface or binding an antithrombogenic reagent such as heparin to the polymer.[1] In order to synthesise the antithrombogenic polymer, it is practical to modify conventional polymers such as poly(vinyl chloride) or polyacrylonitrile because the physical and mechanical properties of the polymers are well known. The photochemical method is reaction-selective and considered to be ideal for the modification of conventional polymers. The author surveyed various photograft and photoblock polymerisation reactions which might be applicable for the present purpose.

2. PHOTOGRAFT AND PHOTOBLOCK REACTIONS

Various photograft and photoblock polymerisation reactions which have been published are shown below.

2.1. Method Using Tetraethylthiuram Disulphide (TETD)[2]

According to the following reaction, a vinyl polymer having photosensitive groups at both ends is obtained.

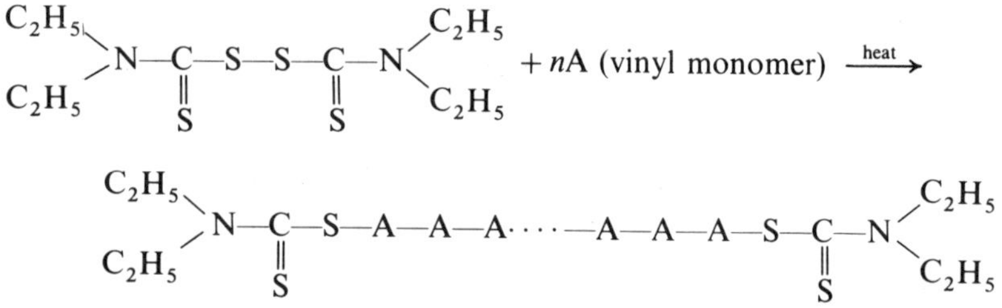

$$(C_2H_5)_2N{-}\underset{\underset{S}{\|}}{C}{-}S{-}S{-}\underset{\underset{S}{\|}}{C}{-}N(C_2H_5)_2 + nA \text{ (vinyl monomer)} \xrightarrow{\text{heat}}$$

$$(C_2H_5)_2N{-}\underset{\underset{S}{\|}}{C}{-}S{-}A{-}A{-}A\cdots{-}A{-}A{-}A{-}S{-}\underset{\underset{S}{\|}}{C}{-}N(C_2H_5)_2$$

Another vinyl monomer is added to the photosensitive polymer and photoblocked by irradiating the mixture.

2.2. Methods Using Chloromethylated Polystyrene[3]

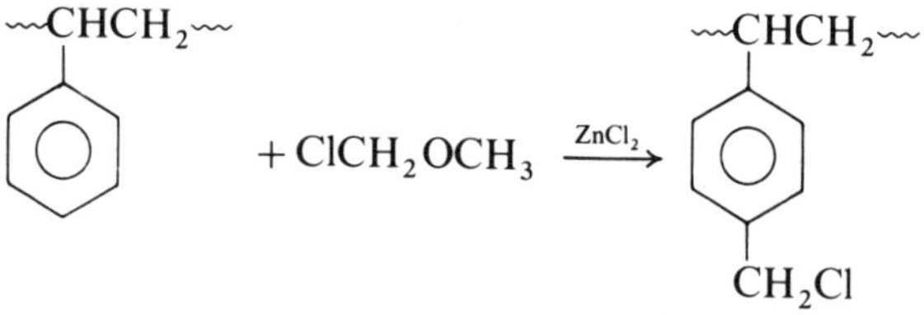

$$\sim CH(C_6H_5)CH_2 \sim + ClCH_2OCH_3 \xrightarrow{ZnCl_2} \sim CH(C_6H_4CH_2Cl)CH_2 \sim$$

To benzene solution of the polymer obtained by the above reaction, another vinyl monomer is added. By irradiating the mixture, graft copolymer is obtained.

2.3. Method Using Carbamated Polystyrene[4]

The chloromethylated polystyrene obtained by the method described above is further carbamated as follows.

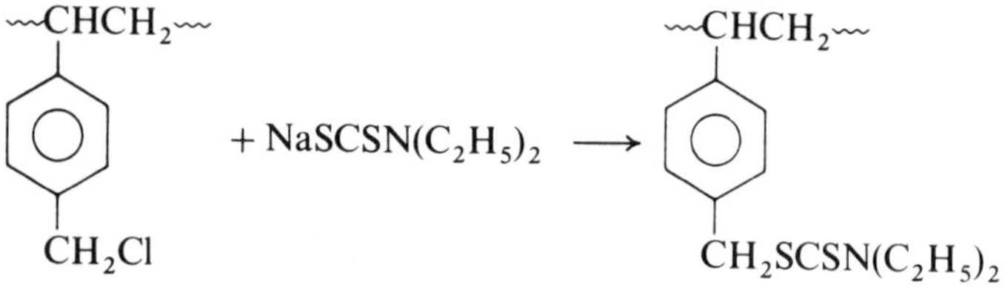

$$\sim CH(C_6H_4CH_2Cl)CH_2 \sim + NaSCSN(C_2H_5)_2 \longrightarrow \sim CH(C_6H_4CH_2SCSN(C_2H_5)_2)CH_2 \sim$$

The carbamated polystyrene is dissolved in benzene or dimethylsulphoamide, to which another vinyl monomer is added. By irradiating the solution, graft copolymer is obtained.

2.4. Method Using Carbamated Poly(vinyl chloride)[5]

By using the same method as described above, graft copolymer of poly(vinyl chloride) is obtained.

2.5. Method Using Chlorine-containing Polydimethyl Siloxane[6]

By applying the method described above to chlorine-containing polydimethylsiloxane, graft polymer is obtained as follows.

$$\overset{\displaystyle CH_2Cl}{\underset{\displaystyle CH_3}{-\!\!(Si}}-O)_m\!\!-\!\!(\overset{\displaystyle CH_3}{\underset{\displaystyle CH_3}{Si}}-O)_n\!\!- \xrightarrow{NaS_2CN(C_2H_5)_2} -\!\!(\overset{\displaystyle CH_2-S\overset{S}{\overset{\|}{C}}N(C_2H_5)_2}{\underset{\displaystyle CH_3}{Si}}-O)_m\!\!-\!\!(\overset{\displaystyle CH_3}{\underset{\displaystyle CH_3}{Si}}-O)_n\!\!-$$

$$+\,pM\text{ (vinyl monomer)} \xrightarrow{h\nu} -\!\!(\overset{\displaystyle CH_2S(M)_p-}{\underset{\displaystyle CH_3}{Si}}-O)_m\!\!-\!\!(\overset{\displaystyle CH_3}{\underset{\displaystyle CH_3}{Si}}-O)_n\!\!-$$

2.6. Method Using Brominated Poly(methyl acrylate)[7]

Brominated poly(methyl acrylate) is synthesised as follows.

$$CH_2{=}CHCOOCH_3 + Br_2 \longrightarrow CH_2Br{-}CHBr_2COOCH_3 \longrightarrow$$

$$CH_2{=}CBr{-}COOCH_3 \xrightarrow{\text{(vinyl monomer)}} \sim\!CH_2-\overset{\displaystyle Br}{\underset{\displaystyle COOCH_3}{\overset{|}{\underset{|}{C}}}}\!\sim$$

By adding another vinyl monomer to the brominated poly(methyl acrylate) and irradiating the mixture, graft polymer is obtained due to the photodissociation of C—Br bond.

2.7. Method Using Bromine-containing Polyacrylonitrile[8-10]

Miller[8] obtained polyacrylonitrile containing bromine atoms at both ends of the polymer by photopolymerisation of acrylonitrile containing bromine atoms at both ends of the polymer, by photopolymerisation of acrylonitrile in dimethylsulphonamide with bromoform or carbon tetrachloride as an initiator. On the other hand,[9,10] Miyama *et al.* obtained polyacrylonitrile containing more than 10 bromine atoms per polymer by the reaction in dimethyl sulphoxide. By using these bromine-containing polymers, other vinyl polymers can be photoblocked or photografted to the trunk polymer.

2.8. Method Using Aqueous Suspension of Starch or Poly(vinyl alcohol)[11]

By adding acrylate drop by drop to an aqueous suspension of starch or poly(vinyl alcohol) and circulating the mixture to a photoreactor, graft copolymer of starch or poly(vinyl alcohol) is obtained.

2.9. Method Using p-Vinylbenzyl-p-tert-butyl Perbenzoate[12]

A vinyl monomer having the following structure is synthesised.

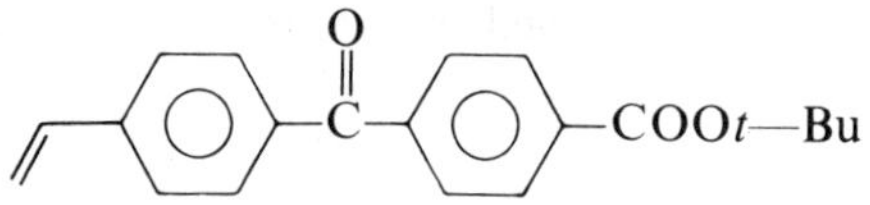

$$CH_2{=}CH{-}C_6H_4{-}\overset{O}{\overset{\|}{C}}{-}C_6H_4{-}COOt{-}Bu$$

A copolymer of this monomer and styrene is mixed with methyl methacrylate. By irradiating the mixture with light of 366 nm, graft polymer is obtained. The mechanism of photo-initiation is proposed as follows.

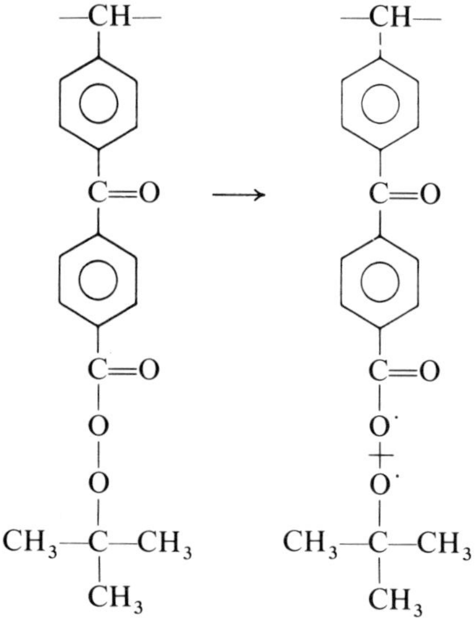

$$-CH(C_6H_4{-}C{=}O{-}C_6H_4{-}C({=}O){-}O{-}O{-}C(CH_3)_3)- \longrightarrow -CH(C_6H_4{-}C{=}O{-}C_6H_4{-}C({=}O){-}O^{\cdot})- + {}^{\cdot}O{-}C(CH_3)_3 \longrightarrow -CH(C_6H_4{-}C{=}O{-}C_6H_4^{\cdot})- + CO_2$$

2.10. Method Using Poly(N-nitroso-acrylamine)[13]

Vinyl monomer is photopolymerised by using poly(N-nitroso-acrylamine) as an initiator, which is synthesised by the nitrosation of Nylon. Block polymer is obtained as follows.

$$\sim H_2C{-}N(NO){-}CO{-}R{-}NH{-}\cdots$$

$$-CO{-}R{-}NH{-}\cdots{-}CO{-}R{-}N(NO){-}CO{-}NH_2\sim \xrightarrow{h\nu}$$

$$\sim CH_2{-}N{=}N{-}O{-}CO{-}R{-}NH{-}\cdots$$

$$-CO{-}R{-}N{-}N{=}N{-}O{-}CO{-}CH_2\sim \xrightarrow[-2N_2]{h\nu}$$

$$\sim H_2C^{\cdot} + {}^{\cdot}OOC{-}R{-}NH{-}\cdots{-}CO{-}(CH_2)_4{-}CH_2{-}{}^{\cdot}OOC{-}CH_2\sim$$

$$\xrightarrow[-CO_2]{\text{(vinyl monomer) }(r+s)M} \sim CH_2{-}M_r{-}OOC{-}RNH{-}\cdots{-}CO{-}R{-}M_s{-}CH_2\sim$$

2.11. Method Using 'Encaged Macroradicals'[14]

By photopolymerising methacrylate in water–ethanol with hydrogene peroxide as an initiator, 'encaged macroradicals' are formed in viscous solution. Another vinyl monomer is added to the solution after the irradiation has ceased. Thus, graft copolymer of poly(methyl methacrylate) is obtained.

2.12. Method Using Iniferter[15]

Otsu *et al.* proposed an idea of Iniferter by developing the method (2.1). According to their definition, Iniferter is explained as follows. If the initiator R—R′ has a high reactivity for chain transfer to the initiator or a part of the radical produced by the initiator easily undergoes primary radical termination, a polymer with two initiator fragments is obtained in the presence of vinyl monomer M. That is, the following successive insertion of monomers into the broken bond of R—R′ occurs.

$$R{-}R' + nM \longrightarrow R{-}\!(M)_n\!{-}R'$$

This type of initiator is defined as Iniferter. By photopolymerising vinyl monomer with the above described dithiocarbamated polymer as an Iniferter, a block copolymer of two components is obtained. By repeating the same procedure with this block copolymer as new Iniferter, a block copolymer of multi-components is obtained.

Examination of all of the methods by considering physical or mechanical properties of trunk polymer and simplicity and yield of reactions indicates that the methods (2.4), (2.5) and (2.7) are suitable for the photochemical synthesis of antithrombogenic polymer.

3. PHOTOCHEMICAL SYNTHESIS OF ANTITHROMBOGENIC POLYMER

3.1. Binding of Chicago Acid to Polymer Surface

This is not an application of photopolymerisation, but that of photochemical binding of small molecule to polymer. To the surface of polyethylene and latex rubber, Chicago Acid, is covalently attached as

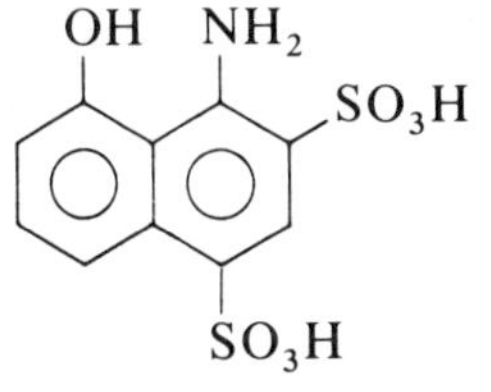

OH NH_2
SO_3H
SO_3H

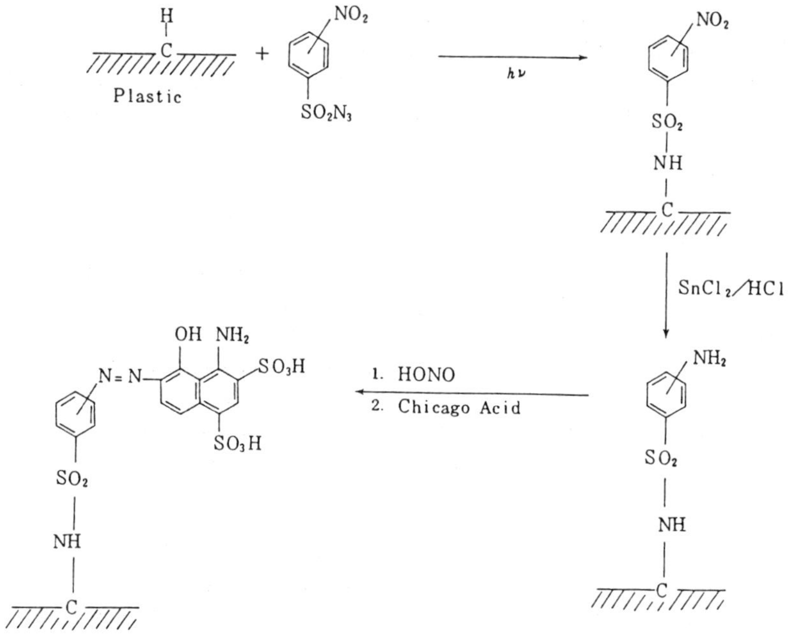

FIG. 1. Scheme of covalent attachment of Chicago Acid to Polymer surface.

shown in Fig. 1. Photolysis of azide generates nitrene which is inserted into C—H bond. The nitro group is then reduced to the amine and in a successive step converted to the diazonium salt. Binding of the antithrombogenic Chicago Acid to the aromatic ring is then accomplished by diazonium coupling. The polymer treated in this way is tested as a extracorporal shunt connecting the carotid artery to the jugular vein in a dog. The polymer showed significantly improved thromboresistance.

3.2. Photografting of Hydrophilic Monomers onto Diethyldithiocarbamated Polydimethylsiloxane[16]

By using method (2.5), hydrophilic vinyl monomers such as 2-hydroxyethylmethacrylate, acrylamide, 2-methylaminoethyl methacrylate, sodium styrene sulphonate, N-vinyl-2-pyrrolidone and methacrylic acid were photografted to diethyldithiocarbamated polydimethylsiloxane. These graft polymers were found to have hydrophilicity, necessary for medical application. However, their antithrombogenicity has not yet been reported.

3.3. Antithrombogenic Poly(vinyl chloride) Graft Copolymer

Miyama *et al.*[17] carried out the following reaction by using method (2.4).

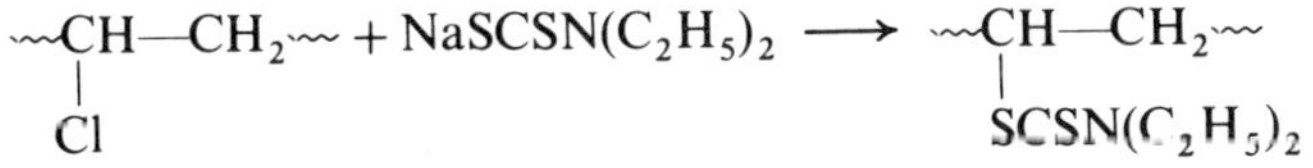

$$\sim\underset{\displaystyle Cl}{\underset{|}{CH}}-CH_2\sim + NaSCSN(C_2H_5)_2 \longrightarrow \sim\underset{\displaystyle SCSN(C_2H_5)_2}{\underset{|}{CH}}-CH_2\sim$$

To the modified polymer, methoxypolyethylene glycol methacrylate (SM),

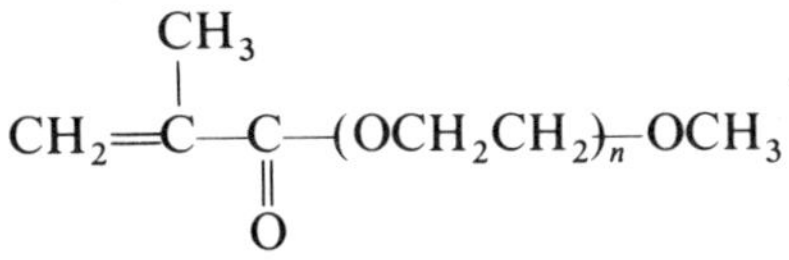

$$CH_2{=}\overset{\displaystyle CH_3}{\overset{|}{C}}-\underset{\displaystyle O}{\underset{\|}{C}}-(OCH_2CH_2)_n-OCH_3$$

and N,N-dimethylaminoethyl methacrylate (DAEM),

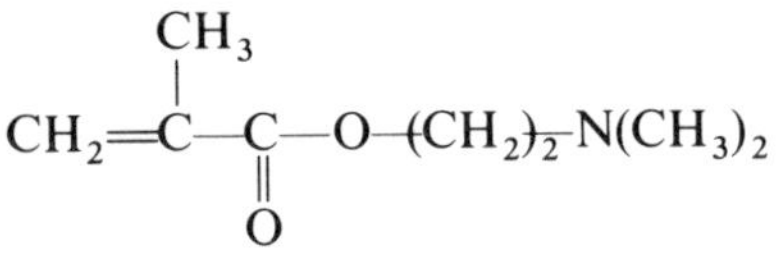

$$CH_2{=}\overset{\displaystyle CH_3}{\overset{|}{C}}-\underset{\displaystyle O}{\underset{\|}{C}}-O-(CH_2)_2-N(CH_3)_2$$

were simultaneously photografted. Dimethylamino groups of the graft polymer are quaternised to give positive charges. To the positive charges of the polymer, antithrombogenic reagent heparin molecules which have negative charges are ionically bounded. That is, the trunk of the graft copolymer takes part of mechanical support, the hydrophilic group (SM) biocompatibility, and heparinised part antithrombogenicity, respectively. In order to avoid contamination by plasticiser, commercial graft copolymer of vinyl chloride, ethylene and vinyl acetate (Graftmer R_3) was used instead of poly(vinyl chloride) with plasticiser. Composition and properties of the heparinised polymer are shown in Table 1. The polymers of Group (II) which have suitable water content show the best *in vivo* results. Here, heparin is not fixed to the polymer, but released slowly from the polymer surface which is effective for antithrombogenicity. The clinical application of this polymer for use in the cardiovascular system was successful. The polymer is now commercialised as an antithrombogenic catheter 'Anthron'. Chemical kinetics for the manufacturing process of the polymer have been well studied.[19]

For use during a short period such as blood bag, poly(vinyl chloride) grafted with SM alone considered to be useful instead of heparinised polymer. It was found[20] that the longer the polyethylene oxide part (*n* of the chemical formula of SM) the less the adhesion of platelets to the polymers surface.

TABLE 1
Chemical Compositions and Physical Properties of H-RSD

	Sample No.	*Graft* (%)	*SM* (%)	*DAEM* (%)	*Ad. Hep.* (%)	*W* (%)	*N*⊕ (*meq./g*)	*SMP* (*m Volt*)	*Smoothness*	*In vivo results*
(I)	103	15	0	15	15	27	0·84	−33	○	×
	102	22	0	22	15	28	0·88	−32	○	×
	113	17	2	15	12	27	0·76	−32	○	×
	107	22	5	17	13	(48)	0·83	−28	○	△
(II)	110	30	12	18	13	37	0·76	−25	○	○
	23	32	16	16	7	49	0·80	−19	○	○
(III)	109	42	23	19	12	76	0·77	−13	○	△
	111	55	20	35	16	80	1·04	−21	×	×

SM: methoxypolyethylene glycol methacrylate.
DAEM: 2-dimethylaminoethyl methacrylate.
Ad. Hep.: heparin content/dry polymer.
W: water content/dry polymer.
SMP: standard membrane potential.
Smoothness: surface condition.

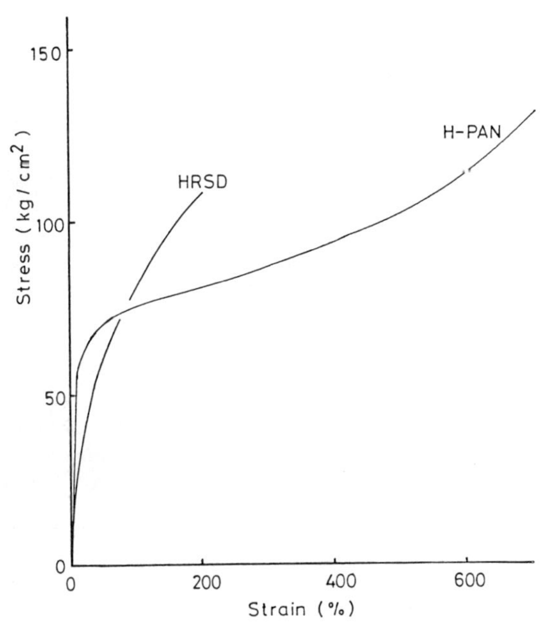

FIG. 2. Stress–strain curves of heparinised polyacrylonitrile graft polymer (H-PAN) and heparinised poly(vinyl chloride) graft copolymer (HRSD).

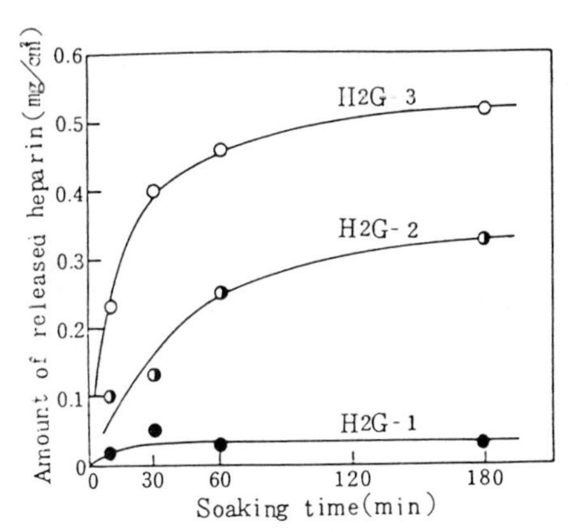

FIG. 3. Change of amount of heparin released with soaking time. Water contents of H2G-1, H2G-2 and H2G-3 are 15·4, 30·4 and 50·4 wt %, respectively.

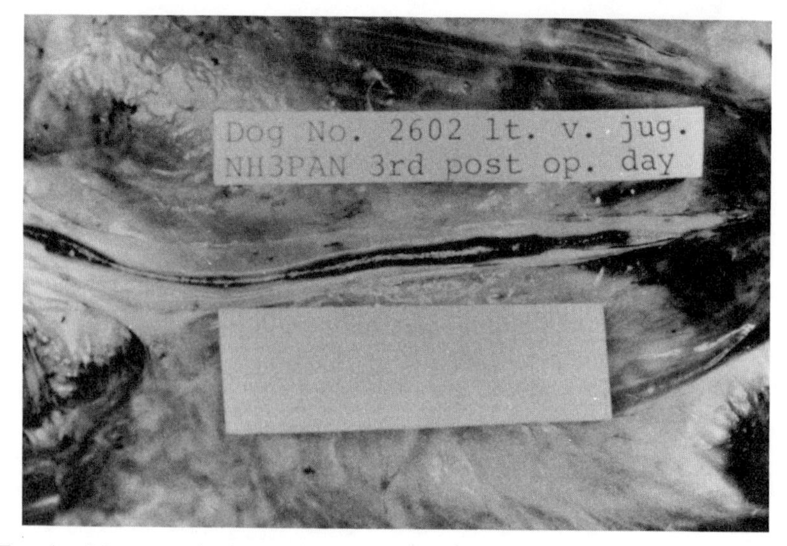

FIG. 4. Macroscopic observation of sutures coated with polymer in vein. H5PAN (top) and NH3PAN (bottom) are heparinised and non-heparinised graft copolymers, respectively.

3.4. Antithrombogenic Polyacrylonitrile

Since polyacrylonitrile is more flexible and tenacious than poly(vinyl chloride), the former is easier to process into membrane or fibre. Therefore, synthesis of heparinised polyacrylonitrile was carried out.[21] By using method (2.7), SM and DAEM were simultaneously photografted to bromine-containing polyacrylonitrile. The grafted copolymer was quaternised and heparinised similarly as described above. The mechanical properties of the heparinised polyacrylonitrile are better than that of heparinised poly(vinyl chloride) as shown in Fig. 2. Elution of heparin from the heparinised polyacrylonitrile is shown in Fig. 3. The larger the water content of the polymer the higher the release rate of heparin. This behaviour is very similar to that of heparinised poly(vinyl chloride). *In vivo* results of the heparinised polyacrylonitrile by Noissiki's method[22] are shown in Fig. 4. A suture coated with the polymer is inserted into jugular and femoral vein of a dog. Three days after the insertion, the animal was sacrificed. The vein in which the suture was inserted was opened and the observation was carried out macroscopically. As shown in the pictures, the

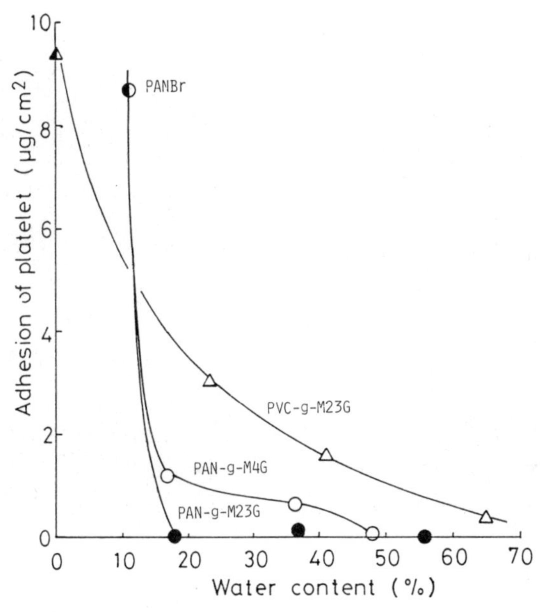

FIG. 5. Change of platelets adhered with water content. PANBr, PAN-g-M4G, PAN-g-M23G and PVC-g-M23G are polyacrylonitrile trunk polymer, polyacrylonitrile grafted with SM of $n = 4$, with that of $n = 23$ and poly(vinyl chloride) grafted with SM of $n = 23$, respectively.

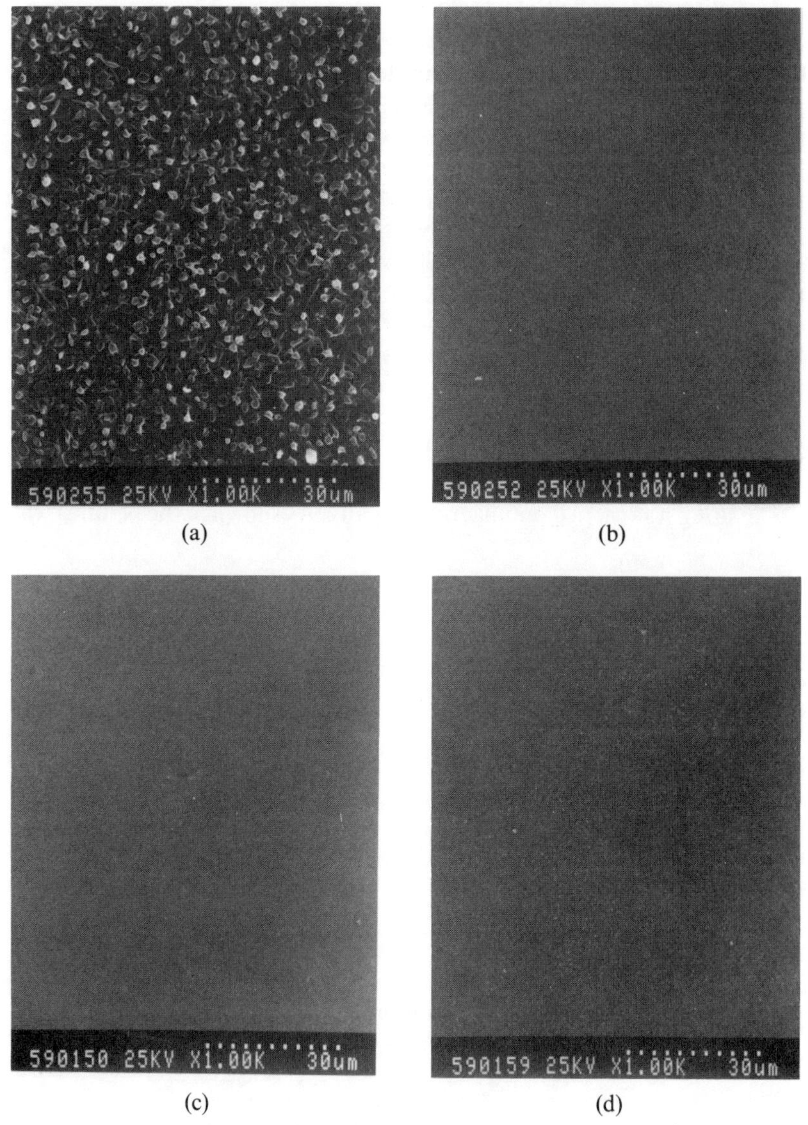

FIG. 6. Scanning-electron-microscopic pictures of platelets adhered on polymer surface. (a), (b), (c) and (d) are PANBr, PAN-g-M4G, PAN-g-M23G and PAN-g-M110G, respectively, which are defined in Fig. 5.

heparinised polymer shows no blood clotting, while non-heparinised polymer shows marked clotting. Kinetic studies of photografting process for this case has been carried out in detail.[23]

Polyacrylonitrile grafted with SM alone[24] showed marked decrease of platelet as shown in Fig. 5. The decrease of platelets adhesion was more pronounced than that for poly(vinyl chloride) grafted with SM as shown in Fig. 6. Medical application of this polymer is now under study.

Photochemical synthesis of medical polymer is not limited to antithrombogenic material, but applicable to various fields of medical use. I believe that successful results will be obtained for other medical use in future.

REFERENCES

1. Kim, S. W., Eberet, C. D., Lin, J. Y. and McRea, J. C., *Am. Soc. Artif. Int. Organs*, **6**, 76–87 (1983); Tsuruta, T. and Sakurai, Y., *Biomed. Sci.*, **2**, Nankodo, Tokyo (1982).
2. Otsu, T., *Kogyo Kagaku Zasshi*, **62** (9), 1461–6 (1959).
3. Okawara, M., Yoneda, Y. and Imoto, E., *Kobunshi Kagaku*, **17** (177), 30–6 (1960).
4. Okawara, M., Yamashita, N., Ishiyama, K. and Imoto, E., *Kogyo Kagaku Zasshi*, **66** (9), 1383–9 (1963).
5. Okawara, M., Morishita, K. and Imoto, E., *Kogyo Kagaku Zasshi*, **69** (4), 761–5 (1966).
6. Inoue, H. and Kohama, S., *J. appl. Polym. Sci.*, **29**, 877–89 (1984).
7. Marvel, C. S., Dec, J., Cooke, H. G. Jr and Cowan, J. C., *J. Am. chem. Soc.*, **62**, 3495–8 (1940).
8. Miller, M. L., *Can. J. Chem.*, **36**, 309–14 (1958).
9. Miyama, H. and Sato, T., *J. Polym. Sci.*, *A-1*, **10**, 2469–71 (1972).
10. Miyama, H. and Harumiya, N., *Kobunshi Ronbunshu*, **36** (5), 351–3 (1979).
11. Trimnell, D. and Stout, E. I., *J. appl. Polym. Sci.*, **25**, 2431–4 (1980).
12. Gupta, S. N., Thijs, L. and Neckers, D. C., *J. Polym. Sci., Polym. Chem. Ed.*, **19**, 855–68 (1981).
13. Grauner, H., *J. Polym. Sci., Polym. Chem. Ed.*, **20**, 1935–9 (1982).
14. Brosse, Jean-Claude and Lenain, Jean-Claude, *J. appl. Polym. Symp.*, **36**, 177–84 (1981).
15. Otsu, T. and Yoshida, M., *Makromol. Chem., Rapid Commun.*, **3**, 127–32 (1982).
16. Nishizawa, E. E., Wynalda, D. J. and Lednicer, D., *Trans. Am. Soc., Artif. Int. Organs*, **XIX**, 13–18 (1973).
17. Miyama, H., Harumiya, N., Mori, Y. and Tanzawa, H., *J. biomed. Mater. Res.*, **11**, 251–65 (1977).
18. Tanzawa, H., Mori, Y., Harumiya, N., Miyama, H., Hori, I., Oshima, N. and Idezuki, Y., *Trans. Am. Soc. Artif. Int. Organs*, 188–94 (1973).

19. Miyama, H., Fujii, N., Shimazaki, Y. and Ikeda, K., *Polym. Photochem.*, **3**, 445–61 (1983).
20. Nagaoka, S., Takiuchi, H., Yokota, K., Mori, Y., Tanzawa, H. and Kikuchi, T., *Kobunshi Ronbushu*, **39** (4), 173–8 (1982); Mori, Y., Nagaoka, S., Takeuchi, H., Nishiumi, S., Kuwano, A. and Miyama, H., *Japan J. Artif. Organs*, **13** (3), 1143–6 (1984).
21. Miyama, H., Fujii, N., Nakamura, T., Nagaoka, S. and Mori, Y., *Kobunshi Ronbunshu*, **40** (10), 691–6 (1983).
22. Noissiki, Y., *Japan J. Artif. Organs*, **11** (3), 794–7 (1982).
23. Miyama, H., Fujii, N. and Ikeda, K., *Polym. Photochem.*, in press (1985).
24. Miyama, H., Fujii, N., Kuwano, A., Nagaoka, S. and Mori, Y., *1st SPSJ International Conference*, Aug. 23 (1984) in Kyoto.

Index